Multiple-Valued Computing in Quantum Molecular Biology

This book focuses on the design methodologies of various quantum circuits, DNA circuits, DNA-quantum circuits and quantum-DNA circuits. It considers the merits and challenges of multi-valued logic circuits in quantum, DNA, quantum-DNA and DNA-quantum computing. *Multiple-Valued Computing in Quantum Molecular Biology: Arithmetic and Combinational Circuits* is Volume 1 of a two-volume set.

From fundamentals to advanced levels, this book discusses different multiple-valued logic DNA-quantum and quantum-DNA circuits. The text consists of four parts. Part I introduces multiple-valued quantum computing and DNA computing. It contains the basic understandings of multiple-valued quantum computing, multiple-valued DNA computing, multiple-valued quantum-DNA computing and multiple-valued DNA-quantum computing. Part II examines heat calculation, speed calculation, heat transfer, data conversion and data management in multi-valued quantum, DNA, quantum-DNA and DNA-quantum computing. Part III discusses multiple-valued logic operations in quantum and DNA computing such as ternary AND, NAND, OR, NOR, XOR, XNOR and multiple-valued arithmetic operations such as adder, multiplier, divider and more. Quantum-DNA and DNA-quantum multiple-valued arithmetic operations are also explained in this section. Part IV explains multiple-valued quantum and DNA combinational circuits such as multiple-valued DNA-quantum and quantum-DNA multiplexer, demultiplexer, encoder and decoder.

This book will be of great help to researchers and students in quantum computing, DNA computing, quantum-DNA computing and DNA-quantum computing researchers.

Multiple-Valued Computing in Quantum Molecular Biology

Arithmetic and Combinational Circuits, Volume 1

Hafiz Md. Hasan Babu

CRC Press

Taylor & Francis Group

Boca Raton London New York

CRC Press is an imprint of the
Taylor & Francis Group, an **informa** business

First edition published 2024
by CRC Press
6000 Broken Sound Parkway NW, Suite 300, Boca Raton, FL 33487-2742

and by CRC Press
4 Park Square, Milton Park, Abingdon, Oxon, OX14 4RN

CRC Press is an imprint of Taylor & Francis Group, LLC

© 2024 Hafiz Md. Hasan Babu

Library of Congress Control Number: 2023942121

ISBN: 978-1-032-46486-2 (hbk)
ISBN: 978-1-032-46491-6 (pbk)
ISBN: 978-1-003-38193-8 (ebk)

DOI: 10.1201/9781003381938

Typeset in Nimbus Roman
by KnowledgeWorks Global Ltd.

To my respected great parents and also to my lovely wife, daughter, and son, who made it possible to write this book

Contents

IV   Multiple-Valued Combinational Circuits in Quantum Molecular Biology   387

Overview   389

16   Multiple-Valued Quantum Combinational Circuits   391

Author

Dr. Hafiz Md. Hasan Babu is currently working as a professor in the Department of Computer Science and Engineering, University of Dhaka, Bangladesh as well as Dean in the Faculty of Engineering and Technology of the University of Dhaka. In addition, at present, he is a member (part-time) of Bangladesh Accreditation Council, Ministry of Education of the Government of the People's Republic of Bangladesh. He is also the Chairman of the Board of Directors at Dhaka Stock Exchange Limited, Bangladesh. Dr. Hasan Babu was Director of the Board of Directors of Bangladesh Submarine Cable Company Limited. He was the Chairman of the Department of Computer Science and Engineering at the University of Dhaka from 2003–2006 and Pro-Vice-Chancellor of the National University of Bangladesh from 2016–2020. He was also a professor and Founding Chairman of the Department of Robotics and Mechatronics Engineering, University of Dhaka. Dr. Hasan Babu obtained his PhD in electronics and computer science from Japan under the Japanese Government Scholarship and received his MSc in computer science and engineering from the Czech Republic under the Czech Government Scholarship. He also received the DAAD Research Fellowship from Germany.

Dr. Hasan Babu was awarded the Dr. M.O. Ghani Memorial Gold Medal by the Bangladesh Academy of Sciences in 2017 for his excellent research in the progress of physical sciences in Bangladesh. In addition, he was awarded the University Grants Commission of Bangladesh (UGC) Gold Medal Award-2017 in the mathematics, statistics, and computer science categories for his research on quantum multiplier-accumulator devices. He is currently an associate editor of the research journal *IET Computers and Digital Techniques* published by the Institution of

Engineering and Technology of the United Kingdom. He was a member of the Prime Minister's ICT Task Force in Bangladesh. Dr. Hasan Babu was also President of the Bangladesh Computer Society from 2017–2020. Presently, he is President of the International Internet Society, Bangladesh chapter.

Professor Dr. Hafiz Babu has published more than 100 research papers. Three of his research papers have received the best research awards in international conferences.

In addition, he has published the following four textbooks by three famous UK and US publishers for graduate and post-graduate students:

1. Hafiz Md. Hasan Babu, *Quantum Computing: A Pathway to Quantum Logic Design*, IOP (Institute of Physics) Publishing, 2020, Bristol, UK

2. Hafiz Md. Hasan Babu, *Reversible and DNA Computing*, Wiley Publishers, 2021, UK

3. Hafiz Md. Hasan Babu, *VLSI Circuits and Embedded Systems*, CRC Press/Taylor & Francis, 2022, USA

4. Md. Jahangir Alam, Guoqing Hu, Hafiz Md. Hasan Babu, and Huazhong Xu, *Control Engineering Theory and Applications*, CRC Press/Taylor & Francis, 2022, USA

Preface

Quantum computing is a type of computation that harnesses the collective properties of quantum states, such as superposition, interference, and entanglement, to perform calculations. The devices that perform quantum computations are known as quantum computers. Quantum computers have the potential to revolutionize computation by making certain types of classically intractable problems solvable.

DNA (deoxyribonucleic acid) encodes biological organisms' genetic information. It is constituted of polymer chains, which are commonly referred to as DNA strands. Nucleotides, or bases, are attached to a sugar phosphate "backbone" in each strand. The four DNA nucleotides are adenine, guanine, cytosine, and thymine, denoted as A, G, C, and T, respectively.

Multiple-valued logic (MVL) is the non-binary-valued system in which more than two levels of information content are available, i.e., L>2. In modern technologies, the dual-level binary logic circuits are normally used.

MVL, quantum computing, and DNA computing have caught the attention of researchers working with physics, biochemistry, molecular biology, and computer science. The high-speed computation capability has already proved that quantum computing is the fastest system to solve the computational problem. And the different approaches to bimolecular coding of DNA computing made the revolution in the modern world of computation. Though these computing techniques are completely different in nature and there are no similarities between them, the new concepts can be evolved from these two. This book introduces "multiple-valued quantum-DNA computing" and "multiple-valued DNA-quantum computing," a noble combination of quantum physics and molecular biology, which is also called the quantum molecular biology for multiple-valued computation.

This book, *Multiple-Valued Computing in Quantum Molecular Biology*, Volume 1 is divided into four parts. It starts with the basics of multiple-valued quantum computing, DNA computing, quantum-DNA computing, and DNA-quantum computing.

Part 1 of this book considers the merits, demerits, challenges and motivations of MVL circuits in quantum, DNA, quantum-DNA, and DNA-quantum computings.

Part 2 covers the heat calculation, speed calculation, heat transfer, data conversion, and data management in multi-valued quantum, DNA, quantum-DNA, and DNA-quantum computings.

In Part 3, MVL operations are described in quantum and DNA computings, such as ternary AND, NAND, OR, NOR, XOR, XNOR, and multiple-valued arithmetic operations, such as adder, subtractor, multiplier, divider, and more, are described with their operational procedures and design architectures. At the end of this part, quantum-DNA and DNA-quantum multiple-valued arithmetic operations are also explained.

Part 4 contains the details of multiple-valued quantum and DNA combinational circuits such as multiple-valued DNA-quantum and quantum-DNA multiplexer, demultiplexer, encoder, decoder, and more.

This book is a great resource for the MVL, DNA, quantum-DNA, and DNA-quantum computing researchers, students, and academicians. This is a novel approach to a book in this field. This book will quench the thirst of beginners to advanced-level readers.

Acknowledgments

I would like to express my sincerest gratitude and special appreciation to the various researchers in the field of multiple-valued computing in quantum molecular biology. The contents in this book have been compiled from a wide variety of research works, where the researchers are pioneers in their respective fields. All the research articles related to the contents are listed at the end of each chapter.

I am grateful to my great parents and dear family members for their endless support. Most of all, I want to thank my lovely wife, Mrs. Sitara Roshan, sweet daughter, Ms. Fariha Tasnim, and sweet son, Md. Tahsin Hasan, for their invaluable cooperation so that I could complete this book.

Finally, I am thankful to all of those, especially beloved students Nitish Biswas, Md. Tareq Hasan, and Rownak Borhan Himel, who have provided their immense support and valuable time to help me finish this book.

Acronyms

AI	artificial intelligence
ALU	arithmetic logic unit
BP	block propagate
CLB	configurable logic block
CPU	central processing unit
CU	control unit
DNA	deoxyribonucleic acid
DNase	deoxyribounuclease
EMR	electron magnetic resonance
ML	machine learning
MUX	multiplexer
MVL	multiple-valued logic
NMR	nuclear magnetic resonance
NTI	negative ternary inversion
PCR	polymerase chain reaction
PROM	programmable read-only memory
PTI	positive ternary inversion
QIT	quandrupole ion trap
RAM	random access memory
RF	radio frequency
RNA	ribonucleic acid
STI	standard ternary inversion
TG	Toffoli gate
XNOR	exclusive NOR
XOR	exclusive OR

Introduction

Modern computers use binary logic in which everything is represented by the numbers, 0 and 1. Despite the fact that binary logic only has two logics: true or false, the logic gates can sometimes cause ambiguity and imprecision. To solve these issues, multi-valued logic is used, which allows switching between more than two states. Furthermore, multi-valued logic with an expanding number of distinct states allows for increased information densities. Quantum computing achieves tremendous processing power, low-energy consumption, and exponential speed above traditional computers by regulating the behavior of minuscule physical things such as atoms, electrons, photons, and other microscopic particles. With the advent of nanotechnology, quantum computing vibrates an incredibly immense role in developing more compact and less power-consuming computers. The quantum computer is a completely new notion from regular computing, and it does not employ binary logic.

The DNA molecule's properties help in the induction of quantum properties including superposition, tunneling, coherence, and entanglement. The collection of quantum component states in which a particle can exist in either a solo or mixed state is known as superposition. Quantum computing relies on quantum bits, sometimes called "qubits," which may also denote either |0> or |1>. It's intriguing that qubits can achieve a mixed state called superposition in which they can be both |1> and |0> at the same time. Think about this scenario: It's just a piece of paper with a number on it. When a coin is tossed, it starts spinning randomly. It has a probability of being the head, tail, or both at the same time throughout rotation since it spins at random. The idea of superpositioning, which is important to quantum physics and quantum computing, is referred to as quantum coherence. Quantum coherence explores a scenario in which the wave property of an item is split in two and the two waves coherently interact with one another. Quantum coherence is based on the idea that everything has wave-like properties. It's similar to quantum entanglement in that it involves two quantum particles sharing states rather than two quantum waves from a single particle. Entanglement is a concept that describes a relationship between two or more particles that interact in such a way that characterizing each particle independently is difficult.

The study of the make-up, structure, and interactions of cellular molecules, such as proteins and nucleic acids, which carry out the biological processes necessary for a cell's maintenance and function, is known as molecular biology. Multiple-valued computing in quantum molecular biology can be introduced to get the advantages of both multi-valued quantum and DNA computing. Multi-valued quantum molecular biology means multiple-valued quantum-DNA and multiple-valued DNA-quantum computing; it can be obtained by merging multiple-valued quantum and multiple-valued DNA computing and have the capacity to execute parallel operations, which is one of the finest advantages of classical multiple-valued computing. Multiple-valued quantum computing is quicker than traditional multiple-valued computing, but the data created by multiple-valued quantum processes must be stored. Furthermore, multiple-valued DNA has the potential to store a significant amount of data. As a result, multiple-valued quantum-DNA may be able to solve the data storage challenges of multiple-valued quantum computing. The advantage of multiple-valued quantum-DNA and DNA-quantum computing is a balanced system, which is rounding a number that is the same as truncating it.

Part I

Multiple-Valued Quantum and DNA Computing

Overview

The non-binary-valued system known as multiple-valued logic (MVL) allows for the availability of information content at levels greater than two. Dual-level binary logic circuits have typically been used in modern technologies. Multi-valued means ternary, quarternary, and more.

Quantum computing is a field of study that focuses on the creation of computer-based technologies based on quantum-theoretical principles. On the quantum (atomic and subatomic) level, quantum theory describes the nature and behavior of energy and matter. To execute certain computational tasks, quantum computing employs a combination of qubits. All of this is done at a far higher rate than their traditional computing equipment. Quantum computers represent a significant advancement in computing capability, with enormous performance benefits for specific use cases. The ability of qubits to be in several states at the same time gives the quantum computer a lot of computing capability. They can accomplish jobs with a mix of $| 1 >, | 0 >$, and both $| 1 >$ and $| 0 >$ at the same time. So, quantum computing can be defined as an area of computing that is focused on the development of computer technology based on the principles of quantum theory. In addition, quantum computing is much faster than classical bit-wise classical computing.

On the other hand, instead of using typical silicon chips, DNA computing uses biological molecules to do computations. The four-character genetic alphabets (A-adenine, G-guanine, C-cytosine, and T-thymine) are used in DNA computing instead of the binary alphabet (1 and 0) utilized by standard computers. This is possible due to the ability to create small DNA molecules with any arbitrary sequence. The input of any DNA operation can be represented by DNA molecules with specific sequences. The instructions are carried out by laboratory operations on the molecules, and the result is defined as some property of the final set of molecules. DNA computing promises significant linkages between computers and life systems, as well as massively parallel computations. DNA computing can carry out millions of operations at the same time.

The advantages of quantum computing and DNA computing can be achieved by combining these two technologies. These two computing systems can be merged to form two new computing processes, which can be called quantum molecular biology. This quantum molecular biology consists of quantum-DNA computing and DNA-quantum computing. It is already established that the combination of these two is beneficial for a two-valued system. So, it is possible to combine these two processes in a multiple-valued system also. This part will discuss multiple-valued quantum computing, multiple-valued DNA computing, multiple-valued quantum-DNA computing, and multiple-valued DNA-quantum computing.

1

Multiple-Valued Quantum Computing

1.1 Introduction

The ability of a quantum computer to solve traditional NP problems in polynomial time has drawn a lot of attention. Quantum computing originated from the ideas of classical information theory, computer science, mathematics, and quantum physics. Most of the data are uncertain in this world. Those data are presented in a classical computer system where the operations are treated in only two ways – on or off, or zero or one. This property of the classical binary-based system limits the overall performance while analyzing the uncertain world phenomena. However, this predicting game (0's, and 1's) changes completely, with quantum computers. In this quantum realm, the processing and storage of 1's and 0's of classical systems give way to qubits or quantum bits as the fundamental architecture block of quantum information, encountered as a two-state quantum-mechanical system. Quantum computers execute computations based on the likelihood of an object's state before it is measured – instead of just 1's or 0's – which implies they have the potential to process exponentially more innumerable data compared to classical computers.

Although numerous works have been accomplished in quantum computing systems over the years, they were mostly focused on binary logic systems (i.e. they perform computations with binary data). However, the usage of digital data is expanding at an exponential rate; therefore, the size of digital data is skyrocketing by the second. So data manipulation and maintenance become a tough row to hoe. In these circumstances, researchers suggest working with multiple-valued computing systems rather than binary systems because more data can be manipulated with less effort in a multiple-valued system. Designing with multiple-valued logic (MVL) has got a lot of attention in the previous three decades. MVL emerged as a separate study in the early 1920s, thanks to a Polish philosopher named Lukasiewicz. His goal was to add a third value to the binary system. The Lukasiewicz system is the result of this investigation. Emil Post, an American mathematician, invented multiple-valued algebra, sometimes known as post algebra, in response to this technique. The design of multi-valued logic gates, as well as the fundamental operations in quantum computing, is currently in high demand. In the context of quantum computing, this chapter will provide an overview of ternary logic (multiple-valued logic) in quantum computing.

DOI: 10.1201/9781003381938-1

FIGURE 1.1
Danish Physicist Niels Henrik David Bohr

1.2 Quantum Physics and Quantum Computing

The study of matter and energy at the most fundamental level is known as quantum physics. Its goal is to learn more about the properties and behaviors of nature's fundamental building blocks. Put simply, it's the physics that explains how everything works: the best description of the nature of the particles that make up matter and the forces with which they interact. Figure 1.1 shows the picture of the great Danish physicist Niels Henrik David Bohr.

Quantum physics underlies how atoms work, and so why chemistry and biology work as they do. You, me, and the gatepost – at some level at least – all human are dancing to the quantum tune. If it needs to explain how electrons move through a computer chip, how photons of light get turned into electrical current in a solar panel or amplify themselves in a laser, or even just how the sun keeps burning, it'll need to use quantum physics.

In the late 1800s and early 1900s, a sequence of experimental findings of atoms that didn't make intuitive sense in the context of classical physics spawned the discipline of quantum physics. The insight that matters and energy may be thought of as discrete packets, or quanta, with a minimum value associated with them was one of the most fundamental discoveries. Light with a set frequency, for example, will provide energy in quanta known as "photons." At this frequency, each photon has the same amount of energy, and this energy cannot be divided into smaller pieces. The word "quantum" comes from Latin and means "how much." Figure 1.2 shows

FIGURE 1.2
German Physicist Werner Heisenberg

the picture of German physicist Werner Heisenberg who is a scientist of quantum mechanics.

Quantum mechanics, the basic mathematical framework that underpins it all, was first developed in the 1920s by Niels Bohr, Werner Heisenberg, Erwin Schrödinger, as shown in Figure 1.3 and others.

It characterizes simple things such as how the position or momentum of a single particle or group of few particles' changes over time. But to understand how things work in the real world, quantum mechanics must be combined with other elements of physics – principally, Albert Einstein's special theory of relativity, which explains what happens when things move very fast – to create quantum field theories. Figure 1.4 displays the photo of Albert Einstein.

Knowledge of quantum principles transformed the conceptualization of the atom, which consists of a nucleus surrounded by electrons. Early models depicted electrons as particles that orbited the nucleus, much like the way satellites orbit Earth. Modern quantum physics instead understands electrons as being distributed within orbitals, mathematical descriptions that represent the probability of the electrons' existence in more than one location within a given range at any given time. Electrons can jump from one orbital to another as they gain or lose energy, but they cannot be found between orbitals.

Other central concepts that help to establish the foundations of quantum physics are – Wave-Particle duality, Superposition, Uncertainty principle, Entanglement, etc.

FIGURE 1.3
Austrian-Irish Physicist Erwin Schrödinger

FIGURE 1.4
German-Born Theoretical Physicist Albert Einstein

Quantum discoveries have been incorporated into foundational understanding of materials, chemistry, biology, and astronomy. These discoveries are a valuable resource for innovation, giving rise to devices such as lasers and transistors, and enabling real progress on technologies once considered purely speculative, such as quantum computers. Physicists are exploring the potential of quantum science to transform the view of gravity and its connection to space and time. Quantum science may even reveal how everything in the universe (or in multiple universes) is connected to everything else through higher dimensions that our senses cannot comprehend.

1.2.1 Applications of Quantum Physics

Any science concerned with systems that display noticeable quantum-mechanical effects, where waves have particle qualities and particles behave like waves, is referred to as quantum physics. Quantum mechanics has applications in both explaining natural events and designing technology that relies on quantum effects, such as integrated circuits and lasers. Quantum mechanics is also crucial for understanding how covalent bonds connect individual atoms to form molecules. Quantum chemistry is the application of quantum mechanics to chemistry. Quantum mechanics may also demonstrate which molecules are energetically favorable to which others and the magnitudes of the energy involved in ionic and covalent bonding processes.

Modern technology operates on a scale where quantum effects are significant in many ways. Quantum chemistry, quantum optics, quantum computing, superconducting magnets, light-emitting diodes, the optical amplifier and laser, the transistor and semiconductors such as the microprocessor, and medical and research imaging such as magnetic resonance imaging and electron microscopy are all important applications of quantum theory. Many biological and physical phenomena, most notably the macromolecule DNA, have explanations based on the nature of chemical bonds.

Electronics: Quantum mechanics is used in the design of many modern electronic gadgets. Lasers, transistors (and consequently microchips), electron microscopes, and magnetic resonance imaging are all examples (MRI). The diode and transistor, which are essential components of modern electronics systems, computers, and telecommunication equipment, were invented as a result of semiconductor research. Another application is the production of high-efficiency light sources such as laser diodes and light-emitting diodes.

Cryptography: Researchers are currently looking for reliable methods of manipulating quantum states directly. Quantum cryptography is being further developed, which will theoretically allow for guaranteed secure information transmission. The detection of passive eavesdropping is an intrinsic advantage of quantum cryptography over classical cryptography. This is a natural consequence of quantum bit behavior; if a bit in a superposition state is viewed, the superposition state collapses into an eigenstate due to the observer effect. Because the intended recipient expected to receive the bit in a superposition state, the attack would be detected because the bit's state would no longer be in a superposition.

Quantum Computing: The development of quantum computers, which are projected to execute certain computing tasks exponentially faster than classical computers, is another goal. Quantum computers use qubits, which can be in a superposition of states, instead of classical bits. Quantum programmers can exploit the superposition of qubits to tackle issues that traditional computers can't, such as searching unsorted databases or integer factorization. According to IBM, quantum computing could help advance medical, logistics, financial services, artificial intelligence, and cloud security.

Quantum teleportation, which deals with strategies for transmitting quantum information over arbitrary distances, is another hot research area.

Macroscale Quantum Effects: While quantum mechanics is mostly concerned with the lower atomic scales of matter and energy, some systems display quantum mechanical effects on a larger scale. One well-known example is super fluidity, which is defined as the frictionless flow of a liquid at temperatures close to absolute zero. Superconductivity, which is the frictionless movement of an electron gas in a conducting material (an electric current) at sufficiently low temperatures, is also a closely related phenomenon. A topologically ordered state corresponding to patterns of long-range quantum entanglement is the fractional quantum Hall effect. Without a phase transition, states with distinct topological ordering (or different patterns of long-range entanglements) cannot change into one another.

Other Phenomena: Many previously inexplicable phenomena, like black-body radiation and the stability of electron orbitals in atoms, are now accurately described by quantum theory. It has also revealed the inner workings of a variety of biological systems, such as scent receptors and protein architectures. Recent research on photosynthesis has revealed that quantum correlations are crucial in this fundamental mechanism in plants and other creatures. Even Nevertheless, in situations involving huge numbers of particles or big quantum numbers, classical physics can often provide good approximations to conclusions acquired by quantum physics.

Classical approximations are utilized and favored when the system is large enough to render the effects of quantum mechanics unimportant because classical formulas are much simpler and easier to compute than quantum formulations.

1.3   What is Quantum Computing?

Quantum Computing is a new kind of computing based on Quantum mechanics that deals with the physical world that is probabilistic and unpredictable in nature. *Quantum* is the discrete small amount of a physical quantity of energy that can exist independently and that is proportional in magnitude to the frequency of the radiation it represents. It is the least measure of any physical entity that is involved in an interaction. The quantum computer is based on the characteristics of quantum. Richard Feynman in Figure 1.5 and Yuri Manin hypothesized quantum computers in the 1980s. The idea for quantum computing came from what was once considered

FIGURE 1.5
Richard Phillips Feynman

one of physics' greatest embarrassments: amazing scientific progress met with an inability to simulate even simple systems.

Richard Feynman addressed the following conundrum at MIT in 1981: classical computers cannot efficiently model the evolution of quantum systems. As a result, he suggested a fundamental model for a quantum computer capable of doing such simulations. He outlined the possibility of massively outpacing traditional computers with this. However, it took more than a decade for the *Shor algorithm*, a unique method, to shift people's minds about quantum computing.

In 1994, Peter Shor developed his algorithm allowing quantum computers to efficiently factorize large integers exponentially quicker than the best classical algorithm on traditional machines. The latter takes millions of years to factor 300-digit numbers. The Shor algorithm has the potential to break many of today's cryptosystems in theory. Quantum computers' ability to break cryptosystems in hours rather than millions of years sparked interest in quantum computing and its applications. Lov Grover developed a quantum database search technique in 1996 that provided a quadratic speedup for a wide range of problems. Any problem that required a random or brute-force search could now be solved four times faster. The first quantum algorithms, such as Grover's algorithm, were solved with a working 2-qubit (Quantum bits) quantum computer in 1998. The competition to usher in a new era of computing power began, and more and more applications were created. Twenty years later, in 2017, IBM unveiled the first commercially viable quantum computer as shown in Figure 1.6, escalating the competition to a new level.

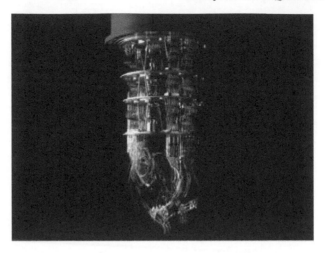

FIGURE 1.6
A Cryostat at Google's Quantum Computing Lab Near Santa Barbara, California
Designed to Keep a Quantum Chip at Temperatures Close to Absolute Zero

1.3.1   Quantum Computing Properties

As classical computers are based on bits, a quantum computer is based on *quantum
bits*, called **qubits** for short. Qubits are physically derived from small quantum objects, such as electrons or photons, where a pure quantum mechanical state such as
the spin indicates the ones and zeros, as is used in classical computers. The behaviors of quantum are the main properties of quantum computing. The key properties
of quantum computers are – superposition, entanglement, and interference.

Quantum Superposition

A qubit's ability to be in *superposition* is one of the features that distinguish it
from a classical bit. One of quantum physics' fundamental principles is superposition. In classical physics, a wave expressing a musical tone can be regarded as a
superposition of many waves with various frequencies. A quantum state in superposition can be thought of as a linear combination of different quantum states. This
quantum state in superposition generates a new quantum state that is legitimate.

The feature of a quantum system is that it exists in several separate quantum states
at the same time. For example, electrons possess a quantum feature called spin, a type
of intrinsic angular momentum. In the presence of a magnetic field, the electron may
exist in two possible spin states, usually referred to as spin up and spin down. Each
electron, until it is measured, will have a finite chance of being in either state. Only
when measured is it observed to be in a specific spin state. In common experience,
a coin facing up has a definite value: it is a head or a tail. In quantum experience,
the situation is more unsettling: material properties of things do not exist until they
are measured. Until "look" (measure the particular property) at the coin, as it were,
it has no fixed face up.

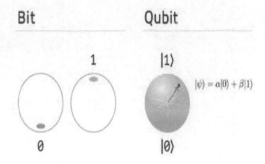

FIGURE 1.7

A Comparison of Classical Bits and Quantum Bits Where Qubit Is in Superposition

Qubits can be in a superposition of both the basis states $|0\rangle$ and $|1\rangle$. When a qubit is measured (to be more precise: only observables can be measured), the qubit will collapse to one of its eigenstates and the measured value will reflect that state. For example, when a qubit is in a superposition state of equal weights, a measurement will make it collapse to one of its two basis states $|0\rangle$ and $|1\rangle$ with an equal probability of 50%. $|0\rangle$ is the state that when measured, and therefore collapsed, will always give the result 0. Similarly, $|1\rangle$ will always convert to 1. Figure 1.7 shows a comparison of classical bits and quantum bits where qubit is in superposition.

Quantum Entanglement

Entanglement is another counter-intuitive phenomenon in quantum physics. When each particle's quantum state cannot be characterized independently of the quantum state of the other particle, the pair or set of particles is said to be entangled. The quantum state of the system as a whole can be characterized; it is in a definite state, even though the individual components are not.

When two qubits are entangled, there exists a special connection between them (Figure 1.8). The entanglement will become clear from the results of the measurements. The outcome of the measurements on the individual qubits could be 0 or 1. However, the outcome of the measurement on one qubit will always be correlated to the measurement on the other qubit. This is always the case, even if the particles are separated from each other by a large distance.

For example, two particles are created in such a way that the total spin of the system is zero. If the spin of one of the particles is measured on a certain axis and found to be counterclockwise, then it is guaranteed that a measurement of the spin of the other particle (along the same axis) will show the spin to be clockwise. This seems strange because it appears that one of the entangled particles "feels" that measurement is performed on the other entangled particle and "knows" what the outcome should be, but this is not the case. This happens, without any information exchange

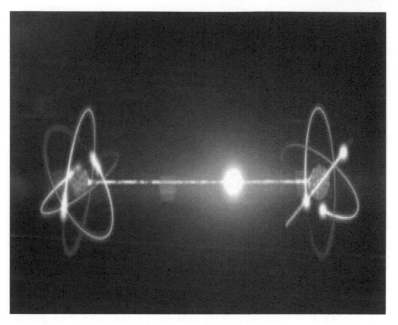

FIGURE 1.8
Quantum Entanglement; Two Quantum Particles with Opposite Spin are Entangled
Together

between the entangled particles. They could even be billions of miles away from each
other and this entanglement would still be present.

Quantum Interference

Quantum interference is the intrinsic behavior of a qubit on account of superposition to influence the likelihood of it collapsing one way or another. Essentially, the
concept states that the elementary particles cannot only be in more than one place
at any given time (through superposition), but that an individual particle, such as a
photon (light particles), can cross its trajectory and interfere with the direction of its
path. Figure 1.9 shows the quantum interference phenomenon identified that occurs
through time.

1.3.2   How Does Quantum Computer Work?

Quantum Computing in the binary logic system is performed using qubits. A qubit is
defined as the basic unit to store information in a quantum computer. The two basic
states of qubits are conventionally written as: |0>, and |1> (it can be pronounced as
'ket 0', and 'ket 1'). A *pure qubit state* is a linear quantum superposition of those
two states. This means that each qubit is represented as a linear combination of $|0\rangle$
and $|1\rangle$, such as:

$$|\Psi\rangle = \alpha|0\rangle + \beta|1\rangle$$

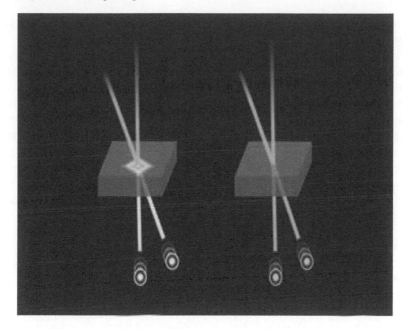

FIGURE 1.9
Quantum Interference Phenomenon Identified That Occurs through Time

where α and β are *complex probability amplitudes*. α and β are constrained by the following equation

$$|\alpha|^2 + |\beta|^2 = 1$$

This implies that the probability that the qubit will be measured in the state $|0\rangle$ is $|\alpha|^2$, and the probability that it will be measured in the state $|1\rangle$ is $|\beta|^2$. Therefore, the total probability of the system being observed in either state $|0\rangle$ or $|1\rangle$ is 1. This can be represented in a 2×2 homogeneous Hilbert space as – $|0\rangle = \begin{bmatrix} 1 \\ 0 \end{bmatrix}$, and $|1\rangle = \begin{bmatrix} 0 \\ 1 \end{bmatrix}$, so, a general state can be written as

$$|\Psi\rangle = \alpha|0\rangle + \beta|1\rangle = \begin{bmatrix} \alpha \\ \beta \end{bmatrix}$$

For two qubits, that can form four qubits pattern $\{|00\rangle, |01\rangle, |10\rangle, |11\rangle\}$, this will be represented as

$$|\Psi\rangle = \begin{bmatrix} \alpha \\ \beta \\ \gamma \\ \varepsilon \end{bmatrix}$$

The above equation is the representation of a general vector in 4-dimensional Hilbert space.

This is significantly different from the state of a classical bit, which can only take the value 0 or 1. Quantum computing holds this phenomenon to gain the power of having multiple states at a time, which is called superposition. Several qubits taken together are a quantum register. Quantum computers perform calculations by manipulating qubits. To perform quantum computing, first, put some individual qubits into superposition. For this, each qubit is either initialized as |0> or |1>. Then mathematically, a transformation is used called *Hadamard transformation*, which rotates a little sphere called the block sphere and rotates it through 90 degrees, so it takes that |0> or |1>, whatever it is, put in a state of superposition. So, if 5 qubits are taken then a system is here that is simultaneously in all (2^5) 32 states. There are also entangled bits which make two different qubits dependent on each other.

1.4 Multiple-Valued Logic

Digital circuits are prevalent in current technology and nearly all of these are based on binary logic. Although vastly used, binary circuits present a few obstacles, and alternatives have been sought, namely, Multiple-Valued Logic (MVL). Many-valued logics are non-classical logics. They are similar to classical logic because they accept the principle of truth-functionality, namely, that the truth of a compound sentence is determined by the truth values of its component sentences (and so remains unaffected when one of its component sentences is replaced by another sentence with the same truth value). But they differ from classical logic by the fundamental fact that they do not restrict the number of truth values to only two: they allow for a larger set **W** of truth degrees.

Multiple-valued logic displays phenomena, it is not possible to see in the binary case, where the only two values available are null and unit elements of Boolean algebra, possessing very specific properties. Reflecting on the two-valued scale, these phenomena give us a new, deeper understanding of the matter. Working with MVL is like painting a picture having all colors available, whereas binary logic can be considered as a painting with only two colors (black and white).

1.4.1 History of Multiple-Valued Logic

Until the beginning of the 20th century, logic was almost exclusively viewed as binary, but mathematicians, logicians, and philosophers needed to represent uncertainty and error, so they started investigating systems that allowed more options than just "true" or "false". In 1920, Polish logician and philosopher Jan Łukasiewicz in Figure 1.10, looking for a third truth value for solving problems and paradoxes, developed one of the earliest MVL systems.

FIGURE 1.10
Jan Łukasiewicz

Meanwhile, American mathematician Emil L. Post introduced in 1921 the basic idea of additional truth degrees with $n \geq 2$, where n is the number of truth values, and applied it to problems related to the representability of functions. After this, several mathematicians, logicians, and philosophers, namely, Godel, Bochvar, and Kleene, in the 1930s, started working with this concept and developed several forms of MVL. Aside from logicians and philosophers trying to create ways of solving their problems, MVL has also been studied, and been an area of great interest, to engineers for several years. Researchers from various areas of engineering have been looking for improvements in technology and performance. In electronic device structure and circuit design, interconnection cost/delay has become a major concern. Higher-radix systems could provide benefits to this problem and also a better relation between circuits and natural representation. Similar advantages can also be found in information and signal processing, discrete system algorithms and data structures, and so on. Figure 1.11, Figure 1.12, and Figure 1.13 show the pictures of three pioneers of the multi-valued logic systems.

1.4.2   Advantages of Multiple-Valued Quantum Computing over Binary Logic

In quantum computation, multi-valued logic (MVL) has various advantages over binary logic, including improved security for quantum cryptography and higher processing capacity for quantum data. A ternary computer (sometimes known as a trinary computer) is a computer that does computations using ternary logic (three possible values) rather than binary logic (two possible values). In ternary quantum computing, the superposition will be formed in the range of |0> to |2>. Ternary computing has many basic benefits over binary computing. These include:

FIGURE 1.11
Pioneer of Multi-Valued Logic Systems Kurt Gödel

FIGURE 1.12
Pioneer of Multi-Valued Logic Systems Stephen Cole Kleene

FIGURE 1.13
Pioneer of Multi-Valued Logic Systems Anatoli Mikhailovich Bochvar

1. Higher data throughput
2. Access to additional instructions
3. Back-compatibility with legacy binary codes
4. Preventing malware and viruses
5. Providing more security

However, the usual aim of ternary computers has not met with overwhelming success thus far, because they are not as efficient as binary computers in computing binary codes, which are widely used and appear to be nearly ubiquitous. Besides, constructing the ternary operational circuits is much more difficult than the binary operational circuit.

1.5 Ternary Logic in Quantum Computing

In binary quantum computing, the basic unit of information is qubits, which have two states: |0> and |1>. The information unit in a three-valued quantum system (ternary quantum system) is termed *qutrit* . The ternary quantum system represents one type of three-dimensional quantum system with the basis states |0>, |1>, and |2>. These basis states are called qutrit states and can be represented by 3×1 vectors:

$$|0> = \begin{bmatrix} 1 \\ 0 \\ 0 \end{bmatrix}, |1> = \begin{bmatrix} 0 \\ 1 \\ 0 \end{bmatrix}, \text{ and } |2> = \begin{bmatrix} 0 \\ 0 \\ 1 \end{bmatrix}$$

In a ternary quantum system, a qutrit can be defined as a linear superposition of the above-mentioned basis states with the following equation:

$$\Psi = \alpha \, |0> + \beta \, |1> + \gamma \, |2>;$$

where α, β, and γ are the complex quantities to represent the probability amplitudes of the basis states and Ψ is the wave function.

1.5.1   Quantum Ternary Fundamental Logic Operations

Quantum ternary logic functions are those functions that have significance if a third value is acquainted with the quantum binary logic. Quantum computing in the ternary logic system is quite interesting. Here, $|0>$, $|1>$, and $|2>$ denote the ternary levels for basic logic gates to represent false, undefined, and true, respectively. The basic operations of quantum ternary logic can be defined as follows:

$$y_{OR} = \max(x, y), y_{NOR} = \overline{\max(x, y)}, y_{AND} = \min(x, y), \ y_{NAND} = \overline{\min(x, y)};$$

and

$$y_{XOR} = \text{sum}(x, y), \quad y_{XNOR} = \overline{\text{sum}(x, y)}, \text{where } x, y = \{|0>, |1>, |2>\}.$$

1.6   Applications of Multiple-Valued Quantum Computing

With the exponential growth in computing power, quantum computing is getting ready for its close-up. Quantum computers are ideally suited to solving complex problems, which are hard for classical computers but easy to factor on a quantum computer. Such an advancement creates a world of opportunities across almost every aspect of modern life. Quantum computing can include practically all applications of quantum physics, computer science, and mathematics because it is a combination of those areas. Besides, multiple-valued quantum computing enhanced the computing ability of quantum computers, which was also mentioned in the previous section. Some common applications of quantum computing are enlisted below.

Artificial Intelligence and Machine Learning: Artificial intelligence and machine learning are some of the prominent areas right now, as the emerging technologies have penetrated almost every aspect of humans' lives. Some of the widespread applications are seen every day in voice, image, and handwriting recognition. However, as the number of applications increased, it becomes a challenging task for traditional computers, to match up the accuracy and speed. And, that's where quantum computing can help in processing complex problems in very less time, which would have taken traditional computers thousands of years.

Computational Chemistry: IBM, once said, one of the most promising quantum computing applications will be in the field of computational chemistry. It is believed that the number of quantum states, even in a tiniest of a molecule, is extremely vast, and therefore difficult for conventional computing memory to process. The ability for quantum computers to focus on the existence of both 1 and 0 simultaneously could provide immense power to the machine to successfully map the molecules which, in turn, potentially opens opportunities for pharmaceutical research. Some of the critical problems that could be solved via quantum computing are — improving the nitrogen-fixation process for creating ammonia-based fertilizer; creating a room-temperature superconductor; removing carbon dioxide for a better climate; and creating solid-state batteries.

Drug Design and Development: Designing and developing a drug is the most challenging problem in quantum computing. Usually, drugs are being developed via the trial and error method, which is not only very expensive but also a risky and challenging task to complete. Researchers believe quantum computing can be an effective way of understanding the drugs and their reactions to humans which, in turn, can save a ton of money and time for drug companies. These advancements in computing could enhance efficiency dramatically, by allowing companies to carry out more drug discoveries to uncover new medical treatments for the better pharmaceutical industry.

Cybersecurity and Cryptography: The online security space currently has been quite vulnerable due to the increasing number of cyber-attacks occurring across the globe, daily. Although companies are establishing necessary security frameworks in their organizations, the process becomes daunting and impractical for classical digital computers. And, therefore, cybersecurity has continued to be an essential concern around the world. With the increasing dependency on digitization, humans are becoming even more vulnerable to these threats. Quantum computing with the help of machine learning can help in developing various techniques to combat these cybersecurity threats. Additionally, quantum computing can help in creating encryption methods, also known as, quantum cryptography.

Financial Modeling: For a finance industry to find the right mix for fruitful investments based on expected returns, the risk associated, and other factors are important to survive in the market. To achieve that, the technique of 'Monte Carlo' simulations is continually being run on conventional computers, which, in turn, consumes an enormous amount of computer time. However, by applying quantum technology to perform these massive and complex calculations, companies can not only improve the quality of the solutions but also reduce the time to develop them. Because financial leaders are in the business of handling billions of dollars, even a tiny improvement in the expected return can be worth a lot for them. Algorithmic trading is another potential application where the machine uses complex algorithms to automatically trigger share dealings by analyzing the market variables, which is an advantage, especially for high-volume transactions.

Logistics Optimization: Improved data analysis and robust modeling will indeed enable a wide range of industries to optimize their logistics and scheduling workflows associated with their supply-chain management. The operating models need to continuously calculate and recalculate optimal routes of traffic management, fleet operations, air traffic control, freight, and distribution, and that could have a severe impact on applications. Usually, to do these tasks, conventional computing is used; however, some of them could turn into more complex for an ideal computing solution, whereas a quantum approach may be able to do it. Two common quantum approaches that can be used to solve such problems are — quantum annealing and universal quantum computers. Quantum annealing is an advanced optimization technique that is expected to surpass traditional computers. In contrast, universal quantum computers are capable of solving all types of computational problems, not yet commercially available.

Weather Forecasting: Currently, the process of analyzing weather conditions by traditional computers can sometimes take longer than the weather itself does to change. But a quantum computer's ability to crunch vast amounts of data, in a short period, could indeed lead to enhancing weather system modeling allowing scientists to predict the changing weather patterns in no time and with excellent accuracy — something which can be essential for the current time when the world is going under a climate change.

Weather forecasting includes several variables to consider, such as air pressure, temperature, and air density, which makes it difficult for it to be predicted accurately. The application of quantum machine learning can help in improving pattern recognition, which, in turn, will make it easier for scientists to predict extreme weather events and potentially save thousands of lives a year. With quantum computers, meteorologists will also be able to generate and analyze more detailed climate models, which will provide greater insight into climate change and ways to mitigate it.

The list could go on and on. With the availability of the quantum computer, it may eventually replace the use of conventional computers in all research centers and computational labs.

1.7 Demerits of Multiple-Valued Quantum Computing

Quantum computers are exceedingly difficult to engineer, build and program. As a result, they are crippled by errors in the form of noise, faults, and loss of quantum coherence, which is crucial to their operation and yet falls apart before any nontrivial program has a chance to run to completion.

This loss of coherence (called *decoherence*), caused by vibrations, temperature fluctuations, electromagnetic waves, and other interactions with the outside environment, ultimately destroys the exotic quantum properties of the computer. Given the current pervasiveness of decoherence and other errors, contemporary quantum computers are unlikely to return correct answers for programs of even modest execution time.

There are certain disadvantages to using the quantum computing system to perform operations. The most typical drawbacks are given below.

Availability: The main disadvantage of Quantum computing is the technology required to implement a quantum computer is not available at present days.

Energy Requirements: The minimum energy requirement for quantum logical operations is five times that of classical computers.

Decoherence: When a measurement of any type is made to a quantum system, decoherence is broken down and the wave function collapses into a single state.

Heating Issue: The fact that a quantum computer's CPU creates tremendous heat from itself is the other major issue, as this problem can induce chaos across the system.

Error Correction Problem: Because qubits or qutrits aren't the same as today's digital bits, they can't be used for error correction in the traditional sense. Error correction of the multiple-valued quantum computing system is much more difficult.

Computer Architecture: A quantum computer's architecture is, in general, quite complicated. The system design becomes substantially more complicated when multiple-valued systems (such as quantum ternary computing systems) are considered. The circuits in multiple-valued quantum computing systems become larger, and as a result, the size of the computer also becomes larger.

High Cost: Cost is another major issue of quantum computing. By most estimates, a single qubit costs around \$10K and needs to be supported by a host of microwave controller electronics, coaxial cabling, and other materials that require large controlled rooms to function. In hardware alone, a useful quantum computer costs tens of billions of dollars to build.

Besides those problems, there are risks. If an advanced quantum computer was created, all the security of the current Internet of Things would collapse like a house of cards. British economist The Economist says quantum computers will be able to do whatever they want in the company. The personal information of billions of Internet users can be taken into their bags. The government database can be hacked. Undue control can be imposed on the banking system. The state defense system can be turned off if desired. Considering these aspects, many do not even hesitate to call it a "terrible" computer.

1.7.1 Challenges in Multiple-Valued Quantum Computing

Quantum computers are currently available on a small scale with a small number of qubits. To be competitive with the classical system, quantum computers need to scale up. Although this issue has improved dramatically over the last decades, further progress in the development of quantum computers requires addressing other technical issues such as quantum error correction. It is fundamentally difficult to scale quantum computer chips because quantum information cannot be copied and subsystems are not independent, leading to design trade-offs that are global by nature. However, superconducting qubits are more and more showing the potential to overcome these hurdles and are starting to evolve as commercially usable in quantum computers. The critical challenges of this are described below.

Quality of Qubit: There are several challenges in building a large-scale quantum computer such as fabrication, verification, and architecture. The power of quantum computing comes from storing a complex state in a single bit. This complex state makes quantum systems challenging to build, verify, and design. The first challenge is to make qubits effectively that will generate valuable instructions. The famous quantum computing algorithm is Shor's algorithm that can quickly factor a large number. He discovered a quantum error-correcting code that can store the information of one qubit onto a highly entangled state of nine qubits.

Qubit Maintenance: Multiple qubits are needed to be controlled, including error detection schemes applying complex algorithms. That control must have low quiescence on the order of 10's, and it has to come from CMOS-grounded adaptive feedback control circuits.

Numerous Cables/Wires: Another challenge is to scale up the number of qubits within a quantum chip. Today multiple control wires or cables are required to create each qubit. These too many wires create anarchy within a computer system which is needed to be organized properly.

Moreover, there are more challenges such as algorithm creation, and internet security problems in quantum computing each time it needs to write a new algorithm. They cannot work as a classical computer. They need unique algorithms to perform tasks in their environment, which is tough to implement each time.

Another challenge is that they need a negative temperature of 460 degrees which is a shallow temperature and difficult to maintain. If a quantum computer is implemented in the best way, then complete internet security breaks because these computers are good at decrypting all the codes on the internet.

1.8   Summary

In this chapter, some basic information has been shown about multiple-valued quantum computing. This chapter only focuses on the fundamental concept. In the first section, some basic information and the history of this quantum physics are discussed. Quantum computation promises the ability to compute solutions to any problem. Though it has potential advantages, it is still underdeveloped. Therefore, researchers should emphasize the quantum field and work to reduce disadvantages. This chapter covers the advantages, disadvantages, and challenges of multiple-valued quantum computing. After reading this chapter, the reader must be able to achieve the fundamental concept and the background of this field. In the near future, the quantum computer will be accessible for all people.

Bibliography

[1] Dubrova, E. (1999, November). Multiple-valued logic in VLSI: challenges and opportunities. In Proceedings of NORCHIP (Vol. 99, No. 1999, pp. 340-350).

[2] Haghparast, M., Wille, R., & Monfared, A. T. (2017). Towards quantum reversible ternary coded decimal adder. Quantum Information Processing, 16(11), 1-25.

[3] Mandal, S. B., Chakrabarti, A., & Sur-Kolay, S. (2011, May). Synthesis techniques for ternary quantum logic. In 2011 41st IEEE International Symposium on Multiple-Valued Logic (pp. 218-223). IEEE.

[4] Anderson, M. (2009). Is quantum mechanics controlling your thoughts? Discover Magazine.

[5] Chen, X., Gu, Z. C., & Wen, X. G. (2010). Local unitary transformation, long-range quantum entanglement, wave function renormalization, and topological order. Physical Review B, 82(15), 155138.

[6] Marella, S. T., & Parisa, H. S. K. Introduction to Quantum Computing.

[7] Perkowski, M., Al-Rabadi, A., & Kerttopf, P. (2002). Multiple-Valued Quantum Logic Synthesis.

2

Multiple-Valued DNA Computing

2.1 Introduction

Molecular computing, or DNA (Deoxyribose Nucleic Acid) computing, is a new discipline of computer science. It's a novel method for massively parallel computation that can tackle NP-complete or non-deterministic polynomial-time complete problems in a fraction of the time. Combinatorial issues are another area where DNA computing excels. It's a great mix of biochemistry, molecular biology, and computer science that allows researchers to carry out arithmetic and logic operations. It does computations utilizing biological molecules rather than typical silicon processors.

Every cell in a living organism has information for a variety of functions that are necessary for the cell's survival. In each cell, nucleic acids are molecules that contain genetic information. The most stable nucleic acid is deoxyribonucleic acid (DNA). Helixes are long polymers made up of millions of nucleotides joined together that form from each DNA strand. These nucleotides are made up of one of four nitrogen bases, a five-carbon sugar, and a phosphate group.

2.2 What is DNA Computing?

The basic idea behind DNA computing is to represent data using a biological (wet) technique, with DNA strands serving as an efficient computing vehicle. Even though the cycle time of a DNA reaction is far slower than that of a silicon-based computer, the DNA process' intrinsic parallel processing is critical. This massive parallelism of DNA processing is very important for solving NP-complete or NP-hard issues. Figure 2.1 shows the DNA base pairs in double helix model.

The genetic information is encoded by the nitrogen bases A (Adenine), T (Thymine), G (Guanine), and C (Cytosine), while the others offer structural stability. T with A and C with G are the base-pairing rules that connect the strands. The order in which these nucleotides are arranged is crucial since it determines how different genes function.

DOI: 10.1201/9781003381938-2

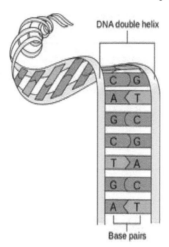

FIGURE 2.1
DNA Base Pairs in Double Helix Model

2.2.1 History of DNA Computing

The concept of DNA computing was introduced in 1994 by USC Professor, Leonard Adleman in Figure 2.2, in the November 1994 Science article, Molecular Computations of Solutions to Combinatorial Problems. Adleman demonstrated that DNA might be utilized to store data and even do massively parallel computations. Adleman encoded a classic "hard" problem (one that exhibits exponential growth with each additional input parameter) known as the Traveling Salesman Problem into strands of DNA and used biological properties of DNA to find the answer using the four bases of DNA (adenine, thymine, cytosine, and guanine). When Adleman realized how DNA replication was eerily similar to an early theoretical computer established by Alan Turing in the 1930s, he came up with the idea of DNA computing.

During replication, DNA polymerase slides down a single DNA strand, reading each base and writing its complement on the next strand, whereas in one version of the Turing Machine, a mechanism moved along a pair of tapes, reading instructions from one and recording the result on the other. Alan Turing's modest machine, interestingly, was shown to have the same computing capabilities as any current computer. Adleman now began to wonder: if Turing's simple machine has such great computational ability, would similarly operating DNA also can do computations? It did, as Adleman's first experiment proved. Even though his experiment required a lot of slow, manual labor to separate the correct answers, had a high chance of error, and was unscalable for larger problems, DNA computing promised enormously high-density storage, unparalleled energy efficiency, and a level of parallelism that digital computers couldn't match. There was the birth of a new field.

FIGURE 2.2
American Computer Scientist Leonard Adleman

2.3 Some Related Terminologies

There are a few things to understand before understanding the working methodology of DNA computing.

2.3.1 Molecular Biology

Molecular biology is a branch of biology that studies the chemical structures and processes of biological phenomena involving molecules, the basic units of life. The study of molecular biology is concerned with nucleic acids (such as DNA and RNA) and proteins—macromolecules that are critical to biological processes—and how they interact and behave within cells. In the 1930s, molecular biology emerged from the associated disciplines of biochemistry, genetics, and biophysics, and it is still strongly tied with those fields today.

For molecular biology , various methodologies have been established, while researchers in the discipline may also use procedures and techniques that are native to genetics and other closely related fields. Molecular biology, in particular, uses techniques like *X-ray diffraction* and *electron microscopy* to better understand the three-dimensional structure of biological macromolecules. Molecular biologists study the molecular basis of genetic processes, mapping the location of genes on specific chromosomes, associating these genes with specific characteristics of an organism, and isolating, sequencing, and modifying specific genes using genetic engineering

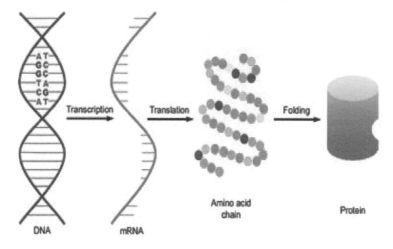

FIGURE 2.3
The Central Dogma of Molecular Biology; the Genetic Information Encoded in the DNA is First Transcribed into the mRNA, Which is Then Translated into the Protein

(recombinant DNA technology). Techniques including *Polymerase Chain Reaction* (PCR), *western blotting, and microarray analysis* can be used in these approaches. Figure 2.3 shows the central dogma of molecular biology.

The field of molecular biology began in the 1940s intending to reveal the basic three-dimensional structure of proteins. In the early 1950s, as knowledge of protein structure grew, the structure of deoxyribonucleic acid (DNA), the genetic blueprint found in all living things, as described in 1953. With more investigation, scientists were able to learn more about not only DNA and ribonucleic acid (RNA), but also the chemical sequences within these substances that tell cells and viruses how to build proteins. Molecular biology remained a pure science with few practical applications until the 1970s when enzymes that could cut and reassemble portions of DNA in bacteria's chromosomes were found. Because it allows manipulation of the genetic sequences that determine the basic features of organisms, recombinant DNA technology became one of the most active disciplines of molecular biology.

2.3.2   DNase Enzyme

Deoxyribonuclease (DNase) is an enzyme that breaks up extracellular DNA found in the purulent sputum during respiratory infections. DNase degrades DNA by catalyzing the hydrolytic breakage of phosphodiester links in the DNA backbone. Nucleases are enzymes that hydrolyze phosphodiester linkages between nucleotides. Deoxyribonucleases are one form of nuclease. There are several different types of deoxyribonucleases, each with its substrate specificities, chemical processes, and biological activities. Deoxyribonuclease (DNase) enzymes perform a variety of important cellular roles by degrading DNA via hydrolysis of its phosphodiester backbone.

Deoxyribonuclease I (DNase I) enzymes cleave single or double-stranded DNA and require divalent metal ions to hydrolyze DNA yielding 3-hydroxyl and 5-phosphorylated products.

In DNA computing, the DNase I enzyme plays a very significant. The base pair bond can be detected using the DNase enzyme role (which will be explained in the next section). Besides, DNase I is used in a range of molecular biology applications. Some of its uses include:

1. Degradation of contaminating DNA after RNA isolation,

2. "Clean-up" of RNA prior to RT-PCR and after in vitro transcription,

3. Identification of protein binding sequences on DNA (DNase I foot printing),

4. Prevention of clumping when handling cultured cells, and

5. Creation of a fragmented library of DNA sequences for in vitro recombination reactions.

2.3.3 Fluorescence Detection

The term *fluorescence* refers to a type of luminescence. When a substance is irradiated with light of a specific wavelength (excitation wavelength), the substance emits light with a longer wavelength (emission wavelength), which is referred to as *fluorescence*. After a molecule absorbs energy in the form of light, it fluoresces. When a photon is absorbed by a fluorescent molecule, it promotes the molecule to an excited electronic state. It subsequently emits a lower-energy photon and decays to the ground (unexcited) state. Because the molecule produces heat in addition to the photon, the released photon is always lower in energy than the absorbed light. The wavelength of emitted radiation is always longer than the wavelength of the excitation light because the wavelength of radiation is inversely proportional to the energy.

Fluorescence detection is generally used for analysis when sensitivity and selectivity are required, especially when the analyte has little or no UV absorbance and can be derivatized to produce fluorescence . The light released from the sample owing to fluorescence is separated into a spectrum by a monochromator, and the intensity of light at a specific emission wavelength is measured, unlike an ultraviolet (UV) detector, which detects the amount of light absorbed by a substance at a specific wavelength. The following are the characteristics of fluorescence detection:

1. Selectivity is high because the measurement is conducted using specific excitation and emission wavelengths specific to the target substance.

2. Fluorescent substances can be detected with high sensitivity.

3. Sensitivity and selectivity can be improved by using derivatization techniques.

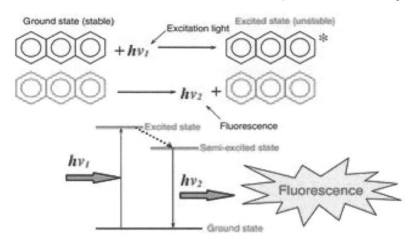

FIGURE 2.4
Mechanism of Fluorescence from a Substance

Principle of Fluorescence

Fluorescent substances follow the following fluorescence principle. When a substance is irradiated with excitation light, it enters a state of excitation. The substance swiftly returns to its original condition due to the instability of this excited state (ground state). Fluorescence is the energy emitted at this moment, and a light detector assesses the intensity of the emitted fluorescence. Selectivity is higher with fluorescence detectors than with absorbency and other types of detectors since excitation and measurement are both done at a specific wavelength. The mechanism of fluorescence from a substance is shown in Figure 2.4.

Applications of Fluorescence Detection

In Multiple-valued DNA computing, fluorescence detection is a must to detect the DNA base pair construction. In general, fluorescence detection has numerous applications in both the research and industrial sector. Some of them are given here.

1. High-Sensitivity Analysis of Enrofloxacin,

2. Conventional and Ultra-Fast Analysis of Anionic Surfactant (LAS),

3. Simultaneous Analysis of Poly-cyclic Aromatic Hydrocarbons (PAHs),

4. Amino Acid Analysis,

5. Repeatability in Carbamate Analysis,

6. Analysis of Reducing Sugars, and so on.

2.4   How Does a DNA Computing Work?

A computation may be thought of as the execution of an algorithm, which itself may be defined as a step-by-step list of well-defined instructions that takes some input, processes it, and produces a result. In DNA computing (Figure 2.5), information is represented using the four-character genetic alphabet (**A** [adenine], **G** [guanine], **C** [cytosine], and **T** [thymine]), rather than the binary alphabet (1 and 0) used by traditional computers. This is achievable because short DNA molecules of any arbitrary sequence may be synthesized to order. An algorithm's input is therefore represented (in the simplest case) by DNA molecules with specific sequences, the instructions are carried out by laboratory operations on the molecules (such as sorting them according to length or chopping strands containing a certain subsequence), and the result is defined as some property of the final set of molecules (such as the presence or absence of a specific sequence).

DNA has cutting, copying, pasting, repairing, and many other operations, just like a CPU has addition, bit-shifting, logical operators (AND, OR, NOT, NOR), and so on which will be explained in Chapter 4 that allows it to accomplish even the most complex computations. The appropriate sequences are sorted out using genetic engineering methods in a DNA computer, which computes in test tubes or on a glass slide coated in 24K gold.

What Adleman was able to demonstrate is that DNA can be assembled in such a way that a test tube full of DNA blocks could assemble themselves to encode all of the possible paths in the traveling salesman problem at the same time.

In DNA, genetic coding is represented by four different molecules, called A, T, C, and G. These four "bits", when chained together, can hold an incredible amount

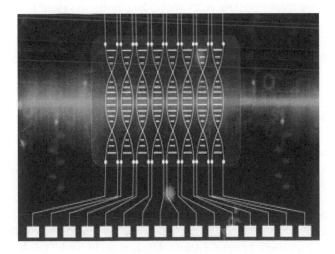

FIGURE 2.5
DNA Computing [Source: Dipositpohotos]

of data. After all, the human genome is encoded in something that can be packed into a single nucleus of a cell.

By mixing these four molecules into a test tube, the molecules naturally assembled themselves into strands of DNA. If some combination of these molecules represents a city and a flight path, each strand of DNA could represent a different flight path for the salesman, all being calculated at once in the synthesis of the DNA strands assembling themselves in parallel.

Then, it would simply be a matter of filtering out the longer paths until only the shortest path left. In his paper, he showed how this could be done with 7 cities and the solution to the problem would be encoded as soon as the DNA strands were synthesized. DNA structures are inexpensive, relatively simple to make, and scalable, which sparked an interest. DNA computing has no theoretical limit because its power grows as more molecules are added to the equation, and unlike silicon transistors, which can only perform a single logical operation at a time, these DNA structures can theoretically perform as many calculations as needed to solve a problem all at once.

However, there is a speed issue. Although Adleman's answer to the traveling salesman's problem was encoded into his DNA strands in the test tube in a matter of seconds, it took days of weeding out faulty options before he found the ideal solution he was seeking—after rigorous preparation for this single computation. Despite this, the principle was valid, and the potential for massive increases in storage capacity and processing speed was clear. This sparked a two-decade effort to figure out how to make practical DNA computing a reality.

2.4.1   DNA Computing for Binary Logic System

Although this book is aimed at the multiple-valued computing system, it is necessary to understand the binary operations to understand multiple-valued DNA computing. Because the working procedures are much identical except that – DNA computing for binary system utilizes the capability of DNase polymerase enzyme to detect the DNA base bond pairs whereas computing multiple-valued DNA computing utilizes fluorescence detection techniques to detect the DNA base bond pairs.

Figure 2.6 shows a typical OR and XOR operation in DNA Computing in a test tube with the help of DNase enzyme – Deoxyribonuclease I and Deoxyribonuclease II, base sequence, and a temperature which is approximately in 60° Celsius for DNA-OR and more than 60° Celsius for DNA-XOR operation. Here, the base sequence ACCTAG and TGGATC are equivalent to binary values 1 and 0, respectively.

For DNA-OR operation, the base sequence in the test tube is TGGATC. Now if the input sequences are TGGATC and ACCTAG, one base pair bond will be created and one sequence (TGGATC for this case) will remain in the test tube. DNase enzyme will detect the remaining sequences which did not create the bond, if found any it will destroy the remained sequences. If the final mixture contains any bond pair then the output will be ACCTAG (equivalent to 1), and TGGATC otherwise. Therefore, for this case, the output will be ACCTAG. Similarly, for each input pattern, the expected output can be found.

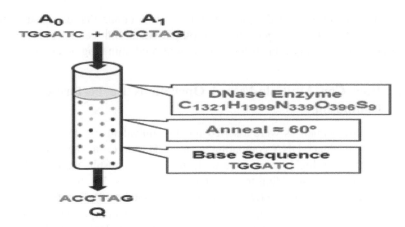

FIGURE 2.6
Performing DNA Operations in Test Tubes: DNA OR Operation

For DNA-XOR operation, there is no base sequence in the test tube. Therefore, if the input pattern contains ACCTAG and TGGATC it will produce output ACCTAG (as the bond will be created). Otherwise, the output will be TGGATC. Similarly, all the fundamental binary operations can be performed in DNA computing.

2.5 Multiple-Valued DNA Computing

Multiple-Valued Logic (MVL) can be used in logic circuits to minimize the number of logic elements and interconnections that connect different areas of the circuit. Delays, area, and energy usage can all be minimized by reducing interconnections. With the development of biological computing and quantum computing, the dominance of binary logic is challenged and multi-valued logic gates attract intense interest in research. Boolean binary logic has only two states that are true and false, or on and off. However, quite often, the logic gates suffer a certain degree of uncertainty and imprecision. In these cases, it is difficult to process information based on binary logic. Multi-Valued logic is likely to play an increasingly important role in the conception and design of a molecular computer since it would be able to deal with uncertain information. Multi-Valued logic is defined as a nonbinary logic and involves the switching between more than two states. The number of distinguishable states increases, which allows for higher information densities. The common application of a multi-valued variable is coding. The Morse code, which uses three different symbols, is probably the best known. The more different symbols there are, the shorter the code words.

A ternary or three-valued logic function has two inputs that can assume three states (say 0, 1, and 2) and generates one output signal that can have one of these three states. Thus ternary logic is the extension of predominating binary logic.

2.5.1 How Can Multiple-Valued Operations Be Performed in DNA Computing?

Two DNA base sequences were asserted for the equivalent two binary values zero and one in the binary DNA computing system. The digits in the multiple-valued system (i.e. ternary logic system) are 0, 1, and 2. As a result, three DNA strands must be created corresponding to the three ternary digits. For the multiple-valued DNA computing, consider the DNA sequences ACCTAG = '0', CAAGCT = '1', and TGGATC = '2'. The operations are performed in DNA computing by conducting chemical reactions. And in binary logic, the bond between DNA strands to perform operations were detected using the DNase Enzyme (if the bond was not created the DNase enzyme would destroy the base sequences).

The working principle of multi-valued DNA computing is easier to comprehend in the ternary system, but the underlying architecture is complex. The *fluorescence level* is utilized to detect the DNA sequence in Ternary DNA computing. Fluorescence is defined as fluorescent molecules temporarily absorbing electromagnetic wavelengths from the visible light spectrum and then emitting light at a lower energy level. Biofluorescence is the term used when it occurs in a live thing. The light that is emitted is a different color than the light that is absorbed as a result of this. An electron is excited by stimulating light, causing its energy to rise to an unstable level. The fluorescence level will be different for each operation, and it will affect the operation's output.

DNA ternary logic functions are those functions that have significance if a third value is acquainted with the DNA binary logic. Here, ACCTAG, CAAGCT, and TGGATC will represent 0, 1, and 2 in ternary logic, which denote the ternary levels for basic logic operations to represent false, undefined, and true, respectively. The basic operations of DNA ternary logic can be defined as follows:

$$y_{OR} = \max(x, y);$$
$$y_{NOR} = \overline{\max(x, y)};$$
$$y_{AND} = \min(x, y)$$
$$y_{NAND} = \overline{\min(x, y)};$$
$$y_{XOR} = \operatorname{sum}(x, y); \text{ and}$$
$$y_{XNOR} = \overline{\operatorname{sum}(x, y)},$$
$$\text{where } x, y = \{0, \ 1, \ 2\}$$

But in this case, 0, 1, and 2 are used for simplicity. The actual values are the DNA strands which are ACCTAG, CAAGCT, and TGGATC as mentioned before.

The following truth table as shown in Table 2.1 shows the input and output mapping of the fundamental logic operations in the ternary logic system.

The output value of the ternary AND logic operation is determined by the minimum value of its inputs. Similarly, the output value of the ternary OR logic operation

TABLE 2.1

Truth Table for Ternary AND, NAND, OR, NOR, XOR, XNOR

Input 1	Input 2	AND	NAND	OR	NOR	XOR	XNOR
0	0	0	2	0	2	0	2
0	1	0	2	1	1	1	1
0	2	0	2	2	0	2	0
1	0	0	2	1	1	1	1
1	1	1	1	1	1	2	0
1	2	1	1	2	0	0	2
2	0	0	2	2	0	2	0
2	1	1	1	2	0	0	2
2	2	2	0	2	0	1	1

is determined by the maximum value of the inputs. The output value of the ternary XOR operation is the sum of its input values. As a result, the outputs of DNA Ternary NAND, NOR, and XNOR logic operations are the quantum ternary AND, OR, and XNOR logic operations reversed. Figure 2.8 shows the circuit architecture of the DNA ternary XOR operation as an example.

DNA ternary XOR operation as shown in Figure 2.8 is defined as $Y_{DTXOR} = $ sum(X, Y), where X and Y are the input from {0, 1, 2}. In Figure 2.7, two input sequences will mix into the test tube and the fluorescence level will produce the output based on the input sequence. The annealing temperature is more than 60°C for the DNA ternary XOR operations. Table 2.2 shows the truth table of DNA ternary XOR operations.

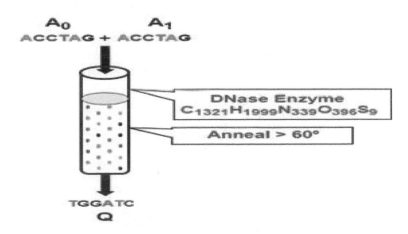

FIGURE 2.7

Performing DNA Operations in Test Tubes: DNA XOR Operation

TABLE 2.2
Truth Table of DNA Ternary XOR Operations

	ACCTAG	CAAGCT	TGGATC
ACCTAG	ACCTAG	CAAGCT	TGGATC
CAAGCT	CAAGCT	TGGATC	ACCTAG
TGGATC	TGGATC	ACCTAG	CAAGCT

The above circuit will produce output according to the above truth table with the help of fluorescent detection. Other circuits can be constructed in the same way.

2.6   Advantages of Multiple-Valued DNA Computing

The DNA computer has clear advantages over conventional computers when applied to problems that can be divided into separate, non-sequential tasks. The reason is that DNA strands can hold so much data in memory and conduct multiple operations at once, thus solving decomposable problems much faster. On the other hand, non-decomposable problems, those that require many sequential operations are much more efficient on a conventional computer due to the length of time required to conduct the biochemical operations.

1. **Parallel Processing:** DNA computers' massively parallel processing capabilities have the potential to speed up huge, but otherwise solvable, polynomial-time tasks requiring only a few operations. For instance, a mix of 1,018 strands of DNA could operate at 10,000 times the speed of today's advanced supercomputers.

2. **Performance Rate:** Performing millions of operations simultaneously allows the performance rate of DNA strands to increase exponentially.

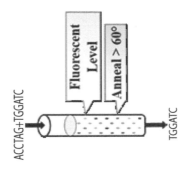

FIGURE 2.8
Multiple-Valued DNA XOR Operation

Adleman's experiment was executed at 1,014 operations per second, a rate of 100 Teraflops (100 trillion floating-point operations per second). The world's fastest supercomputer runs at just 35.8 Teraflops.

3. **Ability to hold tremendous amounts of info in very small spaces:** DNA has an incredible amount of info density. A strand of DNA is encoded with the four bases (nucleotides) – as it is known already, which are represented by the letters A, T, C, and G, in the same way, that binary data is encoded with 1's and 0's. DNA has a phenomenal data density of about 18 Mbits per inch since these bases are separated every 0.35 nanometers along the molecule. In two dimensions, the data density is over one million Gbits per square inch if one base per square nanometer is assumed. When compared to a normal high-performance hard drive, which has a data density of roughly 7 Gbits per square inch, this is a factor of over 100,000 smaller.

4. **Low Power Consumption:** DNA computers can perform 2×10^{19} irreversible operations per joule, whereas supercomputers can execute a maximum of 109 operations per joule. DNA computers need power only to prevent DNA from denaturation. Energy is going to be very valuable shortly. So, it's one of the best advantages.

2.7 Limitations of Multiple-Valued DNA Computing

Overall, many technical challenges remain before DNA computing can be widely used. New techniques must be developed to reduce the number of computational errors produced by unwanted chemical reactions with the DNA strands, and steps in processing DNA need to be eliminated, combined, or accelerated.

1. **Requires exponential resource in terms of memory:** Generating solution sets, even for some relatively simple problems, may require impractically large amounts of memory (Adams). Although DNA can store a trillion times more information than current storage media, how the information is processed necessitates a massive amount of DNA if larger-scale problems are to be solved.

2. **Accuracy:** DNA synthesis is liable to errors, such as mismatching pairs, and is highly dependent on the accuracy of the enzymes involved. In addition, the chance of errors increases exponentially, limiting the number of operations to do successively before the probability becomes greater than producing the correct result.

3. **Resource-intensive:** i) Each stage of parallel operations requires time measured in hours or days, with extensive human or mechanical intervention between steps; ii) Since a set of DNA strands is tailored to a specific

problem, a new set would have to be made for each new problem; and iii) Algorithms can be executed in polynomial time due to the massive parallelism inherent in DNA computation, but they are limited in applicability to small instances of these problems because they require the generation of an unrestricted solution space. For example, the DNA encoding of all paths of a Traveling Salesman problem with 200 cities would weigh more than the earth.

4. **Not Programmable:** DNA computers could not replace traditional computers as they are not programmable, and the average dunce cannot sit down at a familiar keyboard and get to work.

2.8   Applications of Multiple-Valued DNA Computing

DNA computing has been applied to various fields, including nanotechnology, combinatorial optimization, Boolean circuit development, and of particular relevance to the present section, scheduling. Computation using DNA molecules piques the interest of computer and biology researchers alike. DNA computing technologies are used to address various combinatorial issues due to their high parallelism and storage density. However, because the exponential expansion of the solution space makes it impossible to apply an exhaustive search strategy to problem cases of practical magnitude, artificial intelligence models are utilized to develop more efficient ways. DNA has also been investigated as good material and a fundamental building component for manufacturing large-scale nanostructures, individual Nanomechanical devices, and calculations.

In 1994, Leonard Adleman proposed the first DNA computation theory. He put his experimental theory to the test with a seven-point Hamiltonian path problem, also known as the traveling salesman problem. The salesman must find the shortest route between seven cities with known distances, passing through none of them twice and returning to the starting point. Even though his DNA computing solution to this problem was not flawless, this demonstration opened up a world of possibilities and applications. The following are a handful of the applications that are currently being developed.

1. **Security:** The use of DNA algorithms in cryptography to build an intrusion detection model is the most recent advancement. With the ability to store 108 terabytes of data in a single gram of DNA, a gigantic one-time pad may be created. Another example is DNA steganography, which uses a new method to hide messages in microdots. Each letter was represented by three chemical bases rather than binary encoding, such as CGA for the letter "A." These signals are then encoded into DNA sequences and concealed in a tube with a large amount of sonicated random human DNA.

2. **Scheduling:** To solve the work scheduling challenge, Zhixing et al proposed a DNA computing-based approach. To demonstrate the model with

six tasks, he depicted the working processes, replicating the technique used for the Hamiltonian Path issue. In the early 2000s, Watada used DNA algorithms to devise elevator schedule systems and restructure the Flexible Manufacturing System. However, due to a lack of theoretical underpinning, only medium-sized tasks were considered.

3. **Empirical Research:** Multiple-valued DNA computing has become one of the most emerging fields of research in nanotechnology.

4. **Clustering:** Clustering is the process of constructing a structure from a complicated collection of data to derive highly meaningful relationships. It involves a variety of concepts and methods. Edges and vertices are assigned using strands in DNA-based clustering. To increase quality, iterative calculations are conducted for each created cluster. When dealing with vast amounts of heterogeneous data with an uncertain number of clusters, this strategy is particularly useful. Due to DNA's great parallelism, it aids in lowering time complexity.

5. **Medicine:** Many areas can benefit from DNA computing, but one particularly important one is medicine. Currently in development is a DNA computer that operates within human cells. As mentioned, the hope is that this technology will eventually allow for the DNA computer to select diseased cells and then exclusively treat diseased cells while leaving healthy cells intact. Using a mechanism known as RNA interference, little molecules of RNA stop a gene from creating protein.

Apart from that, DNA computing has been used in a variety of sectors, such as nanotechnology, combinatorial optimization, Boolean circuit construction, genetic programming, pharmaceutical applications, and so on.

2.9 Challenges in Multiple-Valued DNA Computing

To work with a third state value is always difficult. More resources are necessary to establish the multiple-valued system. Besides, DNA computing itself has some challenges in implementing the computing system.

Unfortunately, there are numerous obstacles in the way of achieving the goal of multiple-valued DNA computing. The problem for scientists is to figure out how to make a molecular computer capable of handling complex yet critical decision-making processes. Although DNA computers have been created to play a rudimentary game of tic-tac-toe or solve relatively simple logical tasks, they are still a long way from being able to work in human cells. It's also far too simple to apply to the more complex computations and applications required for decision-making. While DNA's potential exceeds current computing capabilities, putting that promise into practical applications is still a complex and challenging task – one that is at best many years away.

Another issue with the creation and use of DNA computing is that researchers have spent many years trying to figure out how to make DNA in such a way that it can solve issues as well as the existing silicon-based computers. The nano-sized particles in DNA make this a difficult problem to solve. Furthermore, DNA computing is still slower than its silicon-chip competitors, but if its full potential is realized, it will outperform traditional computing in terms of speed and economy.

DNA computing is a fascinating notion, straddling the line between science and biology. It has not only proven to be a promising technology for data analysis, but it has also demonstrated the potential and power to transport data in nanotechnology and other intriguing applications. DNA computing will presumably overcome its current hurdles with sustained study and development, paving the way for efficient and successful computer applications in a wide range of sectors.

2.10 Motivations toward Multiple-Valued DNA Computing

The field of DNA computing is an emerging concept still in its infancy and its applications are still being understood. DNA computing can be harnessed to act along with the living cells to provide new detection methods in medical devices. With the flexible molecular algorithms on the rise, one might be able to assemble a complex entity on the nanoscale with the reprogrammable tileset. Though replacing silicon chips-based computers seems highly unlikely shortly, the concept of solving problems beyond the scope of conventional computers gives rise to unfathomable applications.

According to American engineer Gordan Moore, the number of transistors per silicon increases every two years, while computer costs are cut in half. As a result, the scale is shrinking and the performance is increasing at a rapid pace. After this researcher began to consider other possibilities, they began looking for new ways to analyze data. Finally, they discovered several fascinating new directions in atypical computing, including quantum computing and molecular computing research. Molecule computing is a branch of computing that makes use of DNA, molecular biology hardware, and DNA in general. It is feasible to cram far more circuitry onto a microchip utilizing molecular computing rather than silicon-based molecular computing. They are only a few nanometers in size, allowing for the production of devices with billions and trillions of switches and components. Boolean logic gates, memory units, and arithmetic functions have all been implemented in synthetic molecular systems. Although these systems can do fundamental Boolean operations and simple computations, their complexity is limited, making them difficult to use.

Furthermore, these systems remain far off from natural information science in cells. If a biocomputer could be built, it would be useful for a variety of applications, including tackling complicated combinatorial problems faster than typical silicon-based computers due to parallel processing. In general, researchers have been inspired to pursue novel computing applications. This field's study leads to new sensing and switchable materials controlled by bioelectronics devices, process signals regulated by external signals, and signal controlled-release processes

2.11 Summary

This chapter mainly focuses on the introduction of DNA computing, its pros and cons, challenges, and motivation for multiple-valued DNA computing. It is now the most exciting area to be explored by researchers. There are many opportunities for expanding and manipulating DNA characteristics, and it is capable of solving real applications, mainly industrial engineering and management engineering problems. If it is possible to implement a DNA computer perfectly, it will become an alternative way to solve the difficulties faced by current silicon computers. Although DNA computing is a more popular and understood topic, it is still theoretical. There are still some obstacles that are yet to be solved for building up a DNA computer. It is difficult to predict what directions the researchers will follow and what applications will be more efficient for DNA computing. Its computing model always depends on one molecular technique to solve various problems. Therefore, it is tough to implement and cannot construct actual intimidation to cryptography security.

Bibliography

[1] Adleman, L. M. (1994). Molecular computation of solutions to combinatorial problems. Science, 266(5187), 1021-1024.

[2] Yin, Z., Yang, J., Yang, Y., & Ma, Y. (2007). DNA Computing Model of the 0-1 Programming Problem. International Journal of Algebra, 1(2), 71-79.

[3] Gangadharan, S., & Raman, K. (2021). The art of molecular computing: whence and whither. BioEssays, 43(8), 2100051.

[4] Levesque, J., & Wagenbreth, G. (2010). High performance computing: programming and applications. CRC Press.

[5] Ten Berge, M., Brinkhorst, G., Kroon, A. A., & de Jongste, J. C. (1999). DNase treatment in primary ciliary dyskinesia: assessment by nocturnal pulse oximetry. Pediatric Pulmonology, 27(1), 59-61.

[6] LEE, Y. K., Lu, H., & Powers, J. M. (2005). Fluorescence of layered resin composites. Journal of Esthetic and Restorative Dentistry, 17(2), 93-100.

[7] Watada, J., & binti abu Bakar, R. (2008, November). DNA computing and its applications. In 2008 Eighth International Conference on Intelligent Systems Design and Applications (Vol. 2, pp. 288-294). IEEE.

3

Multiple-Valued Quantum-DNA Computing

3.1 Introduction

Quantum and DNA computing are both distributed and parallel types of computing. They're useful for tasks that require high-complexity computations and/or large data sets, such as searching, sorting, merging, pattern recognition, image processing, and encryption. Quantum and DNA algorithms cannot be efficiently simulated on classical computers because they are incapable of coping with parallelism.

Quantum Computing's superpower is the ability to solve complex problems at super high speed, and a DNA computer's superpower is its high speed along with a massive storage system in a very small amount of DNA molecule. Both DNA and quantum computers have the potential to outperform traditional digital computers, but significant technical challenges must be addressed. Quantum computers are more powerful than classical turing machines because of their coherent superposition of states. Biotechnology techniques can be used to evolve DNA computers. Both of these qualities could be captured by combining DNA and quantum computers. Self-assembling quantum logic circuits from gates attached to DNA strands could be achieved using DNA computers. Furthermore, quantum computers might be built directly from the physical properties of the DNA molecule. So, if a cross-platform can be created that connects quantum and DNA computers with the same specific goal (solving complicated problems), will make a revolution in computation. In this chapter, a completely new idea will be introduced, where a cross-platform environment will be established using Quantum and DNA computing systems.

In terms of how inputs are given and outputs are generated, multiple-valued Quantum and DNA computing can be combined in two ways. They can be named as:

1. Multiple-Valued Quantum-DNA Computing, and
2. Multiple-Valued DNA-Quantum Computing

In this chapter, multiple-valued Quantum-DNA Computing is focused.

DOI: 10.1201/9781003381938-3

FIGURE 3.1
German-American Biophysicist Max Delbrück

3.2 Quantum Mechanics and Quantum Molecular Biology

Quantum mechanics is the fundamental theory that explains the properties of sub-atomic particles, atoms, molecules, molecular assemblies, and maybe beyond. Quantum mechanics governs key life functions such as photosynthesis, respiration, and vision at the nanometer and sub-nanometer scales. All things in quantum mechanics have wave-like features, and quantum coherence characterizes the correlations between the physical quantities describing such items due to their wave-like nature when they interact.

The idea that quantum phenomena – like coherence – may play a functional role in macroscopic living systems is not new. In 1932, 10 years after quantum physicist Niels Bohr was awarded the Nobel Prize for his work on the atomic structure, he delivered a lecture entitled 'Light and Life' at the International Congress on Light Therapy in Copenhagen. This raised the question of whether quantum theory could contribute to a scientific understanding of living systems. In attendance was an intrigued Max Delbrück (Figure 3.1), a young physicist who later helped to establish the field of molecular biology and won a Nobel Prize in 1969 for his discoveries in genetics.

All living systems are made up of molecules, and fundamentally all molecules are described by quantum mechanics. Traditionally, however, the vast separation of scales between systems described by quantum mechanics and those studied in biology, as well as the seemingly different properties of inanimate and animate matter, has maintained some separation between the two bodies of knowledge. Recently, developments in experimental techniques such as ultrafast spectroscopy, single-molecule spectroscopy, time-resolved microscopy, and single-particle imaging have enabled us to study biological dynamics on the increasingly small length and time scales, revealing a variety of processes necessary for the function of the living system that depends on a delicate interplay between quantum and classical physical effects.

Quantum biology is the application of quantum theory to aspects of biology for which classical physics fails to give an accurate description. It is the study of applications of quantum mechanics and theoretical chemistry to biological objects and problems.

Fundamentally, all matters – animate or inanimate – are quantum mechanical, being constituted of ions, atoms, and/or molecules whose equilibrium properties are accurately determined by quantum theory. As a result, it could be claimed that all of biology is quantum mechanical. However, this definition does not address the dynamical nature of biological processes, or the fact that a classical description of intermolecular dynamics seems often sufficient. Quantum biology should, therefore, be defined in terms of the physical 'correctness' of the models used and the consistency in the explanatory capabilities of classical versus quantum mechanical models of a particular biological process.

Good examples of biological processes is shown in Figure 3.2 in which quantum effects are visible are the transport of electrons and protons in photosynthesis, respiration, vision, catalysis, olfaction, and in basically every other biological transport process. Further examples include the transfer of electronic and/or vibrational energy, and magnetic field effects in electron transfer and bird migration.

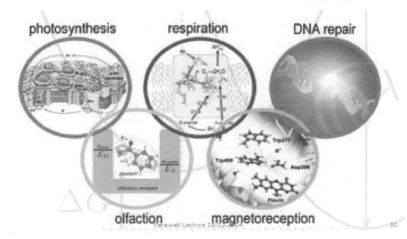

FIGURE 3.2
Quantum Biology in Photosynthesis

Quantum biology promises to give rise to design principles for biologically inspired quantum nanotechnologies, with the ability to perform efficiently at a fundamental level in noisy environments at room temperature and even make use of these 'noisy environments' to preserve or even enhance the quantum properties.

3.2.1 Why Quantum Molecular Biology Is an Important Field?

By definition, electrons, protons, excitations, chemical bonds, and electronic charges are quantum, and comprehending their dynamics necessitates knowledge of quantum mechanics. Furthermore, the properties of the next level of organization in biological systems – biomolecular complexes – are largely determined by these fundamental entities, whose interactions with one another and with their environment are frequently impossible to describe accurately without considering quantum biology laws. In addition, often in biology, the environment plays an essential role in the outcome of a biomolecular process. Photosynthesis and vision are two prominent examples. Thus, to understand biology, and the amazing selectivity of biological processes, quantum molecular biology is needed.

Quantum molecular biology furthermore can potentially have a huge impact on numerous technologies, including sensing, health, the environment, and information technologies. For example, energy technologies might be revolutionized by bioinspired solar cells, and chemical, magnetic, and biological sensing technologies may be taken to a new level when applying the principles found in natural equivalents.

3.3 Relationship between Multi-Valued and Two-Valued Quantum and DNA Computing

It has been already discussed that in quantum computing, the basic unit of information is qubits. It has two states – $|0>$ and $|1>$. But in a ternary quantum system or multiple-valued quantum system, the basic information unit is qutrit. It represents three – a dimensional quantum system with three basic states $|0>$, $|1>$ and, $|2>$. This ternary computing system performs any operations using qutrit and it also uses superposition, entanglement, and interference phenomena.

The purpose of multiple-valued logic is to improve the computations power. It provides more security, performs high-speed computations, and provides high storage capacity, etc. over the two-valued quantum computing.

On the other hand, DNA computing also accomplishes significant, complex mathematical problems using genetic molecules. DNA computing performs DNA logic gate operations, input and output are converted in the form of encoded strands of DNA where ACCTAG and TGGATC denote 1 and 0, respectively. And MVL DNA logic is such a ternary logic gate that consists of two inputs and one output where each input or output stays in three different states (0, 1, and 2). Multi-Valued DNA computing has significant advantages such as it can reduce the number of nucleic acid strands, minimize the cross interactions, and enhance the computing rates while increasing complexity. Many researchers are still working on both computing systems. Previously, this has been shown that DNA computing and quantum computing for two-valued is possible to combine with the help of some techniques like NMR and NMR relaxation, etc. As it is possible to combine the two different computing systems, also possible to combine multiple-valued quantum computing and multiple-valued DNA computing using trap ions, NMR, and NMR relaxation because Multiple-valued quantum and multiple-valued DNA computing are nothing but advanced computing techniques. That is why Multiple-valued quantum-DNA computing will be discussed in this chapter which will add a new dimension to the computer computing world.

3.3.1 Multiple-Valued Quantum-DNA Computing: A New Computing Approach

Quantum computing is a type of computation that harnesses the collective properties of quantum states, such as superposition, interference, and entanglement, to perform calculations. Where, *DNA computing* is an emerging branch of computing that uses DNA, biochemistry, and molecular biology hardware, instead of traditional electronic computing. The core concepts and operating principles of those two computing systems are diametrically different. Multiple-valued Quantum computing works with qutrits and multiple-valued DNA computing works with nitrogen base sequences. So, if a cross-platform is needed to be built with those, so many questions may arise and so many difficulties may have to be faced. In this section, the establishment of Quantum-DNA computing will be explained in a nutshell.

In Multiple-valued Quantum-DNA computing, a portion of computation will be performed in the Quantum computing system and the rest of the computing will be performed in the DNA system. The most general outer view is shown in Figure 3.3. But to make it real, a lot of functionalities have to be added to that architecture.

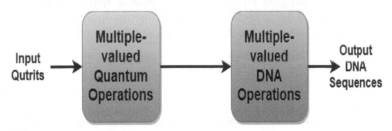

FIGURE 3.3
The General View of the Multiple-Valued Quantum-DNA Computing

3.3.2   Data Conversion Circuits to Convert Qutrits to DNA Base Sequences

Figure 3.4 shows that the multiple-valued Quantum-DNA environment has two separate computing systems such as i) multiple-valued Quantum system, and ii) multiple-valued DNA computing system. Where the outputs of the multiple-valued Quantum system are the input to the multiple-valued DNA system. But in reality, the input of the DNA system is nitrogen base sequences. That implies the DNA system cannot work with the qutrits. Therefore, the data conversion circuits are required to convert qutrits into the equivalent DNA base sequences.

Data conversion circuits are the hardware that can convert Quantum bits to the equivalent DNA information and vice versa.

To establish a cross-platform between multiple-valued quantum and DNA computing systems, there are three types of data conversion circuits available as mentioned below:

1. Nuclear Magnetic Resonance (NMR) technique

2. NMR Relaxation, and

3. Trap ions

But to establish the multiple-valued quantum-DNA system, NMR relaxation and trap ions can be used. Now updated multiple-valued quantum-DNA system with data conversion circuits may look like Figure 3.4.

The data conversion circuit will convert the output qutrits from the multiple-valued quantum system into the equivalent DNA base sequences.

3.3.2.1   NMR and NMR Relaxation

Nuclear Magnetic Resonance is abbreviated as NMR. When positioned in a high magnetic field, an NMR apparatus can study the molecular structure of a substance by viewing and quantifying the interaction of nuclear spins. For the analysis of molecular structure at the atomic level, electron microscopes and X-ray diffraction instruments can also be used, but the advantages of NMR are that sample measure-

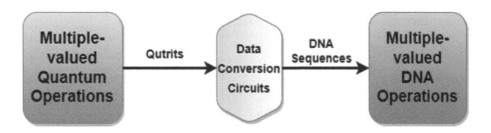

FIGURE 3.4
The General View of the Multiple-Valued Quantum-DNA Computing with Data Conversion Circuits

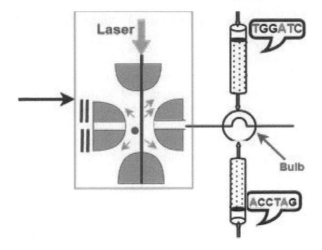

FIGURE 3.5
The General Diagram of Converting Qutrit to DNA Sequence Using Trap Ion

ments are non-destructive and there is less sample preparation required. It is possible to convert a DNA sequence to a quantum-bit using NMR. But when converting quantum-bit to the DNA base sequence, NMR relaxation is needed.

NMR relaxation is the process by which an excited magnetic state returns to its equilibrium distribution. When a molecule drops into the NMR probe as a sample, it goes to an excited state with the help of a magnetic field. When the electromagnetic resonance is not emitted, the magnetic field becomes weak in which the superposition state molecule loses its energy and comes into the ground state and this process is called NMR relaxation.

3.3.2.2 Trap Ions as Data Conversion Circuits

Qutrits can be converted into nitrogen base sequences using another method – Trap ion. A trapped ion quantum computer is an approach for a large-scale quantum computer. Ions, or charged atomic particles, can be confined and suspended in free space using electromagnetic fields. Qutrits are stored in stable electronic states of each ion, and quantum information can be transferred through the collective quantized motion of the ions in a shared trap (interacting through the Coulomb force). Lasers are applied to induce coupling between the qutrit states (for single-qutrit operations) or coupling between the internal qubit states and the external motional states (for entanglement between qutrits). The general diagram to convert the qutrits into the DNA sequence is shown in Figure 3.5.

3.3.3   Intermediatory System to Control Quantum-DNA Data Flow

The data conversion problem has been solved by the data conversion circuits. Another major problem is controlling the data flow. This happens due to the execution speed difference between the two computing systems.

According to IBM, quantum computing is so fast that the machine completed a theoretically defined computation in 200 seconds that would take the world's most powerful supercomputer 10,000 years to complete. Quantum computers are 158 million times quicker than the world's fastest supercomputer as a result of this. Therefore, quantum operations provide output instantaneously. But DNA operations require much time for preparation. Hence, the resulting qubits from the quantum computer cannot enter into the DNA system instantaneously. There must be a time delay, otherwise the qubit properties can be damaged. Eventually, the entire system may become obsolete. Therefore, the necessity of a temporary storage device or system arises where the qubits can be stored for a very short time while working with the Quantum-DNA system.

To overcome this problem an intermediary storage system is proposed namely Quantum Cache Memory to store the qubits or qutrits temporarily.

Cache memory is a supplementary memory system that temporarily stores frequently used instructions and data for quicker processing. It is an extremely fast memory type that holds frequently requested data and instructions so that they are immediately available for further processing. Cache memory can be used in the quantum system due to its speed and reliability. It can pass and get data very frequently. The mapping and swapping techniques in the cache memory are much optimized. But to use a cache memory in the quantum system, it is needed to redesign the cache memory to make it suitable for the quantum system.

The general organization of the multiple-valued Quantum-DNA computing system including multiple-valued quantum cache memory and data conversion circuits is shown in Figure 3.6.

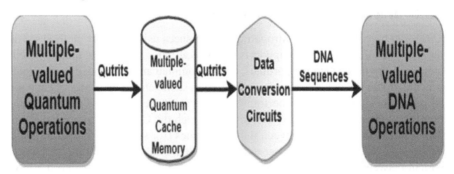

FIGURE 3.6
The General Organization of Multiple-Valued Quantum-DNA Computing with Quantum Cache Memory

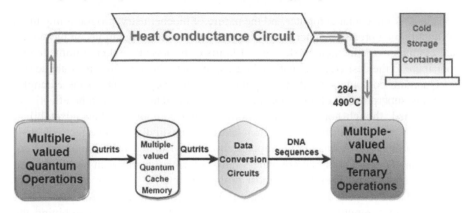

FIGURE 3.7
The General Organization of Multiple-Valued Quantum-DNA Computing with Data Conversion Circuits, Quantum Ternary Cache Memory, and Heat Conductance Circuit

3.3.4 Heat Transfer Circuit

There is another difficulty that must have to be overcome to establish the multiple-valued Quantum-DNA computing system. Quantum computer produces a tremendous amount of heat, and to connect the quantum computer with the DNA computer, this extremely excessive amount of heat must be controlled.

To solve this problem a heat conductance circuit can be added to the system to transfer the excessive amount of heat into cold storage. The Figure 3.7 shows the complete architecture of the multiple-valued Quantum-DNA computing system.

Note that, the DNA computing system requires an amount of 284-490°C heat in its preparation phase. So, the necessary heat can be given from the quantum system.

3.4 Advantages of Multiple-Valued Quantum-DNA Computing

Quantum computing involves performing calculations at a scale where quantum mechanical effects are significant. Biomolecules and biomolecular processes are used to develop computational algorithms in DNA computing. Both of these new computing paradigms have been considered as potential successors to solid-state computers. Both have significant advantages over traditional computing, as well as significant challenges to overcome to achieve successful implementation.

Biomolecular computers can be evolved using molecular biology techniques and enzymes to adapt to changing settings and input. It is the flexibility and robustness of biological systems to environmental change that has inspired evolutionary programs and artificial neural networks. On the other hand, due to the intricacy of the interactions in a biomolecular computer and the nature of biochemistry, programming them to function in a precise manner is difficult. Furthermore, computations on biomolecular computers are inefficient. Because of faults or because they do not immediately contribute to the desired computational outcome, many interactions are squandered.

Quantum computers excel at computational efficiency because of the entanglement and superposition of an exponential number of states. They can be effectively programmed, though not as easily as conventional computers. Their adaptability, however, is nonexistent because of their extreme sensitivity to the effects of environmental changes. A quantum computer should be isolated from the external environment. Traditional digital computers fall somewhere in between biological and quantum computers. They are the most programmable but are less adaptable than biological and less efficient than quantum. Therefore, in what follows, first, DNA and quantum computing will be reviewed. Then, several ideas for combining the adaptability of DNA computers and the efficiency of quantum computers will be presented. The presentation is somewhat speculative, of necessity, but the intent is to suggest possibilities, not supply detailed blueprints for implementation.

Quantum-DNA computing holds all the individual advantages of quantum and DNA computing. While considering multiple-valued Quantum-DNA computing it will also provide the advantages of multi-valued logic. It uses less power, but it has a great computing performance. It will be the world's fastest computing system in the modern era. Parallelism can be gained in computing and the ability to store large amounts of data and information in a small amount of space by employing the DNA portion. The quantum portion can solve highly complicated problems in a matter of seconds, saving a significant amount of time.

The world as a whole could attain the greatest success ever in the future world by integrating these two computing technologies.

3.5   Disadvantages of Multiple-Valued Quantum-DNA Computing

The issue of cost is a major stumbling block. This isn't for all computation. Regular computational problems solved by the multiple-valued Quantum-DNA computing system may not provide mesmerizing computational differences in terms of execution speed compared to the traditional system. Besides due to the scarcity of technologies, it is not viable to implement quantum-DNA circuits in all aspects of computation.

The control of heat is a critical issue here. The heat released by quantum circuits is difficult to control. Another significant disadvantage is the inability to manage it. It is difficult to operate and control without a thorough understanding of quantum computing and DNA computing.

3.6 Summary

Multiple-valued quantum-DNA computing is a completely new idea which has been introduced in this chapter. This chapter has presented the basic organization of multiple-valued Quantum-DNA computing. The basic technologies to build multiple-valued Quantum-DNA circuits like multiple-valued quantum cache memory, NMR and NMR relaxation, trap ion, and heat transfer circuits have also discussed. The merits and demerits of the multiple-valued quantum-DNA computing are also shown to get the advantages of quantum computing and DNA computing together. The concept of quantum biology is explained here and how quantum biology is related to the Quantum-DNA computing system is also explained in this chapter.

Bibliography

[1] Marais, A., Adams, B., Ringsmuth, A. K., Ferretti, M., Gruber, J. M., Hendrikx, R., ... & Van Grondelle, R. (2018). The future of quantum biology. Journal of the Royal Society Interface, 15(148), 20180640.

[2] Dubrova, E. (1999, November). Multiple-valued logic in VLSI: challenges and opportunities. In Proceedings of NORCHIP (Vol. 99, No. 1999, pp. 340-350).

[3] Hobo, F., Takahashi, M., & Maeda, H. (2009). S33 NMR cryogenic probe for taurine detection. Review of Scientific Instruments, 80(3), 036106. doi:10.1063/1.3103573.

[4] Kodibagkar, V. D., & Conradi, M. S. (2000). Remote tuning of NMR probe circuits. Journal of Magnetic Resonance, 144(1), 53-57.

[5] Auguin, D., Catherinot, V., Malliavin, T. E., Pons, J. L., & Delsuc, M. A. (2003). Superposition of chemical shifts in NMR spectra can be overcome to determine automatically the structure of a protein. Spectroscopy, 17(2-3), 559-568.

[6] Ishida, K., Tanaka, M., Ono, T., & Inoue, K. (2016, October). Single-flux-quantum cache memory architecture. In 2016 International SoC Design Conference (ISOCC) (pp. 105-106). IEEE.

[7] Haghparast, M., Wille, R., & Monfared, A. T. (2017). Towards quantum reversible ternary coded decimal adder. Quantum Information Processing, 16(11), 1-25.

4

Multiple-Valued DNA-Quantum Computing

4.1 Introduction

In the previous chapter, a new computing approach was established by merging multiple-valued quantum computing and DNA computing system. It made a relationship between molecular biology and quantum mechanics. The super-speed ability of quantum computers and large storage capability along with massive parallel execution of DNA computers have made the most powerful computing system. Figure 4.1 shows the picture of DNA at atom scale.

There is another possible way to make a cross-platform environment with multiple-valued quantum computing and multiple-valued DNA computing. It is called *Multiple-valued DNA-Quantum computing*. In this chapter, the fundamental concept and general organizations of the multiple-valued DNA-Quantum computing will be discussed.

4.2 Relationship between Multi-Valued DNA Computing and Multi-Valued Quantum Computing with Two-Valued DNA Computing and Quantum Computing

DNA computing performs computations using biological molecules rather than traditional silicon chips. It has more advantages over conventional computers. DNA strands can hold so much data in memory and perform a million operations at once. In two-valued DNA sequence computing, ACCTAG(1) and TGGATC (0) are used for performing any logical operation. In multi-valued DNA computing, ACCTAG, CAAGCT, and TGGATC are expressed 0,1, 2, respectively. The DNA sequence is detected with the help of the fluorescent level that has been already discussed previously. The advantages of multi-valued DNA are parallelism, less power consumption, and best for the combinatorial problem.

On the contrary, Quantum computing is also an advanced system that performs any complex computations very easily and multi-valued quantum computing adds more speciality and advantages to the quantum computing world.

DOI: 10.1201/9781003381938-4

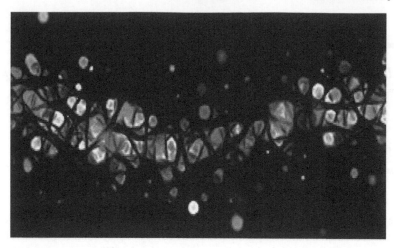

FIGURE 4.1
DNA at Atom Scale

Both DNA and quantum computers have the potential to exceed the power of conventional digital computers. Quantum computers are more powerful than conventional computers because of superposition states. DNA computers are also on the way to development through biotechnology techniques. As a two-valued Quantum computer and two-valued DNA sequence are possible to implement, it is also possible to develop the multi-valued DNA computing and the multi-valued quantum computing. It has been already shown that the two-valued of quantum computing and DNA computing are possible to merge. The two-valued computing systems are possible to combine the multi-valued quantum and DNA computing in order to get the more advantages. By combining these multi-valued DNA and Quantum computings, both of the characteristics of two computing systems might be captured. Some techniques such as trap ions, NMR, and NMR relaxation can be used here which will be discussed later.

4.3   Multiple-Valued DNA-Quantum Computing

Multiple-valued DNA-Quantum computing is an integrated computing system where the first portion of computation is performed in the multiple-valued DNA computing system and the rest of the computation is performed in the multiple-valued quantum computing system. In the previous chapter, the idea of the multiple-valued Quantum-DNA computing system was explained which consists of five components such as: (i) Multiple-valued quantum System, (ii) Multiple-valued quantum cache memory,

(iii) Data Conversion Circuits, (iv) Multiple-valued DNA system, and (v) heat conductance circuit.

Similarly, the multiple-valued DNA-Quantum computing system will also consist of five components as follows:

1. Multiple-valued DNA Computing system,
2. Multiple-valued DNA Cache memory,
3. Data Conversion Circuits,
4. Multiple-valued Quantum Computing System, and
5. Heat Conductance Circuit.

4.3.1 General Organization of Multiple-Valued DNA-Quantum Computing

The general organization of the multiple-valued DNA-Quantum computing system is similar to the Quantum-DNA system. The first component is the DNA computing system and the last component is the quantum computing system. Besides, multiple-valued DNA cache memory is used to store the output DNA information from the DNA computing system. Figure 4.2 shows the general organization of the multiple-valued DNA-Quantum Computing system.

From Figure 4.2, it is understandable that to store the DNA system's outputs (which is DNA information, a multiple-valued DNA cache memory is required. Besides, either the NMR process or trap ions can be used to convert DNA sequences to

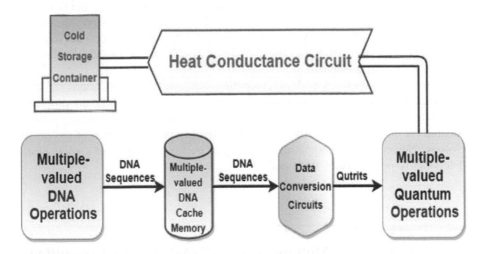

FIGURE 4.2
The General Organization Multiple-Valued DNA-Quantum Computing with Data Conversion Circuits, Multiple-Valued DNA Cache Memory, and Heat Conductance Circuit

the corresponding qutrits. The construction process, circuit diagram, and the working procedure of NMR, NMR relaxation, and trap ions were briefly described in the previous chapter.

And again, to complete the organization of the circuit, a heat conductance circuit must be used through which the tremendous heat can be transferred from the quantum system to a cooler . Now, it is important to note that DNA operations which will be performed in the first place has no the required heat and this heat can't be provided to the DNA system from the quantum system, because the heat is generated in the quantum system only when quantum operations are performed. Therefore, the required heat for the DNA system is not provided by the quantum system. Rather, this heat can be provided from the outside.

4.4   Difference between Multiple-Valued Quantum-DNA and DNA-Quantum Computing

The concept of multiple-valued Quantum-DNA and DNA-Quantum computing is almost identical. But the key differences lie in their general organizations and their applications. The architectures of those computing systems show that

in multiple-valued Quantum-DNA computing, the quantum operations are performed first, then a quantum cache memory temporarily stores the output qutrits of the quantum system. From the quantum cache memory, the qutrits are passed through the data conversion circuits when the multiple-valued DNA system is ready for execution. On the contrary, in multiple-valued DNA-Quantum computing, the DNA operations are performed first, then a DNA cache memory temporarily stores the output DNA information of the DNA system. From the DNA cache memory, the base sequences are passed through the data conversion circuits to the quantum system to perform the rest of the operation.

From the architectures of those computing systems, their applications can also be guessed. Quantum-DNA computing is usually more useful where the problems are related to quantum mechanics directly. But DNA-Quantum computing is most suitable for the problems which are directly related to the information of molecular biology. However, both computing systems can perform any sort of computations that could be performed by them individually.

4.5   Advantages of Multiple-Valued DNA-Quantum Computing

Just like multiple-valued Quantum-DNA computing, multiple-valued DNA-Quantum computing can perform operations super expeditiously along with a massive storage facility. Multiple-valued DNA-Quantum computing can be used for some special purpose that other computing systems may not be handled efficiently. Those special applications include:

Solving NP-Complete and Hard Computational Problems: Much of the work on DNA computing has continued to focus on tackling NP-complete and other difficult computational problems after Adleman and Lipton's original achievements. NP-complete problems are those for which there is no known polynomial-time solution using traditional computer algorithms. That is, as the complexity of these problems grows, the amount of time it takes to solve them grows exponentially. These issues are also described as intractable, but if a tractable solution to one of them can be found within the NP-complete domain, it can be utilized to solve all of the other problems in the set.

Apart from technological issues, real-world issues are also addressed, such as the majority problems of company planning and management science. Many of the cost optimization challenges that managers confront are now tackled via heuristic methods and other approximations since they are NP-complete. These issues include scheduling, routing, and the best utilization of raw materials, and they correspond to problems that have already been handled using multiple-valued DNA-Quantum computation, either theoretically or empirically.

Storage and Associative Memory: DNA could potentially be utilized to replicate and perhaps improve the human brain's associative skills. Baum proposed using DNA to create a huge content-addressed memory.

Baum has also proposed that a memory may be built, in which only content-addressable and associative portions of the data are retained, with other information on an item compactly stored in addresses relative to the associative section of the entry.

Given the brain's limit of around 10^{15} synapses and Feynman's low-end estimate of the brain's ability to discriminate about 10^6 concepts, a DNA-based associative memory system could have certain benefits over the brain. Baum thinks that a huge bathtub of DNA, around 50g in 1000l, could carry over 10^{20} words without accounting for superfluous molecules. This massive storage along with the quantum phenomena can produce a strong computing capability that the world has ever imagined.

DNA2DNA Applications: Another area of DNA-Quantum computation exists where conventional computers are incapable of competing at this time. The use of DNA computers to execute operations on unknown sections of DNA without needing to sequence them first is known as DNA2DNA computations. This is accomplished by re-coding and amplifying unknown strands into a redundant form, allowing them to be processed using techniques similar to those employed in the sticker model of DNA computation. The followings are some of the potential uses of DNA2DNA computations:

1. DNA Sequencing,

2. DNA Fingerprinting,

3. DNA Mutation and Population Screening, and so on.

Uses in Nano Computing: The combination of DNA and nanotechnology provides us with a wide range of benefits. When the structure of DNA is considered, it'll be noticed that it's a double-stranded molecule. This molecule can unzip to generate a branching molecule, but the branches can also be weaved into self-assembling structures. Scientists may unzip a double-stranded DNA molecule partway, and the branches will unite with corresponding sequences on other DNA that is branching out in the same way. With the appropriate supervision, DNA molecules can be connected in such a way that they form useful structures which can be used in a variety of applications. Previously, several DNA Nano mechanical devices were built to exhibit motions including open/close, extension/contraction, and motors/rotation mediated by DNA.

Intelligent systems Based on DNA-Quantum Computing: This new invention which is introduced here as multiple-valued DNA-Quantum computing can also be used to develop intelligent systems.

4.6   Summary

This chapter aims at giving the idea of a concept of establishing a cross-platform environment with multiple-valued DNA and Quantum computing systems. The general architecture of the computing system has been explained which is actually quantum molecular biology. Its components are described with proper figures. The difference between multiple-valued Quantum-DNA computing and multiple-valued DNA-Quantum computing has also been discussed. The advantages and applications of multiple-valued DNA-Quantum computing have discussed elaborately.

Bibliography

[1] Sharma, V. DNA Computing: A Complete Overview.

[2] Haghparast, M., Wille, R., & Monfared, A. T. (2017). Towards quantum reversible ternary coded decimal adder. Quantum Information Processing, 16(11), 1-25.

[3] Kodibagkar, V. D., & Conradi, M. S. (2000). Remote tuning of NMR probe circuits. Journal of Magnetic Resonance, 144(1), 53-57.

[4] Hobo, F., Takahashi, M., & Maeda, H. (2009). S33 NMR cryogenic probe for taurine detection. Review of Scientific Instruments, 80(3), 036106. doi:10.1063/1.3103573

[5] Mostafanasab, H., & Yousefian Darani, A. (2021). On Cyclic DNA Codes Over

$$F_2 + uF_2 + u^2F_2$$

F 2+ u F 2+ u 2 F 2. Communications in Mathematics and Statistics, 9(1), 39-52.

Part II

Heat Measurement, Heat Transfer, Speed Calculation, Data Conversion, and Data Management in Multiple-Valued Quantum and DNA Computing

Overview

Multiple-valued quantum computers are frequently characterized as a revolutionary invention that will influence the outcome of history. The rapid computing behavior of quantum computers is one of its primary advantages and capabilities. Quantum computers are extremely powerful and speedy. They can conduct calculations that today's supercomputers would take decades or perhaps millennia to complete. Quantum supremacy is a term used by specialists to describe this occurrence. This was only a theory for a long time. Google's quantum computer prototype, on the other hand, was able to do such a calculation and demonstrate quantum superiority in practice in 2019. Multiple-valued DNA computing uses DNA molecules to represent data and can store huge amounts of information with the ability of computing in parallelism. For combining the advantages of both DNA and Quantum computation systems, quantum molecular biology is needed where DNA-quantum and Quantum-DNA computation systems can be introduced. Quantum molecular biology is a study where the quantum phenomena and the interactions of molecular DNA have been discussed. Multiple-Valued computing in quantum molecular biology will be a big thing near future. It will be able to compute parallel operations at super-fast speed. Quantum computing produces too much heat during computation. On the other hand, DNA computing needs extra heat to perform computation.

The main objectives of this part are to measure excessive heat of multi-valued quantum computing and required heat for multi-valued DNA computing and the heat transfer process from the quantum part to the DNA part. Another important objective is to calculate the speed or performance of multi-valued quantum and DNA computing. Data conversion via NMR, NMR relaxation, and Trap ion from qubit to DNA sequences and vice versa is also an objective of this part. The last chapter of this part describes the data management process during multi-valued quantum and DNA computing where the multi-valued quantum and DNA cache memory is used during computation. All these necessary matters are discussed in this part within five chapters.

5

Heat Calculation

5.1 Introduction

Multi-Valued quantum computer calculations are especially promising for analyzing or simulating extremely complicated processes with large volumes of data. Multi-Valued quantum computation can be applied in cryptography, data analysis, task scheduling, predicting and forecasting, medical research, and artificial intelligence. Multi-Valued DNA computing is capable of performing these millions of processes in parallel way. A single cubic centimeter can contain over 10 trillion DNA molecules. This cubic centimeter of particles could effectively do 10 trillion calculations at the same time and store up to 10 terabytes of information. In many ways, DNA computing is capable of much of the breathless but erroneous coverage that quantum computing receives. By observing quantum and DNA computing, it is easy to find out the advantages of both computation systems. To make some operations, some circuit designs are always needed which can lead us towards some expected outputs.

Quantum computing produces so much heat during computation and DNA computing needs heat to perform computation. Produced heat from a circuit is always an important topic for each type of computing circuit. Nowadays, produced heat or required heat for quantum computing and DNA computing is an arresting matter for researchers. This chapter presents some ways to find out the amount of heat produced from a multi-valued quantum computing circuit. And further, it is needed to calculate the amount of required heat for multi-valued DNA computation.

5.2 Basic Definitions for Heat Calculation in Quantum Circuit

This section describes the basic information and theory to calculate produced heat from quantum circuits and the required heat to perform a DNA operation in quantum and DNA computing.

In quantum operation, qubit is generating heat when they become isolated and start to compute. In quantum physics, thermodynamics exists and the thermodynamics rule is quite the same for qubits also. When a quantum system is a single qubit, it is easy to write the Hamiltonian matrix as follows:

$$H = \frac{-1}{2}\boldsymbol{\epsilon}\sigma \tag{5.1}$$

DOI: 10.1201/9781003381938-5

This may correspond to an electric spin in a vertical magnetic field where €
is the energy difference between the states $|\uparrow> = |0>$ and $|\downarrow> = |1>$. The same Hamil-
ton matrix may also refer to a two=level atom where it is needed to identify its ground
and excited states as $|0>$ and $|1>$, respectively.

The Gibbs state of the qubit takes the form

$$\rho_\beta = \frac{1}{2cosh\left(\beta\frac{€}{2}\right)}e^{\beta€\frac{\sigma}{2}} = \frac{1}{1 + e^{-\beta€}}\left(|0 > 0| + e^{-\beta€}|1 > 1|\right) \tag{5.2}$$

Researchers have introduced the inverse temperature $\beta = 1 / K_B T$. Then the oc-
cupation number,

$$n_\beta = tr\left(\hat{n}\hat{\rho}_\beta\right) = \frac{1}{1 + exp\left(\beta_\epsilon\right)} \tag{5.3}$$

So, the average energy of thermal qubit,

$$E = \frac{1}{1 + exp\left(\beta€\right)}, 0 < E < \frac{1}{2}€ \tag{5.4}$$

This will be identified as the thermodynamic energy of the thermal qubit. The
von Neumann entropy of the Gibbs state can be calculated, and with one eye on
thermodynamics, the right-hand side is expressed. In terms of the energy E,

$$S\left(E\right) = \frac{-€ - E}{€}log\frac{€ - E}{€} - \frac{E}{€}log\frac{E}{€}, 0 < E < \frac{1}{2}€$$

Then S_{th} Energy will, $S_{th}\left(E\right) = \left(k_B ln2\right)S\left(E\right)$

So, the entropy of a single thermal qubit is,

$$\frac{dS_{th}\left(E\right)}{dE} = \frac{1}{T}$$

And, the entropy for n-thermal qubit is,

$$\frac{dS_{th}\left(E\right)}{dE} = \frac{n}{T} \tag{5.5}$$

It is known that,

$$\beta = \frac{1}{K_B T}$$

Here β is the inverse temperature, kb is the Boltzmann constant, and T is the
room temperature initially.

$$\beta = \frac{1}{K_B T}$$

$$= \frac{1}{8.617 \; X \; 10^{-5} X 300}$$

$$= 39 \; ev^{-1}$$

So, the average energy of thermal qubit, $E = \frac{1}{1 \pm e^{39X\epsilon}}$

$$= \frac{1}{1 \pm e^{39x0.9}}$$

$$= 5.7051 \times 10^{-16}$$

Here, ϵ is emissivity and this value will be 0 to 1 concerning the molecule and assumed that $\epsilon = 0.9$ for ideal purposes.

Now, according to Qubit Thermodynamics, S_{th} energy is,

$$S_{th}(E) = (K_b \ln 2) S(E)$$

$$= -k_B \frac{\epsilon - E}{\epsilon} \ln \frac{\epsilon - E}{\epsilon} - k_B \frac{E}{\epsilon} \ln \frac{E}{\epsilon}$$

$$= 1.9118 \times 10^{-18}$$

and $S_{th}(E)$ is quantum mechanics qubit entropy.

This can be identified as the thermodynamic energy of the thermal qubit. The von Neumann entropy of the Gibbs state is calculated, and with one eye on thermodynamics, the right-hand side is expressed. In terms of the energy E:

$S_{th}(E, N) = N(K_b \ln 2) S(E/N)$

$$= -k_B \frac{\epsilon - E/N}{\epsilon} \ln \frac{\epsilon - E/N}{\epsilon} - k_B \frac{E/N}{\epsilon} \ln \frac{E/N}{\epsilon} \tag{5.6}$$

Here, € = 0.9
E = 5.7051 x 10⁻¹⁶
N = Number of qubits in the operation
K_B = Boltzmann constant, 8.617×10^{-5} evT⁻¹

Now, by using the above equations, it is possible to calculate the produced heat from any quantum circuit. Some basic quantum operational circuits are described below for calculating the produced heat, where N is the number of the quantum qubits.

5.3 Heat Calculation in Multiple-Valued Quantum Circuit

Quantum computing embraces emerging technology areas such as quantum information systems and quantum cryptography. Its behavior is defined by quantum algorithms that exploit the quantum mechanical phenomenon of matter. The advantage of quantum computing is parallelism based on the linear superposition of quantum states. It has been shown that certain problems can be solved in fewer steps by quantum algorithms than by existing classical ones. A 2-state quantum system is usually defined with the two pure states $|0>$ and $|1>$ as basis states, and the unit of information is known as a qubit. Presently, multi-valued quantum computing is gaining importance in the field of quantum information theory and quantum cryptography as it can represent an n-dimensional quantum system, defined by the basic states $|0>, |1>, ..., | -1 >$. The unit of information is called a qubit. For a multi-valued quantum system, this unit is termed as qutrit [6]. A multi-valued quantum system exists in a linear superposition of three basic states, labeled $|0>$, $|1>$, and $|2>$.

Multi-Valued logic is one of the cases that drew the attention of the researchers. Meanwhile, the multi-valued circuits were more accepted because of their advantages over the binary circuits. Multi-Valued quantum computation has high fault tolerance and high speed as compared to the binary and is useful for quantum simulation, quantum tomography, and quantum games.

5.3.1 Quantum Multi-Valued Half Adder

A multi-valued half adder is a type of adder, an electronic circuit that performs the addition of multi-valued numbers. The half adder can add two single multi-valued digits and provide the output plus a carry value. It has two inputs, called A and B, and two outputs S (sum) and C (carry). Figure 5.1 describes the quantum operational circuit of the multi-valued half adder.

Quantum multi-valued half adder will add two qubits from the quantum multi-valued digit $|0>$, $|1>$, and $|2>$ and produce two outputs as quantum sum $|S>$ and quantum carry $|C>$.

Multi-Valued half adder is 16 qubits (considering 14 ancilla qubits and 2 input qubits) quantum operation; and for this quantum operation, heat measurement is calculated using the following formula:

$$\frac{dS_{th}(E)}{dE} = \frac{n}{T}$$

It is known that, for the N-qubit gate, $S_{th}(E, N) = N (Kb \ln 2) S (E/N)$

$$= -k_B \frac{\epsilon - E/N}{\epsilon} \ln \frac{\epsilon - E/N}{\epsilon} - k_B \frac{E/N}{\epsilon} \ln \frac{E/N}{\epsilon}$$

Quantum multi-valued half adder has 16 qubits as shown in Figure 5.1. So, N = 16.
Thus, $S_{th}(E, N) = 2.0607 x 10^{-18}$, and

$$T = \frac{dE \times N}{dS_{th}(E)}$$

$$= \frac{5.7051 \times 10^{-16} \times 16}{2.0607 \times 10^{-18}}$$

$$= 4429.64 \, k$$

Therefore, the produced heat from Quantum multi-valued half adder is 4429.64 Kelvin. Figure 5.1 shows the circuit diagram of quantum multi-valued half adder.

Quantum Ternary 1-to-3 Decoders

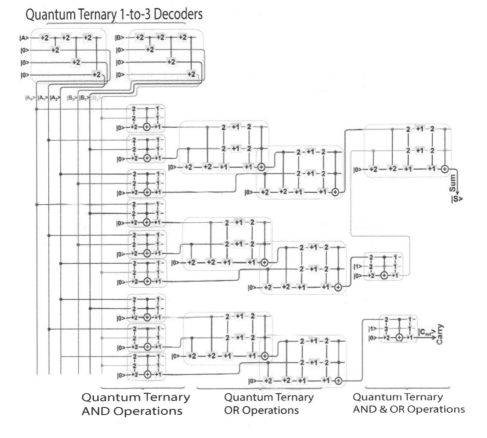

Quantum Ternary AND Operations

Quantum Ternary OR Operations

Quantum Ternary AND & OR Operations

FIGURE 5.1

The Circuit Diagram of Quantum Multi-Valued Half Adder

5.3.2 Quantum Multi-Valued Half Subtractor

A multi-valued half subtractor is a type of subtractor, an electronic circuit that performs the subtractions of multi-valued numbers. The half subtractor can subtract two single multi-valued digits and provide the output plus a borrow value. It has two inputs, called A and B, and two outputs D (difference) and B (borrow). Quantum multi-valued half subtractor will subtract two bits from the quantum multi-valued digit |0>, |1>, and |2>, and produce two outputs as quantum difference |D> and quantum borrow bit |B>. Multi-Valued half subtractor is 16 qubits (considering 14 ancilla qubits and 2 input qubits) quantum operation; and for this quantum operation, the heat measurement is calculated using the following formula:

$$\frac{dS_{th}(E)}{dE} = \frac{n}{T}$$

It is known that, for N qubit gate, S_{th} (E, N) = N (Kb ln 2) S (E/N)

$$= -k_B \frac{\epsilon - E/N}{\epsilon} ln \frac{\epsilon - E/N}{\epsilon} - k_B \frac{E/N}{\epsilon} ln \frac{E/N}{\epsilon}$$

Quantum multi-valued half subtractor has 16 qubits in the figure. So, N = 16. Thus, S_{th} (E, N) = 2.0607 × 10^{-18}, and

$$T = \frac{dE \times N}{dS_{th}(E)}$$

$$= \frac{5.7051 \times 10^{-16} \times 16}{2.0607 \times 10^{-18}}$$

$$= 4429.64 \text{ k}$$

Thus, the produced heat from the quantum multi-valued half subtractor is 4429.64 Kelvin. The quantum circuit for the multi-valued half subtractor is shown in Figure 5.2.

5.3.3 Quantum Multi-Valued 3-to-1 Multiplexer

A multiplexer (MUX) is a device that can receive multiple input signals and synthesize a single output signal in a recoverable manner for each input signal. It is also an integrated system that usually contains a certain number of data inputs and a single output. A multi-valued multiplexer of 3^n inputs have n select lines. And a selected line is used as the selection input that is sent to the output.

A multi-valued multiplexer is 10 qubits (considering 6 ancilla qubits and 4 input qubits) quantum operation; and for this quantum operation, the heat measurement is calculated using the following formula:

$$\frac{dS_{th}(E)}{dE} = \frac{n}{T}$$

It is known that, for N qubit gate, S_{th} (E, N) = N (Kb ln 2) S (E/N)

$$= -k_B \frac{\epsilon - E/N}{\epsilon} ln \frac{\epsilon - E/N}{\epsilon} - k_B \frac{E/N}{\epsilon} ln \frac{E/N}{\epsilon}$$

The quantum multi-valued multiplexer has 10 qubits as shown in Figure 5.3. So, N = 10.
Thus, S_{th} (E, N) = 2.034 × 10^{-18}, and

$$T = \frac{dE \times N}{dS_{th}(E)}$$

$$= \frac{5.7051 \times 10^{-16} \times 16}{2.034 \times 10^{-18}}$$

$$= 2802.35 \text{ k}$$

Quantum Ternary 1-to-3 Decoders

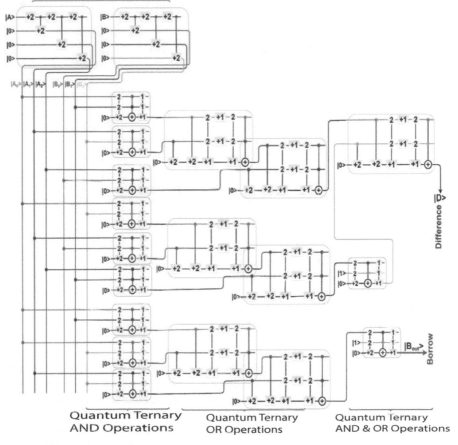

FIGURE 5.2

The Circuit Diagram of Quantum Multi-Valued Half Subtractor

Thus, the produced heat from the quantum multi-valued multiplexer is 2802.35 Kelvin. The quantum circuit for the multi-valued quantum 3-to-1multiplexer is shown in Figure 5.3.

5.4 Basic Definitions for Heat Calculation in DNA Circuit

DNA has the characteristics of enabling classical logical operation using DNA sequence. DNA prefers to be in double-stranded form, while single-stranded DNA naturally migrates towards complementary sequences to form double-stranded com-

Quantum Ternary 1-to-3 Decoder

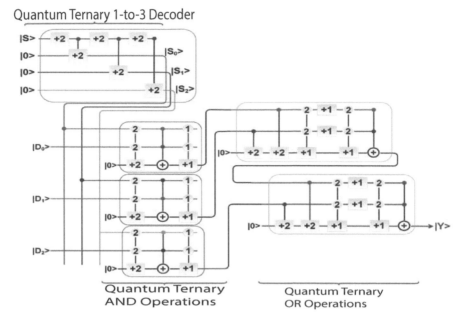

FIGURE 5.3
The Circuit Diagram of Quantum Multi-Valued 3-to-1 Multiplexer

plexes. Complementary sequences pair the bases adenine (A) with thymine (T) and cytosine (C) with guanine (G). DNA sequences pair in an antiparallel manner, with the 5′ end of one sequence pairing with the 3′ end of the complementary sequence.

The input of each DNA operation is the single standard sequence. If it is assumed true then the complementary DNA sequence is false. Suppose ACTCGT is the input sequence and then the complementary sequence TGAGCA. In DNA computing, when the logic gate is designed, a predetermined single strand sequence can be supplied to induce an appropriate chemical reaction. This sequence also helps to evaluate the output value whether it is true or false.

When the mixing step appears, it is needed to mix the two sequences to achieve a union of DNA sequences. In the mixing step, it is needed to give some heat to mix these. Then the annealing appears and in annealing, it is needed to cool this little and make a double sequence bond. After annealing, the step appears as melting. In melting step, it is needed to heat the double-strand DNA sequence to make them a single strand complementary sequence and this sequence will be used in the DNA logic operation after some steps. So, it is needed to know the DNA melting temperature.

Depending on the nature of the sequence, it is needed to use one of two methods to calculate the melting temperature, T_m:

1. Nearest Neighbors, and

2. Basic

1. Nearest Neighbors:

$$T_m = \frac{\Delta H}{A + \Delta S + Rln\frac{C}{4}} - 273.15 + 16.6log\,[Na^+] \tag{5.7}$$

where

1. Tm = melting temperature in°C,

2. ΔH = enthalpy change in kcal mol^{-1} (accounts for the energy change during annealing / melting),

3. A = constant of -0.0108 kcal K^{-1}mol^{-1} (accounts for helix initiation during annealing / melting),

4. ΔS = entropy change in kcal K^{-1}mol^{-1} (accounts for energy unable to do work, i.e., disorder),

5. R = gas constant of 0.00199 kcal K^{-1}mol^{-1} (constant that scales energy to temperature),

6. C = oligonucleotide concentration in M or mol L-1 (Here 0.0000005, i.e., 0.5 μ M),

7. -273.15 = conversion factor to change the expected temperature in Kelvins to°C, and

8. $Na+$ = sodium ion concentration in M or mol L-1 (Here 0.05, i.e., 50 mM).

This example will demonstrate the manual calculation of the Tm for the following sequence:
5'-AAAAACCCCCGGGGGTTTTT-3'
The above sequence paired is with its reverse complement as follows:
5'-AAAAACCCCCGGGGGTTTTT-3'
3'-TTTTTGGGGGCCCCCAAAAA-5'

$$T_m - \frac{\Delta H}{A + \Delta S + Rln\frac{C}{4}} - 273.15 + 16.6log\,[Na^+] \tag{5.8}$$

T_m = 69.6 degree Celsius.

2. Basic Method:

A secondary method which is used to calculate T_m as the basic method. It considers a modified Marmur Doty formula, which is used for oligonucleotides with short sequences lengths, (those that are 14 bases or less):

$$T_m = 2(A + T) + 4(C + G) - 7$$

1. T_m = melting temperature in°C,

2. A = number of adenosine nucleotides in the sequence,

3. T = number of thymidine nucleotides in the sequence,

4. C = number of cytidine nucleotides in the sequence,

5. G = number of guanosine nucleotides in the sequence, and

6. *-7* = correction factor accounting for in solution.

5.5   Heat Calculation in Multi-Valued DNA Circuit

DNA computing embraces emerging biology-dependent technological areas such as DNA security, DNA computing systems, and scheduling systems. The advantage is parallelism based on DNA strands, which are composed of Adenine, Thymine, Guanine, and Cytosine. It has been shown that certain problems can be solved in fewer steps and with less memory by DNA computing algorithms than by existing classical computing systems. A 2-strand base DNA system is usually defined with the two pure strands as CAAGCT and ACCTAG. At the same time, the multi-valued DNA computing can be used as an important area in the field of DNA computation systems, DNA scheduling, and security system as it can represent multiple DNA strands in the system for carrying information. For a multi-valued DNA computation system, the basic three strands can be TGGATC, CAAGCT, and ACCTAG.

Multi-Valued DNA computation systems can draw the attention of researchers because of their advantages over binary computation systems. For ensuring security as intrusion detection, job scheduling, and huge data clustering can be solved using a multi-valued DNA computing system.

5.5.1   DNA Multi-Valued Decoder

A Decoder is a combinational circuit that has 'n' input lines and a maximum of 2^n output lines. One of these outputs will be active High based on the combination of inputs present when the decoder is enabled. That means the decoder detects a particular code. In Figure 5.4, the decoder is the unary function for input variable A as A0, A1, and A2 which is used for multi-valued function implementation. Figure 5.4 shows the DNA multi-valued decoder.

Here, one input sequence "A" as follows:

1. A = ACCTAG

The calculation of the melting temperature of a specific DNA sequence is as follows:
For Input, A = ACCTAG
$$T_{m(A)} = 2(A + T) + 4(C + G) - 7$$
$$= 2(2 + 1) + 4(2 + 1) - 7$$
$$= 11.0°C$$

For each test tube, the heat required for specific steps of DNA multi-Valued decoder is as follows:

1. Basic gate operation preparing (98°C – 94°C),
2. Synthesizing (98°C – 94°C),
3. Mixing (95°C – 22°C),
4. Annealing (70°C – 20°C),
5. Melting (11°C),
6. Amplifying,
7. Separating,
8. Extracting,
9. Cutting   (20°C,
10. Ligating,
11. Substituting,
12. Marking,
13. Destroying, and
14. Detecting and Reading (98°C – 25°C).

So, in a DNA multi-valued decoder,
the maximum required heat = (98 + 98 + 95 + 70 + 11 + 20 + 98)°C − 490°C,

And the minimum required heat = (94 + 94 + 22 + 20 + 11 + 20 + 25)°C = 286 °C.

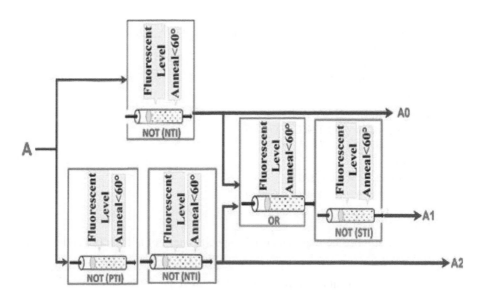

FIGURE 5.4
DNA Multi-Valued Decoder

Again, in a specific basic DNA operation, all the processes are happening in the test tube after mixing is completed. Here in specific cases, it is needed to keep the temperature high and sometimes need to keep the temperature low for several steps. So, the temperature is at a maximum of 94 – 98°C. When a DNA operation applies, it is needed to keep the temperature around 20°C and during the detection time, it is needed at around 25°C.

5.5.2   DNA Multi-Valued 3-to-1 Multiplexer

The multiplexer is one of the most important designs of a binary digital system. It is such a device that allows only one input from several input signals and the selection input helps the multiplexer to transmit data into a single medium. Multiplexers aid to improve the efficiency of the communication system. It allows the transmission of data such as audio, video, etc. from different channels via cables. And here, it will be implemented using MVL which helps to simulate the circuit easier way than the digital binary multiplexer. It is possible to construct a DNA 3-to-1 multi-valued multiplexer circuit based on a conventional 3-to-1 multi-valued circuit. Figure 5.5 illustrates the DNA multi-valued 3-to-1 multiplexer.

Here, the four input sequences are as follows:

1. I_0 = TGGATC
2. I_1 = CAAGCT
3. I_2 = ACCTAG
4. S = TGGATC

Calculation of melting temperature of a specific DNA sequence is as follows:
For Inputs,, I_0 and S = TGGATC
$T_{m\ (I0\ or\ S)} = 2(A + T) + 4(C + G) - 7$
$= 2(1 + 2) + 4(1 + 2) - 7$
$= 11.0°C$
Again, Input, I_1 = CAAGCT
$So,\ T_{m\ (I1)} = 2(A + T) + 4(C + G) - 7$
$= 2(2 + 1) + 4(2 + 1) - 7$
$= 11.0°C$
Again, Input, I_2 = ACCTAG
$So,\ T_{m\ (I2)} = 2(A + T) + 4(C + G) - 7$
$= 2(2 + 1) + 4(2 + 1) - 7$
$= 11.0°C$
For each test tube, the required heat for specific steps of a DNA full adder circuit is as follows:

1. basic gate operation preparing (98 °C – 94 °C)
2. Synthesising (98 °C – 94 °C)
3. Mixing (95°C – 22 °C)

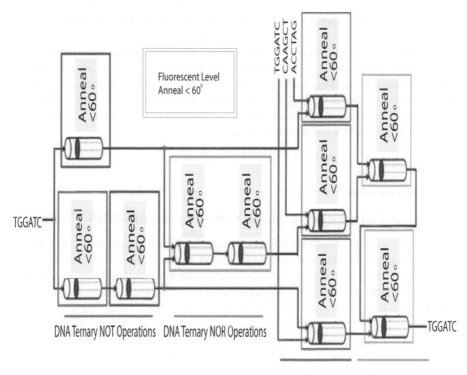

FIGURE 5.5
DNA Multi-Valued 3-to-1 Multiplexer

4. Annealing (70°C – 20 °C)
5. Melting. (11 °C)
6. Amplifying
7. Separating
8. Extracting
9. Cutting   (20 °C)
10. Ligating
11. Substituting
12. Marking
13. Destroying
14. Detecting and Reading. (98°C – 25°C)

So, for a DNA multi-valued multiplexer, the overall maximum required heat
= (98 + 98 + 95 + 70 + 11 + 20 + 98) °C = 490 °C,

and the minimum required heat = (94 + 94 + 22 + 20 + 11 + 20 + 25) °C
= 286 °C.

Again, for a specific basic DNA operation, all the processes are completed in the test tube after mixing is finished. Here in specific cases, it is needed to keep the temperature high and sometimes to keep the temperature low for several steps. So, it is necessary to keep the temperature at a maximum of 94 – 98°C. When a DNA operation occurs, it is needed to keep the temperature around 20°C and also, at detection time it must keep at around 25°C.

5.6   Heat Calculation in Multi-Valued Quantum-DNA Circuit

According to multi-valued quantum computing, multi-valued quantum computation has more advantages than classical computation systems . Quantum computers are also more powerful in terms of computation than supercomputers. They process data 1000 times faster than normal computers and supercomputers. Quantum computers can perform computations in a fraction of a second that would take a traditional computer 1000 years to finish. Parallelism based on linear superposition of quantum states is a benefit of multi-valued quantum computing. It has been demonstrated that multi-valued quantum algorithms can solve some problems in fewer steps than conventional classical algorithms.

The use of multi-valued (more than two DNA strands) DNA strands has high parallel computation which compensates the slow processing of the chips. When compared to traditional storage systems, DNA requires just about 1 bit per cubic nanometer of the memory space. The chemical interactions in DNA provide energy to make or repair new strands. Therefore, there is essentially no power use. As a result, a multi-valued quantum-DNA computation system can be created to find a super-fast computation system with a lot of memory. The advantages of multi-valued quantum computing and multi-valued DNA computing can be combined together in a multi-valued quantum-DNA computation system.

In a multi-valued quantum-DNA computing system, input is received as a qubit, which is then converted into DNA sequences by NMR relaxation after passing through a specific number of quantum gates.

5.6.1   Multi-Valued Quantum-DNA Half Adder

A multi-valued half adder is an electronic circuit that performs the addition of ternary numbers. The half adder is able to add two single ternary digits and provide the output plus a carry value. It has two inputs, called A and B, and two outputs S (sum) and C (carry). Multi-Valued quantum-DNA half adder adds two qubits from the quantum ternary digit |0>, |1>, and |2> and produces two outputs as quantum sum |S> and quantum carry bit |C> in a DNA sequence.

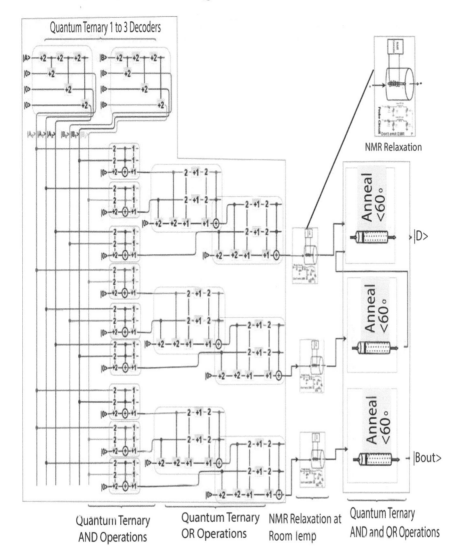

FIGURE 5.6

Multi-Valued Quantum-DNA Half Adder at Room Temperature

To create a half adder, seven OR, eleven AND, and two decoder operations are needed. Figure 5.6 describes the multi-valued quantum-DNA circuit of the half adder. From Figure 5.6, it can be assumed that seventeen quantum gates and three DNA operations are needed to find the expected output from multi-valued quantum-DNA half adder.

A multi-valued quantum-DNA half adder has 16 qubits (considering 14 ancilla qubits and 2 input qubits) for the quantum operation. The output qubit from the quantum operation will be converted into a DNA sequence by using NMR relaxation later. The produced heat generated from the quantum operation will be calculated using the following formula:

$$\frac{dS_{th}(E)}{dE} = \frac{n}{T}$$

It is known that, for N-qubit gate, $S_{th}(E, N) = N\,(Kb \ln 2)\,S\,(E/N)$

$$= -k_B \frac{\epsilon - E/N}{\epsilon} \ln \frac{\epsilon - E/N}{\epsilon} - k_B \frac{E/N}{\epsilon} \ln \frac{E/N}{\epsilon}$$

Quantum multi-valued half adder has 16 qubits in the figure. So, $N = 16$. Thus, $S_{th}(E, N) = 2.0607 \times 10^{-18}$, and

$$T = \frac{dE \times N}{dS_{th}(E)}$$

$$= \frac{5.7051 \times 10^{-16} \times 16}{2.0607 \times 10^{-18}}$$

$$= 4429.64\ k$$

Therefore, the produced heat from the quantum-DNA multi-valued half adder is 4429.64 Kelvin.

Now, the DNA operation requires at most two DNA sequences in the quantum-DNA full adder which is as follows:

1. TGGATC

2. ACCTAG

The melting temperature of a specific DNA sequence can be calculated as follows:

For TGGATC
$T_m = 2(A + T) + 4(C + G) - 7$
$= 2(1 + 2) + 4(1 + 2) - 7$
$= 11.0°C$
Again, For ACCTAG
$T_m = 2(A + T) + 4(C + G) - 7$
$= 2(2 + 1) + 4(2 + 1) - 7$
$= 11.0°C$

For each test tube, the heat generated from specific steps of a multi-valued quantum-DNA full adder during DNA operation is as follows:

1. Preparation of basic operations (98 °C – 94 °C)

2. Synthesizing (98 °C – 94 °C)

3. Mixing (95°C – 22 °C)

4. Annealing (70°C – 20 °C)

5. Melting (Depends on the sequence)

6. Amplifying

7. Separating

8. Extracting

9. Cutting   (20 °C)

10. Ligating

11. Substituting

12. Marking

13. Destroying

14. Detecting and Reading. (98°C − 25°C)

So, the overall maximum required heat during a DNA operation
= (98 + 98 + 95 + 70 + 11 + 20 + 98) °C = 490 °C;

And the minimum required heat = (94 + 94 + 22 + 20 + 11 + 20 + 25) °C = 286 °C.

Again, all the processes occur in the test tube during DNA basic operations after mixing is completed. Here in specific cases, it is needed to keep the temperature high and sometimes the temperature low for several steps. So, it is necessary to keep the temperature at a maximum of 94 − 98°C. The temperature is around 20°C when DNA basic operations occur and the detection time is around 25°C.

5.7   Heat Calculation in Multi-Valued DNA-Quantum Circuit

The use of multi-valued (more than two DNA strands) DNA strands have high parallel computation, which compensates for the slow processing of the chips. As compared to traditional storage systems, DNA requires just about 1 bit per cubic nanometer of memory space. The chemical interactions in DNA provide energy to make or repair new strands. Therefore, there is essentially no power use. On the other hand, according to multi-valued quantum computing, the multi-valued quantum computation has more advantages than classical computation systems. Quantum computers are also more powerful in terms of computation than supercomputers. They process data thousands of times faster than normal computers and supercomputers. Quantum computers can perform computations in a fraction of a second that would take a traditional computer 1000 years to finish. Parallelism based on linear superposition of quantum states is a benefit of multi-valued quantum computing. It has been demonstrated that multi-valued quantum algorithms can solve some problems in fewer steps than conventional classical algorithms.

As a result, a multi-valued DNA-quantum computation system can be created to find a super-fast computation system with a lot of memory. The advantages of multi-valued quantum computing and multi-valued DNA computing can be combined together in the multi-valued DNA-quantum computation system. In a multi-valued DNA-Quantum computing system, input is received as a DNA sequence, which is then converted into quantum qubits by NMR after passing through a specific number of quantum gates. This process will be discussed in later chapters.

5.7.1   Multi-Valued DNA-Quantum Half Subtractor

A multi-valued half subtractor is an electronic circuit that performs the subtractions of ternary numbers. The half subtractor is able to subtract two single ternary digits and provide the output plus a borrow value. It has two inputs, called A and B, and two outputs D (difference) and B (borrow). DNA-quantum multi-valued half subtractor subtracts two DNA sequences from the ternary inputted DNA sequence (TGGATC, ACCTAG, and CAAGTC) and produces two outputs as difference |D> and borrow |B> in quantum bits.

The multi-valued DNA-Quantum circuit for the multi-valued half subtractor is shown in Figure 5.7.

From the Figure 5.7, it can be assumed that twenty-two DNA operations and six quantum operations are required to get the expected output from the multi-valued DNA-quantum half subtractor.

Here, the operations require at most two DNA sequences in the multi-valued DNA-quantum half subtractor which are as follows:

1.   TGGATC

2.   ACCTAG

The melting temperature of a specific DNA sequence is calculated as follows:
For TGGATC
$$T_m = 2(A + T) + 4(C + G) - 7$$
$$= 2(1 + 2) + 4(1 + 2) - 7$$
$$= 11.0°C$$
Again, For ACCTAG
$$T_m = 2(A + T) + 4(C + G) - 7$$
$$= 2(2 + 1) + 4(2 + 1) - 7$$
$$= 11.0°C$$
For each tube, the heat required for specific steps of the multi-valued DNA-quantum half subtractor during DNA operation is calculated as follows:

1.   Preparation of basic gate operations (98 °C – 94 °C)

2.   Synthesising (98 °C – 94 °C)

3.   Mixing (95°C – 22 °C)

4.   Annealing (70°C – 20 °C)

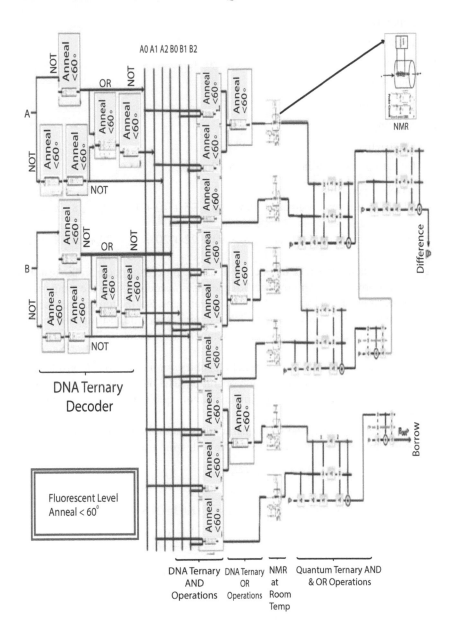

FIGURE 5.7
Multi-Valued DNA-Quantum Half Subtractor at Room Temperature

5. Melting. (Depends on the sequence)
6. Amplifying

7. Separating
8. Extracting
9. Cutting   (20 °C))
10. Ligating
11. Substituting
12. Marking
13. Destroying
14. Detecting and Reading. (98°C – 25°C)

So, in DNA operations,
the overall maximum required heat = (98 + 98 + 95 + 70 + 11 + 20 + 98) °C = 490 °C,
    and the minimum required heat = (94 + 94 + 22 + 20 + 11 + 20 + 25) °C = 286 °C.

Again, in a specific basic DNA operation, all the processes occur in the test tube after mixing is completed. Here in specific cases, it is needed to keep the temperature high and sometimes the temperature low for several steps. So, the temperature should be at a maximum of 94 – 98°C. When a DNA operation occurs, it is needed to keep the temperature around 20°C and also, the temperature should be 25°C during the detection time.

Further, a multi-valued DNA-quantum half subtractor has 9 qubits (considering 3 ancilla qubits and 6 input qubits) quantum operation. The output of a DNA operation is a DNA sequence, which is converted into a quantum qubit after processing through NMR at room temperature.

The produced heat generated from quantum operation in the multi-valued DNA-quantum half subtractor is calculated using the following formula:

$$\frac{dS_{th}(E)}{dE} = \frac{n}{T}$$

It is known that, for N-qubit gate, $S_{th}(E, N) = N (Kb \ln 2) S (E/N)$

$$= -k_B \frac{\epsilon - E/N}{\epsilon} \ln \frac{\epsilon - E/N}{\epsilon} - k_B \frac{E/N}{\epsilon} \ln \frac{E/N}{\epsilon}$$

In the figure, the multi-valued DNA-quantum half subtractor has 9 qubits quantum operation.
So, $N = 9$.
    Thus, $S_{th}(E, N) = 1.996 \times 10^{-18}$

$$T = \frac{dE \times N}{dS_{th}(E)}$$

$$= \frac{5.7051 \times 10^{-16} \times 9}{1.996 \times 10^{-18}}$$

$$= 2572.43 \text{ k}$$

Therefore, the produced heat from the multi-valued DNA-quantum half subtractor is 2572.43 Kelvin.

5.8 Summary

Multiple-valued quantum computing focuses on speedy technology based on quantum-theoretical principles, which is the behavior of energy and matter of a qubit. A combination of qubits is used to perform any specific task in quantum computing. Multiple-valued DNA computing promises significant and meaningful linkages between computers and life systems, as well as massively parallel computations. The DNA computing can carry out millions of operations at the same time. This chapter has presented the calculations to estimate the heat produced by multi-valued quantum computing and the required heat to perform computation in multi-valued DNA computing.

Bibliography

[1] Diósi, L. (2011). A Short Course in Quantum Information Theory: An Approach from Theoretical Physics (Vol. 827). Springer.

[2] Freier, S. M., Kierzek, R., Jaeger, J. A., Sugimoto, N., Caruthers, M. H., Neilson, T., & Turner, D. H. (1986). Improved free-energy parameters for predictions of RNA duplex stability. Proceedings of the National Academy of Sciences, 83(24), 9373-9377.

[3] Kosloff, R. (2013). Quantum thermodynamics: A dynamical viewpoint. Entropy, 15(6), 2100-2128.

[4] Klimov, A. B., Sánchez-Soto, L. L., de Guise, H., & Björk, G. (2004). Quantum phases of a qutrit. Journal of Physics A: Mathematical and General, 37(13), 4097.

[5] Marky, L. A., & Breslauer, K. J. (1987). Calculating thermodynamic data for transitions of any molecularity from equilibrium melting curves. Biopolymers: Original Research on Biomolecules, 26(9), 1601-1620.

[6] Watada, J. (2008). DNA computing and its application. In Computational Intelligence: A Compendium (pp. 1065-1089). Springer, Berlin, Heidelberg.

[7] Zheng, X., Yang, J., Zhou, C., Zhang, C., Zhang, Q., & Wei, X. (2019). Allosteric DNAzyme-based DNA logic circuit: Operations and dynamic analysis. Nucleic Acids Research, 47(3), 1097-1109.

[8] Adleman, L. M. (1994). Molecular computation of solutions to combinatorial problems. Science, 266(5187), 1021-1024.

[9] Alcoba, D. R., Bochicchio, R. C., Lain, L., & Torre, A. (2010). On the measure of electron correlation and entanglement in quantum chemistry based on the cumulant of the second-order reduced density matrix. The Journal of Chemical Physics, 133(14), 144104.

6

Speed Calculation

6.1 Introduction

Multi-Valued DNA computing uses biological molecules to do computations. The four-character genetic alphabets such as A-adenine, G-guanine, C-cytosine, and T-thymine are used in DNA computing. The input of any DNA operation can be represented by DNA molecules with specific sequences. The instructions are carried out by laboratory operations on the molecules, and the result is defined as some property of the final set of molecules. Quantum computers represent a significant advancement in computing capability with enormous performance benefits for specific use cases. The ability of bits to be in several states at the same time gives the quantum computer a lot of computing capability and it is much faster than classical bitwise computing.

To advance the computation process, the use of DNA-Quantum computing and Quantum-DNA computing systems will merge all the advantages of both DNA computing and Quantum computing. These two new computing processes are established in two-valued system and now in multiple-valued system also.

Besides the accuracy of a system, time or speed is a metric that can be used to measure the performance of a system. In this chapter, the calculated speed or consumed time for different operations will be provided in multi-valued quantum, multi-valued DNA, multi-valued quantum-DNA, and multi-valued DNA-quantum computation systems.

6.2 Speed Calculation for a Quantum Operation

Researchers have proposed the theory and formula to calculate the average required operational time in any quantum computation . Numerous researchers have also applied it in their work and research studies. The average computation time needed for an operation is calculated by Equation 6.1 as follows:

$$\tau = \frac{h}{4E};$$

(6.1)

DOI: 10.1201/9781003381938-6

TABLE 6.1

Execution Times for Basic Quantum Gate
Operations

Operations	Time(μ s) Now (Future)
Single Qubit Gate	**1 (1)**
Double Qubit Gate	**10 (10)**
Movement	**20 (10)**

where τ is the required operational time, h is the plank's constant, and E is the quantum mechanical average energy of the performing system. It has also shown that the minimum operation time of any quantum logic gate in quantum computation is given below:

$$\tau = \frac{h}{4E}\left(1 + 2\frac{\theta}{\pi}\right) \tag{6.2}$$

Here, τ is the required operational time, h is the plank's constant, E is the quantum mechanical average energy of the performing system, and θ is the phase shift modulo π. It considers any basic quantum gate that complements the state of a qubit and then adds to it an arbitrary phase shift.

Numerous researchers have noted that any basic quantum operation needs some basic amount of time which is tabulated in Table 6.1. It describes that a single gate operation as NOT needs 1 s and a double qubit gate as CNOT, V, and V+ needs approximately 10 microseconds. Sometimes one output of a gate operation in quantum computing can be needed for another gate operation to perform. In that case, the movement time of approximately 20 microseconds, and in the future, it can be required more or less 10 microseconds.

As a basic quantum operational gate is composed of a single or double qubit gate, the required operational time for a quantum operation can be calculated by using the information provided in Table 6.1.

6.3   Speed Calculation in Multi-Valued Quantum Circuit

Quantum computing embraces emerging technology areas such as quantum information systems, quantum cryptography. Its behavior is defined by quantum algorithms that exploit the quantum mechanical phenomenon of matter. The advantage of quantum computing is parallelism based on the linear superposition of quantum states. It has been shown that certain problems can be solved in fewer steps by quantum algorithms than by existing classical ones. A 2-state quantum system is usually defined with the two pure states $|0>$ and $|1>$ as basis states, and the unit of information is known as a qubit. Presently, multi-valued quantum computing is gaining importance in the field of quantum information theory and quantum cryptography as it can represent an n-dimensional quantum system, defined by the basic states $|0>, |1>, ...,|-1>$.

The unit of information is called a qubit. For a multi-valued quantum system, this unit is termed qutrit. A multi-valued quantum system exists is a linear superposition of three basic states, labeled $| 0 >$, $| 1 >$, and $| 2 >$.

Multi-Valued logic is one of the cases that drew the attention of the researchers. Meanwhile, the multi-valued circuits were more accepted because of their advantages over the binary circuits. Multi-Valued quantum computation has high fault tolerance and high speed as compared to the binary and it is useful for quantum simulation, quantum tomography, and quantum games.

In multi-valued quantum computing, a single qubit gate (NOT, shift operation as, 1, and 2), double qubit gate (CNOT, controlled shift operation), and triple qubit gate (C^2NOT, controlled shift operation) are there. The calculation of the required time for a quantum gate and the operational gate is described in this section.

A. Speed Calculation in Multi-Valued 1-Qubit Controlled Operation

Here, Figure 6.1 shows the multi-valued Quantum 1-qubit controlled operation, which is an important and basic gate for multi-valued quantum computing. The one qubit-controlled operation or CNOT operational gate takes two input and provide one output (A, B as input and R as output in the figure). It is a two-qubit gate and according to Section 6.2, it can be assumed that it needs 10 microseconds to perform.

According to Section 6.2, all double qubit gates are 1-qubit controlled operations and all 1-qubit shift operations need almost 10 μs to perform.

B. Speed Calculation in C^2NOT Operation

Here, Figure 6.2 illustrates a multi-valued quantum 2-qubit controlled operation, which is named as C^2NOT gate. It can be used to implement all multi-valued quantum operations. It is a three-qubit gate and it can be implemented as two double-qubit operational gates. Further, according to Section 6.2, it can be assumed that each double qubit gate needs 10 μs to perform.

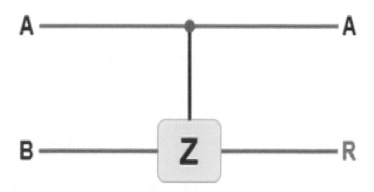

FIGURE 6.1
Multi-Valued 1-Qubit Controlled Operation

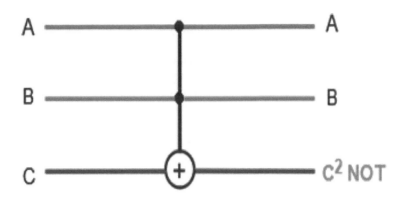

FIGURE 6.2
Multi-Valued 3-qubit C²NOT Gate Operation

According to Section 6.2, the required time for the C²NOT gate operation is (10 + 10) μs.
    = 20 μs.
In addition, all triple qubit gate operations are 2-qubit controlled operations and all 2-qubit shift operations need almost 20 μs to perform operation.

C. Speed Calculation in Quantum Multi-Valued AND Operation
Here, Figure 6.3 shows the multi-valued quantum AND operation. The quantum multi-valued AND operation requires the shifting operations that include (+1) and (+2) shifting, where one (+2) operation is controlled by both inputs and will open only if the value of both inputs is |2>. The other (+1) operation is controlled by both inputs and will open only if the value of both inputs is |1>. And a C²NOT gate is used which will work only if the inputs A and B meet conditions of A! = B and A, B! = 0.

The following steps can help to calculate required time for multi-valued quantum AND operation as follows:

1. The triple qubit (+2) shift operations required time is 20 μs,

2. The triple qubit C²NOT operations required time is 20 μs,

3. The triple qubit (+1) shift operations required time is 20 μs

So, the required time for the AND quantum gate operation = (operational time for (+2) shift operation + operational time for C²NOT gate + operational time for (+1) shift operation) μs.
    = (20 + 20 + 20)μs.
    = 60 microseconds (μs) Thus, the total required time for quantum multi-valued AND operation is 60 microseconds.

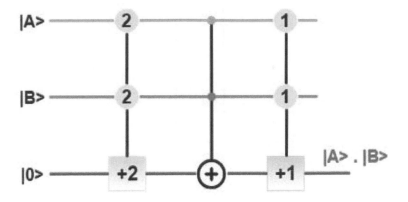

FIGURE 6.3
The Circuit Diagram of Multi-Valued AND Operation

D. Speed Calculation in Quantum Multi-Valued OR

Here, Figure 6.4 describes multi-valued quantum OR operation. The quantum multi-valued OR operation requires the shifting operations that include (+1) and (+2) shifting, where two (+1) operations are not controlled, and two (+1) operations are controlled by two inputs. Further, two (+2) operations are controlled by each input. And, a two-input controlled C^2 NOT gate is also required to get the expected result.

The following steps can help to calculate the required time for a multi-valued quantum OR operation as follows:

1. The required time for double qubit (+2) shift operation is 10 μs,

2. The required time for double qubit (+2) shift operation, whose input depends on the output of first (+2) shift operation is 10 μs,

3. The required time for triple qubit (+1) shift operations, whose input depends on the output of first and second (+2) shift operation is 20 μs,

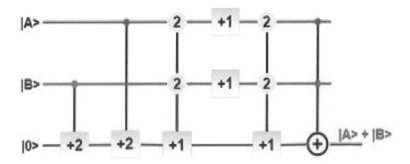

FIGURE 6.4
The Circuit Diagram of Quantum Multi-Valued OR Operation

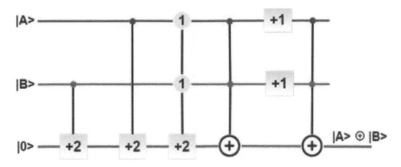

FIGURE 6.5
The Circuit Diagram of Multi-Valued XOR Operation

4. Next, two one qubit operation (+1) shift will perform parallelism. So, it can be avoided.

5. The required time for triple qubit (+1) shift operations, whose input depends on the output of previous operations is 20 μs,

6. The required time for the last triple qubit C^{2}NOT operation, whose input depends on the output of the previous operation is 20 μs.

So, the required time for the quantum AND gate operation is (2 × 10 + 3 × 20) microseconds
= 80 microseconds
Thus, the total required time for quantum multi-valued OR operation is 80 μs.

E. Speed Calculation in Quantum Multi-Valued XOR Operation
Here, Figure 6.5 depicts a multi-valued quantum XOR operation. The quantum multi-valued XOR operation requires the shifting operations that include (+1) and (+2) shifting, where two (+1) operations are not controlled, and two (+2) operations are controlled by respective one input. And one (+2) operation is controlled by both inputs. And, two two-input-controlled C^2 NOT gates are also required to get the expected result.

The following steps can help to calculate the required time for multi-valued quantum OR operation as follows:

1. The required time for double qubit (+2) shift operation is 10 μs,

2. The required time for double qubit (+2) shift operation, whose input depends on the output of first (+2) shift operation is 10 μs,

3. The required time for triple qubit (+2) shift operations, whose input depends on the output of first and second (+2) shift operation is 20 μs,

4. The required time for triple qubit C^{2}NOT operation, whose input depends on the output of the previous operations is 20 μs.

5. Next, two one-qubit operations (+1) shifts will perform in parallel way. So, its operational time can be avoided.

6. The required time for the last triple qubit C^2NOT operation, whose input depends on the output of the previous operations is 20 μs.

So, the required time for the quantum XOR gate operation is $(2 \times 10 + 3 \times 20)$ microseconds

$= 80\ \mu s$.

Thus, the total required time for quantum multi-valued XOR operation is 80 μs.

F. Speed Calculation in Quantum Multi-Valued NAND Operation

The quantum multi-valued NAND operation is jointly quantum AND and quantum NOT operation. Therefore, it is only needed to invert the output of the quantum multi-valued AND operation. The circuit diagram of the quantum multi-valued NAND operation is shown in Figure 6.6.

The following steps can help to calculate the required time for multi-valued quantum NAND operation as follows:

1. The required time for triple qubit (+2) shift operation is 20 microseconds,

2. The required time for triple qubit C^2NOT operation, whose input depends on the output of the previous operations is 20 microseconds.

3. The required time for triple qubit (+1) shift operations, whose input depends on the output of first and second operation is 20 microseconds,

So, the required time for the quantum NAND gate operation is (3×20) microseconds

$= 60$ microseconds

Thus, the total required time for a quantum multi-valued NAND operation is 60 microseconds.

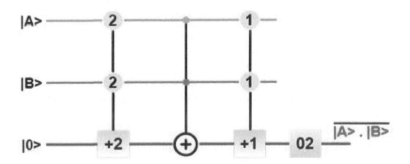

FIGURE 6.6
The Circuit Diagram of Quantum Multi-Valued NAND Operation

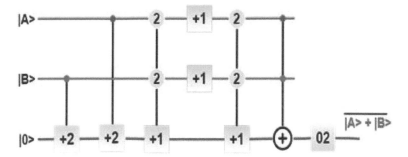

FIGURE 6.7
The Circuit Diagram of Quantum Multi-Valued NOR Operation

G. Speed Calculation in Quantum Multi-Valued NOR Operation
The quantum multi-valued NOR operation is quantum OR and quantum NOT operation together. Therefore, it is only needed to invert the output of the quantum multi-valued OR operation. The circuit diagram of the quantum multi-valued NOR operation is shown in Figure 6.7.

The following steps can help to calculate required time for quantum multi-valued OR operation:

1. The double qubit (+2) shift operation is 10 microseconds,
2. The double qubit (+2) shift operation, whose input depends on the output of first (+2) shift operation is 10 microseconds,
3. The triple qubit (+1) shift operations, whose input depends on the output of first and second (+2) shift operation is 20 microseconds,
4. Next, two one qubit operation (+1) shift will perform in parallel way. So, its operational time can be avoided.
5. The triple qubit (+1) shift operation, whose input depends on the output of the previous operations is 20 microseconds.
6. The triple qubit C^2NOT operation, whose input depends on the output of the previous operations is 20 microseconds.
7. The last single-qubit operation needs 1 microseconds.

So, the required time for the quantum XOR gate operation is $(2 \times 10 + 3 \times 20 + 1)$ microseconds
= 81 microseconds
Thus, the total required time for quantum multi-valued NOR operation is 81 microseconds.

H. Speed Calculation in Quantum Multi-Valued Decoder
The quantum multi-valued 1-to-3 decoder operation in Figure 6.8 requires the shifting operations that include (+2) shifting only, where three (+2) operations are

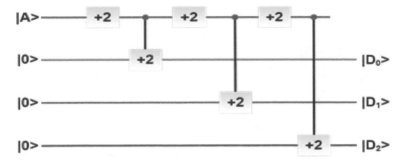

FIGURE 6.8

The Circuit Diagram of a Quantum Multi-Valued Decoder

not controlled and three (+2) operations are controlled by respective one input. For a quantum multi-valued decoder, one input and three ancilla qubits are needed and it provides three outputs as $|D_0>$, $|D_1>$, and $|D_2>$ as shown in Figure 6.8. The circuit of a quantum multi-valued 1-to-3 decoder operation is shown in Figure 6.8.

The following three output pipelines can help to calculate the required time of a quantum multi-valued OR operation as follows:

1. D_0, : It gives the output after performing one single-qubit (+2) shift operation and one double qubit (+2) shift operation which requires total $(1 + 10) = 11 \ \mu s$;

2. D_1, : It gives the output after performing two single-qubit (+2) shift operations and one double qubit (+2) shift operation which requires total $(1 + 1 + 10) = 12 \ \mu s$; and

3. D_2, : It gives the output after performing three single qubit (+2) shift operations and one double qubit (+2) shift operation which requires total $(1 + 1 + 1 + 10) = 13 \ \mu s$.

So, the maximum required time for the quantum decoder operation is 13 microseconds (μs).

6.3.1   Quantum Multi-Valued 3-to-1 Multiplexer

A multiplexer (MUX) is a device that can receive multiple input signals and synthesize a single output signal in a recoverable manner for each input signal. It is also an integrated system that usually contains a certain number of data inputs and a single output. A multi-valued multiplexer of 3^n input has n select lines. And the selected line is used as the selection of inputs for a particular output. The quantum circuit for the quantum multi-valued 3-to-1 multiplexer is depicted in Figure 6.9, where $|D_0>$, $|D_1>$, and $|D_2>$ are input qubits and —Y is the output qubit that is selected by a selection qubit —S.

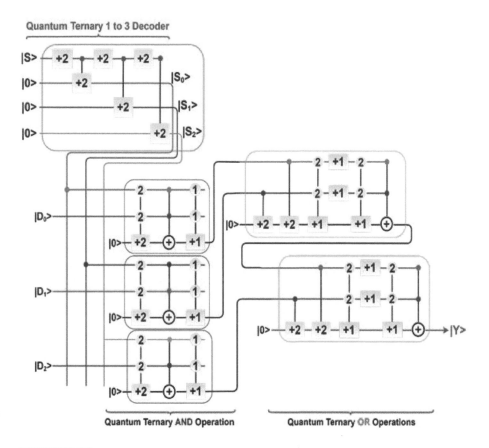

FIGURE 6.9

The Circuit Diagram of Quantum Multi-Valued 3-to-1 Multiplexer

To find the required time for the operation of a quantum multi-valued 3-to-1 multiplexer, divide it into three pipelines as some of the basic quantum multi-valued gate operations are performed in parallel. Three pipelines are as follows:

1. Decoder, AND, OR, and OR;

2. Decoder, AND, OR, and OR; and

3. Decoder, AND, and OR.

AND and OR quantum multi-valued gate operation needs 60 and 80 μs, respectively. Further, a quantum multi-valued decoder needs 13 μs to perform its operation. As both the first and second pipelines are the largest pipelines for providing the output of the quantum multi-valued 3-to-1 multiplexer. On the otherhand, quantum multi-valued basic operations occur within the time in parallel.

The required time for quantum multi-valued 3-to-1 multiplexer is (Decoder + AND + OR + OR) μs.

where, the required time for a quantumm multi-valued decoder operation is 13 μs,
the required time for the quantum mMulti-Valued AND operation is 60 μs,
the required time for the quantum multi-valued OR operation is 80 μs, and
the trquired time for the quantum multi-valued 3-to-1 multiplexer operation is $(13 + 60 + 80 + 80)\mu s = 233 \ \mu$s.

6.3.2 Quantum Multi-Valued Half Adder

A multi-valued half adder is a type of adder, an electronic circuit that performs the addition of ternary numbers. The half adder is able to add two single ternary digits and provide the output plus a carry value. It has two inputs, called A and B, and two outputs S (sum) and C (carry).

Quantum multi-valued half adder adds two qubits from the quantum multi-valued qubits |0>, |1>, and |2> and produces two outputs as quantum sum |S> and quantum carry qubit |C>. Figure 6.10 shows a quantum multi-valued half adder.

To find the required time of a quantum multi-valued half adder operation, it can be divided into multiple pipelines, where just five of them are presented below. As well, these five pipelines of quantum multi-valued operations and some of other basic multi-valued quantum gate operations perform in parallel. Five pipelines are as follows:

1. Decoder, AND, OR, OR, and OR;

2. Decoder, AND, OR, OR, and OR;

3. Decoder, AND, OR, OR, and AND;

4. Decoder, AND, OR, OR, and AND; and

5. Decoder, AND, OR, OR, AND, and OR.

AND and OR quantum multi-valued operations need 60 microseconds and 80 microseconds, respectively. And the decoder needs 13 microseconds to perform its operation.

The last pipeline is the largest pipeline for providing an output of the quantum multi-valued half adder and it is taken for measuring the total required performing time of the adder. Other quantum basic multi-valued operations are performed within this time in parallel.

The required time for a quantum multi-valued half adder operation is (Decoder + AND + OR + OR + AND + OR) microseconds,
where the required time for a quantum multi-valued decoder operation is 13 μs,
the required time for a quantum multi-valued AND operation is 60 μs,
the required time for a quantum multi-valued quantum OR operation is 80 μs, and
the required time for a quantum multi-valued 3-to-1 multiplexer operation is $(13 + 2 \times 60 + 3 \times 80) \ \mu$s
= 373 μs.

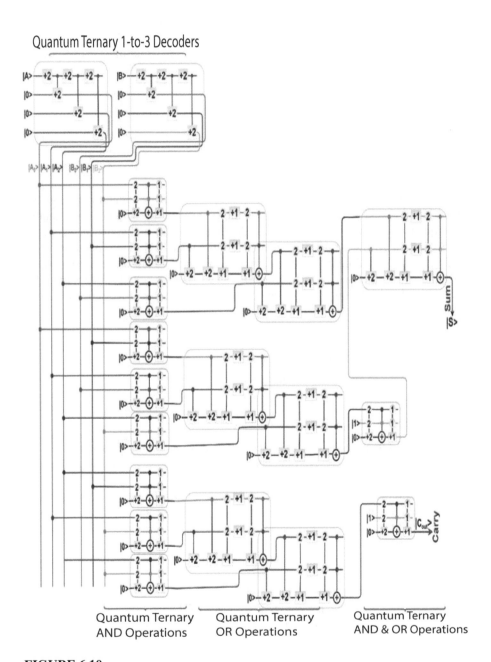

FIGURE 6.10
The Circuit Diagram of Quantum Multi-Valued Half Adder

6.3.3 Quantum Multi-Valued Half Subtractor

A multi-valued half subtractor is a type of subtractor, an electronic circuit that performs the subtractions of ternary numbers. The half subtractor is able to subtract two single ternary digits and provide the output plus a borrow value. It has two inputs, called A and B, and two outputs D (difference) and B (borrow). Quantum multi-valued half subtractor subtracts two qubits from the quantum ternary digit |0>, |1>, and |2> and produce two outputs as quantum difference |D> and quantum borrow qubit |B>. The quantum circuit for the multi-valued half subtractor is shown in Figure 6.11.

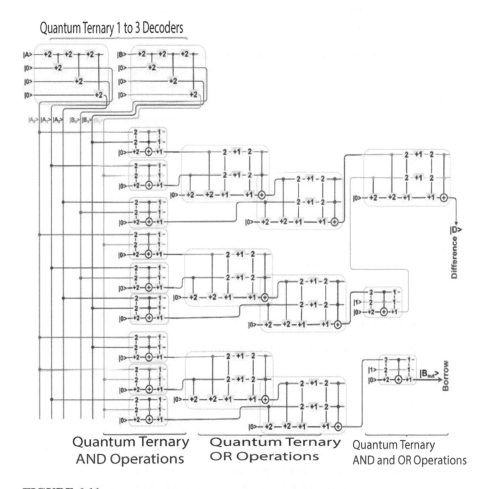

FIGURE 6.11
The Circuit Diagram of Quantum Multi-Valued Half Subtractor

To find the required time for performing operations in a quantum multi-valued half subtractor, it can be divided into multiple pipelines, where just five of them are presented below. As well, these five pipelines of quantum multi-valued operations and some other basic quantum multi-valued operations are performed in parallel. Five pipelines are as follows:

1. Decoder, AND, OR, OR, and OR;

2. Decoder, AND, OR, OR, and OR;

3. Decoder, AND, OR, OR, and AND;

4. Decoder, AND, OR, OR, and AND; and

5. Decoder, AND, OR, OR, AND, and OR.

AND and OR quantum multi-valued operations need 60 microsecond and 80 microseconds, respectively. And the decoder needs 13 microseconds to perform its operation. Among all pipelines input to an output, the last pipeline is the largest pipeline for providing output of the quantum multi-valued half subtractor, it is taken for measuring the total required time for performing the operations. Other quantum multi-valued basic operations are performed within this time in parallel.

The required time for multi-valued half subtractor is (Decoder + AND + OR + OR + AND + OR) microseconds;
where the required time for a quantum multi-valued decoder is 13 microseconds,

the required time for a quantum multi-valued AND operation is 60 microseconds,

the required time for a quantum multi-valued OR operation is 80 microseconds, and

the required time for a quantum multi-valued half subtractor is $(13 + 2 \times 60 + 3 \times 80)$ microseconds = 373 microseconds.

6.4   Speed Calculation for DNA Operation

DNA molecules can be used as information storage media. Usually, DNA sequences of around 8-20 base pairs are used to represent bits, and numerous methods have been developed to manipulate and evaluate these. In order to manipulate a wet technology to perform computations, one or more of the following techniques are used as computational operators for copying, sorting, splitting, or concatenating the information contained within DNA molecules as ligation, hybridization, polymerase chain reaction (PCR), gel electrophoresis, and enzyme reaction.

Allosteric DNAzyme-based DNA logic circuit, described a procedure to make a DNAzyme-based logic circuit. Here, All DNA operations are formed by annealing twice: firstly, the mixture of the inhibitor DNA strands and E6-type DNAzymes in 1× TAE/Mg_2+ buffer (40 mM Tris, 20 mM acetic acid, 1 m MEDTA$_2$Na and 12.5 mM Mg(OAc)$_2$, pH 8.0) is heated at 95°C for 4 min, 65°C for 30 min, 50°C for 30 min, 37°C for 30 min, 22°C for 30 min, and preserved at 20°C; and then the substrates are added into the annealed mixture and incubated at constant temperature 20°C for 4 hours (total 6 hours for preparing the DNA operation).

After that Logic operations are triggered through displacement reaction in 1× TAE/Mg2+ buffer (40 mM Tris, 20 mM acetic acid, 1 mM EDTA$_2$Na, and 12.5 mM Mg (OAc)$_2$, pH 8.0). The input DNA strands are added to a solution containing DNA operations and reacted for > 2 hours at 20°C. Next, the displaced products are stored at 20°C for native PAGE or fluorescence detection. In addition, polyacrylamide gel electrophoresis (PAGE) needs 2 hours and the PCR process for fluorescence detection needs less than 2 hours.

Here a specific biochemical process is described briefly which serves as the basis of the DNA computing approach as Polymerase Chain Reaction (PCR). Polymerases perform several functions, including the repair and duplication of DNA. PCR is a process that quickly amplifies the amount of specific DNA molecules in a given solution using primer extension by the polymerase. Each cycle of the reaction doubles the quantity of this molecule, leading to an exponential growth in the number of sequences. It consists of the following key processes:

1. Initialization: mix a solution of template, primer, dNTP, and the enzyme that is heated to 94 −98°C for 1 − 9 minutes to ensure that most of the DNA template and primers are denatured;

2. Denaturation: heat the solution to 94 − 98°C for 20 − 30 seconds for separation of DNA duplexes;

3. Annealing: lower the temperature enough (usually between 50−64°C) for 20−40 seconds for primers to anneal specifically to the ssDNA template;

4. Elongation/Extension: raise temperature to optimal elongation temperature of *Taq* or similar DNA polymerase (70 − 74°C) for the polymerase adds dNTP's from the direction of 5' to 3' that are complementary to the template; and

5. Final Elongation/Extension: after the last cycle, a 5 − 15 minutes elongation may be performed to ensure that any remaining ssDNA is fully extended.

Steps 2 to 4 are repeated 20−35 times; fewer cycles results in less product, too many cycles increase the fraction of incomplete and erroneous products. PCR is a routine job in the laboratory that can be performed by an apparatus named a *thermal cycler*. According to the PCR process, producing an operational output of DNA computation requires around 2 hours.

So, except for initial preparation and the last phase of fluorescence detection for each operation in a particular test tube, it requires more or less 2 hours.

6.5 Speed Calculation in Multi-Valued DNA Circuit

DNA computing embraces emerging biology-dependent technological areas such as DNA computing systems, DNA security, and scheduling systems. The advantage of DNA computing is parallelism based on DNA strands, which are composed of Adenine, Thymine, Guanine, and Cytosine. It has been shown that certain problems can be solved in fewer steps and with less memory by DNA computing algorithms than by existing classical computing systems. A 2-strand base DNA system is usually defined with the two pure strands as CAAGCT and ACCTAG. At the same time, multi-valued DNA computing can be used as an important matter in the field of DNA computation systems, DNA scheduling, and security system as it can represent multiple DNA strands in the system for carrying information. For a multi-valued DNA computation system, the basic three strands can be TGGATC, CAAGCT, and ACCTAG.

Multi-Valued DNA computation systems can draw the attention of researchers because of their advantages over binary computation systems. For ensuring security as intrusion detection, job scheduling and huge data clustering can be solved using a multi-valued DNA computing system.

6.5.1   DNA Multi-Valued Decoder

A Decoder is a combinational circuit that has 'n' input lines and a maximum of 2^n output lines. One of these outputs will be active High based on the combination of inputs present when the decoder is enabled. That means the decoder detects a particular code. In Figure 6.12, decoder is the unary function for input variable A as A0, A1, and A2 which is used for multi-valued function implementation. So, the DNA multi-valued decoder is shown in Figure 6.12.

To find the required time for performing the operations of the DNA multi-valued decoder, it is divided into four pipelines as some of the basic DNA multi-valued operations are performed in parallel. Four pipelines are as follows:

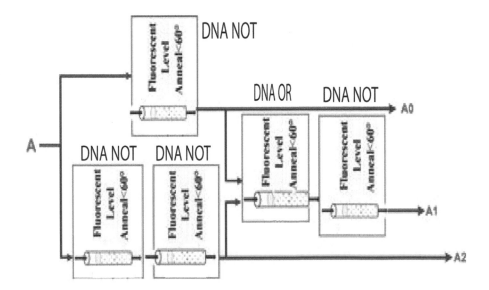

FIGURE 6.12
DNA Multi-Valued Decoder

1. NOT;

2. NOT, OR, and NOT;

3. NOT, NOT, OR, and NOT; and

4. NOT and NOT.

DNA basic (i.e., OR and NOT) operations need more or less 2 hours to perform the operations. In addition, it needs 6 hours for preparing any DNA basic operation and 2 hours for fluorescence detection which is fixed for any DNA multi-valued basic operation.

As the third pipeline is the largest pipeline for processing input to the output of the multi-valued decoder, it is taken for measuring the total required time for performing operations. Other DNA multi-valued basic operations are performed within this time in parallel.

So, the required time for four basic operations in DNA is $= (2 + 2 + 2 + 2)$

$= 8$ hours;

where the required time for DNA (OR and NOT) operations need more or less 2 hours.

The total required time for performing DNA multi-valued decoder operation is the summation of initial preparation time, fluorescence detection time, and multi-valued DNA operation time.

So, the required time for DNA multi-valued full adder operation is

$=$ (Basic operation preparation time + Four basic DNA multi-valued operation time + Fluorescence detection time)

$= (6 + 8 + 2)$ hrs

$= 16$ hours (approximately).

6.5.2   DNA Multi-Valued 3-to-1 Multiplexer

The multiplexer is one of the most important designs of a binary digital system. It is such a device that allows only one input from several input signals and the input which is selected by a multiplexer is transmitted into a single medium. Actually, multiplexers aid to improve the efficiency of the communication system. It allows the transmission of data such as audio, video, etc. from different channels via cables. And here, this will be implemented using MVL to simulate the circuit so easily than the digital binary Multiplexer. Then the designing of a DNA 3-to-1 multi-valued multiplexer circuit is constructed based on a combinational digital 3-to-1 multi-valued circuit.

To find the required time for performing operations of a DNA 3-to-1 multiplexer circuit, it is divided into four pipelines as some of the basic DNA operations are performed in parallel. All these pipelines are as follows:

1. NOT, AND, OR, and OR;

2. NOT, NOR, NOR, AND, OR, and OR;

3. NOT, NOT, NOR, NOR, AND, OR, and OR; and

4. NOT, NOT, AND, and OR.

Any DNA multi-valued basic (i.e. AND, OR, NOT, and NOR) operation needs more or less 2 hours to perform its operation. In addition, it needs 6 hours for preparing any multi-valued DNA basic operation and 2 hours for fluorescence detection which is fixed for any multi-valued basic operation.

As the third pipeline is the longest pipeline for processing input to the output of the 3-to-1 multiplexer circuit, it is taken for measuring the total required time for performing operations. Other multi-valued DNA basic operations are performed within this time in parallel.

So, the required time for seven multi-valued basic operations in DNA is (2×7) = 14 hours;

where the time for performing multi-valued DNA (AND, OR, NOT and NOR) operation needs more or less 2 hours.

The total time for performing multi-valued DNA 3-to-1 multiplexer circuit operation is the summation of initial preparation time, fluorescence detection time and DNA operation time.

So, the required time for DNA 3-to-1 multiplexer circuit operation is

= (Basic operation preparation time + seven basic DNA operations time + fluorescence detection)

= $(6 + 14 + 2)$ hrs = 22 hours (approximately).

The circuit architecture of a DNA multi-valued 3-to-1 multiplexer is shown in Figure 6.13.

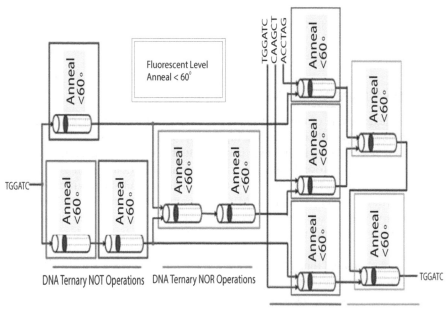

FIGURE 6.13
DNA Multi-Valued 3-to-1 Multiplexer

6.6   Speed Calculation in Multi-Valued Quantum-DNA Circuit

Multi-Valued quantum computation is far more advantageous than classical computation systems. Quantum computers are also more powerful in terms of computation than supercomputers. They process data thousands of times faster than normal computers and supercomputers. Parallelism based on linear superposition of quantum states is a benefit of multi-valued quantum computing. It has been demonstrated that multi-valued quantum algorithms can solve some problems in fewer steps than conventional classical algorithms.

The multi-valued (more than two DNA strands) DNA strands are used to compute different logical and mathematical operations. When compared to traditional storage systems, DNA requires just about 1 bit per cubic nanometer of memory space. The chemical interactions in DNA provide energy to make or repair new strands, therefore there is essentially no power use. As a result, it is possible to create a multi-valued Quantum-DNA computation system to find a super-fast computation system with a lot of memory. The advantages of multi-valued quantum computing and multi-valued DNA computing can be combined together to form the multi-valued quantum-DNA computation system.

In a multi-valued Quantum-DNA computing system, the input is received as a qubit, which is then converted into DNA sequences by NMR relaxation after passing through a specific number of quantum gates.

6.6.1   Multi-Valued Half Adder at 0-Kelvin

A multi-valued half adder is an electronic circuit that performs the addition of ternary numbers. The half adder is able to add two single ternary digits and provide the output plus a carry value. It has two inputs, called A and B, and two outputs S (sum) and C (carry).

Quantum-DNA multi-valued half adder adds two qubits from the quantum ternary qubits |0>, |1>, and |2> and produces two outputs as quantum sum |S> and quantum carry |C>.

From the multi-valued quantum-DNA circuit as shown in Figure 6.14, it is found that 11 AND, 7 OR, and 2 decoder operations perform to provide the output in DNA sequence. In this circuit, nine AND, six OR, and two decoders perform with ternary quantum qubit and the rest perform with DNA sequences. The output of all quantum operations goes through NMR relaxation at 0-kelvin. It produces DNA sequences as an output and it can be used as an input in all DNA operations.

To find the required time of quantum-DNA multi-valued half adder, it can be divided into multiple pipelines, where just five of them are presented below. I addition to these five pipelines of quantum-DNA multi-valued operations, some other basic multi-valued quantum operations are performed in parallel. Five pipelines are as follows:

1.   Quantum (Decoder, AND, OR, OR), and DNA (OR);
2.   Quantum (Decoder, AND, OR, OR), and DNA (OR);

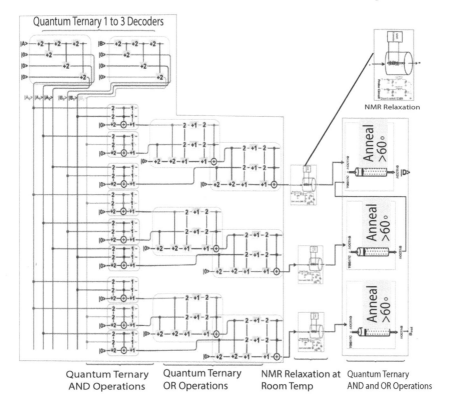

FIGURE 6.14

Quantum-DNA Circuit for Multi-Valued Half Adder

3. Quantum (Decoder, AND, OR, OR), and DNA (AND);
4. Quantum (Decoder, AND, OR, OR), and DNA (AND); and
5. Quantum (Decoder, AND, OR, OR), and DNA (AND, OR).

AND and OR multi-valued quantum operations need 60 μs and 80 μs, respectively. And the decoder needs 13 μs to perform its operation. Among all pipelines from input to output, the last pipeline is the largest pipeline for providing the output of the quantum-DNA multi-valued half adder which is used for measuring the total required performing time. Besides these operations, quantum and DNA multi-valued basic operations are calculated within this time in parallel. The quantum-DNA circuit for the multi-valued half subtractor is shown in Figure 6.15.

The required time for quantum operation in multi-valued half adder operation is (Decoder + AND + OR + OR) μs;

where the required time for multi-valued quantum decoder operation is 13 μs,
the required time for basic multi-valued quantum AND operation is 60 μs,
the required time for basic multi-valued quantum OR operation is 80 μs, and
the required time for the quantum operations is (13+ 60 + 80 +80) μs

= 233 μs. Any DNA basic operation (i.e., AND, OR, NOT, and XOR) needs more or less 2 hours to perform its operation. In addition, it needs 6 hours for preparing any DNA basic operation and 2 hours for fluorescence detection which are fixed for any multi-valued basic operation.

So, the required time for DNA AND and OR operations are (2 + 2) hrs
= 4 hrs.

Thus, the total performing time required for DNA operations is the summation of initial preparation time, fluorescence detection time and DNA operation time.

So, the required time for DNA operations is

= (Basic operation preparation time + DNA basic operation time + Fluorescence detection)

= (6 + 4 + 2) hrs

= 12 hours (approximately).

Therefore, the required time is the summation of quantum operations and DNA operations to find the expected output from the quantum-DNA multi-valued half adder circuit.

So, the total required time for the quantum-DNA multi-valued half adder operation is

= (The required time for quantum operation + The required time for DNA operation)

= (233 microseconds + 12 hours)

=12 hours (approximately).

The quantum-DNA circuit for the multi-valued half adder operation is shown in Figure 6.14.

6.6.2 Multi-Valued Half Subtractor at 0-Kelvin

A multi-valued half subtractor is an electronic circuit that performs the subtractions of ternary numbers. The half subtractor is able to subtract two single ternary digits and provide the output with a borrow value. It has two inputs, called A and B, and two outputs D (difference) and B (borrow). A quantum-DNA multi-valued half subtractor subtractr two qubits from the quantum ternary digit $|0>$, $|1>$, and $|2>$ and produces two outputs as quantum difference $|D>$ and quantum borrow bit $|B>$.

From the multi-valued quantum-DNA half subtractor circuit as shown in Figure 6.15, it is found that 11 AND, 7 OR, and 2 decoder operations are performed to provide the output in a DNA sequence. In this circuit, nine AND, six OR, and two decoder operations are performed with multi-valued quantum qubit and the rest are performed with DNA sequences. The output of all quantum operations goes through NMR relaxation at 0-kelvin. It produces DNA sequences as an output and it can be used as an input in all DNA operations.

To find the required time of a quantum-DNA multi-valued half subtractor operation, it can be divided into multiple pipelines and here, just five of them are presented below. Besides, these five pipelines of quantum-DNA multi-valued operations, some

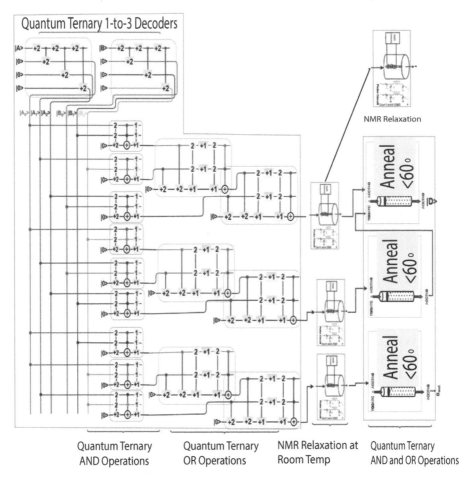

FIGURE 6.15
Quantum-DNA Circuit for Multi-Valued Half Subtractor

other basic Quantum and DNA multi-valued operations are performed in parallel. Five pipelines are as follows:

1. Quantum (Decoder, AND, OR, OR), and DNA (OR);

2. Quantum (Decoder, AND, OR, OR), and DNA (OR);

3. Quantum (Decoder, AND, OR, OR), and DNA (AND);

4. Quantum (Decoder, AND, OR, OR), and DNA (AND); and

5. Quantum (Decoder, AND, OR, OR), and DNA (AND, OR).

AND and OR multi-valued quantum operation needs 60 μs and 80 μs, respectively. And a decoder needs 13 μs to perform its operation. Among all pipelines from input to output, the last pipeline is the largest pipeline for providing the output of the quantum-DNA multi-valued half subtractor which is taken for measuring the total required performing time. Besides these operations, quantum and DNA multi-valued basic operations are performed within this time in parallel. The quantum-DNA circuit for the multi-valued half subtractor is shown as shown in Figure 6.15.

The required time for quantum operation in multi-valued half adder is

= (Decoder + AND + OR + OR) μs; where the required time for multi-valued quantum decoder operation is 13 μs,

the required time for basic multi-valued quantum AND operation is 60 μs,

the required time for basic multi-valued quantum OR operation is 80 μs, and

the required time for quantum operations is $(13 + 60 + 80 + 80)$ μs = 233 μs.

Any DNA basic operation (i.e., AND, OR, NOT, and XOR) needs more or less 2hr to perform its operation. In addition, it needs 6hr for preparing any DNA basic operation and 2hr for fluorescence detection which are fixed for any multi-valued basic operation.

So, the required time for DNA AND and OR operations is $(2 + 2)$ hrs

= 4 hrs.

Therefore, the total performing time required for DNA operations is the summation of initial preparation time, fluorescence detection time and DNA operation time.

Thus, the required time for DNA operations is = (Basic operation preparation time + DNA operation time + Fluorescence detection time)

= $(6 + 4 + 2)$

= 12hr (approximately).

Now, the required time is the summation of Quantum operations and DNA operations to find the expected output of the quantum-DNA multi-valued half subtractor.

So, the total required time for quantum-DNA multi-valued half adder operation is

= (The required time for quantum operation + The required time for DNA operation)

= (233 microseconds + 12hr)

= 12hr (approximately).

6.6.3 Multi-Valued Multiplexer at 0-Kelvin

The multiplexer is one of the most important designs of a multi-valued system. It is such a device that allows only one input from several input signals and the input which is selected by multiplexer is transmitted into a single medium. Multiplexers aid to improve the efficiency of the communication system. It allows the transmission of data such as audio, video, etc. from different channels via cables. And here, it is implemented using quantum-DNA multi-valued system that helps to simulate the circuit as easily as the binary multiplexer. Figure 6.16 shows the quantum-DNA 3-to-1 multi-valued multiplexer circuit based on a digital 3-to-1 multi-valued circuit.

From the multi-valued quantum-DNA 3-to-1 multiplexer circuit as shown in Figure 6.16, it is found that 3 ANDs, 2 ORs, and 1 decoder operation circuits are used

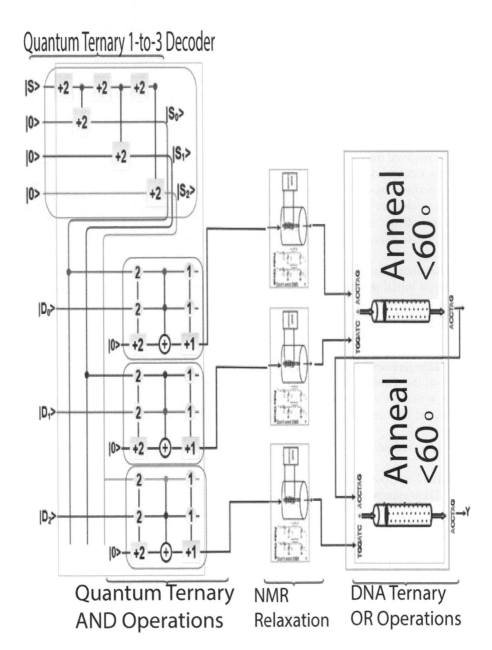

FIGURE 6.16
Quantum-DNA Circuit for Multi-Valued 3-to-1 Multiplexer

to provide the output in a DNA sequence. In this circuit, 3 ANDs and one decoder perform operation with ternary quantum qubit and the rest perform operations with DNA sequences. The output of all quantum gates goes through NMR relaxation at 0-kelvin. It produces DNA sequences as an output and it can be used as an input in all DNA operation circuits.

To find the required time of the quantum-DNA 3-to-1 multiplexer circuit, it is divided into three pipelines, where some of the basic DNA operations are performed in parallel. All these pipelines are as follows:

1. Quantum (Decoder, AND) and DNA (OR, OR);

2. Quantum (Decoder, AND) and DNA (OR, OR); and

3. Quantum (Decoder, AND) and DNA OR.

Multi-Valued quantum AND operation needs 60 μs. And the decoder needs 13 μs to perform its operation. Among all pipelines from input to output, the first pipeline is the largest pipeline for providing an output of the quantum-DNA multi-valued 3-to-1 multiplexer which is taken for measuring the total required time. Besides these operations, other quantum and DNA multi-valued basic operations are performed within this time in parallel.

So, the required time for quantum operation in multi-valued 3-to-1 multiplexer is (Decoder + AND) μs;

where the required time for multi-valued quantum decoder operation is 13 μs,

the required time for basic multi-valued quantum AND operation is 60 μs, and

the required time for quantum operations is $(13 + 60)$ μs

$= 73 \mu s$.

Any multi-valued DNA basic (i.e., AND, OR, NOT, and NOR) operation needs more or less 2 hours to perform its operation. In addition, it needs 6 hours for preparing any DNA multi-valued basic operation and 2 hours for fluorescence detection which are fixed for any multi-valued basic operation.

So, the required time for two multi-valued basic operations in DNA is $(2+2)$ hrs

$= 4$ hours,

where the time for multi-valued DNA (AND, OR, NOT and NOR) operation needs more or less 2 hours.

The total required time for performing DNA operations in a 3-to-1 multiplexer circuit is the summation of initial preparation time, fluorescence detection time, and DNA operation time.

So, the required time for DNA operations in 3-to-1 multiplexer circuit = (Basic operation preparation time + four basic DNA operation time + fluorescence detection time)

$= (6 + 4 + 2)$ hrs

$= 12$ hours (approximately).

Thus, the required time is the summation of quantum operations and DNA operations to find the expected output of a quantum-DNA multi-valued 3-to-1 multiplexer operation.

So, the total required time for a quantum-DNA multi-valued 3-to-1 multiplexer is

$= $ (The required time for quantum operation + The required time for DNA operation)

= (73 microseconds + 12 hours)
= 12 hours (approximately).

6.7 Speed Calculation in Multi-Valued DNA-Quantum Circuit

It is possible to create a multi-valued DNA-Quantum computation system to find a super-fast computation system with a lot of memory. The advantages of multi-valued quantum computing and multi-valued DNA computing can be combined in this multi-valued DNA-Quantum computation system.

In a multi-valued DNA-Quantum computing system, input is received as a DNA sequence, which is then converted into quantum qubits by NMR after passing through a specific number of quantum gates.

6.7.1 Multi-Valued Half Subtractor at room temperature

A multi-valued half subtractor is an electronic circuit that performs the subtractions of ternary numbers. The half subtractor can subtract two single ternary digits and provide the output with a borrow value. It has two inputs, called A and B, and two outputs D (difference) and B (borrow). DNA-quantum multi-valued half subtractor subtracts two qubits as DNA sequence from the quantum ternary qubit |0>, |1>, and |2> and produces two outputs as quantum difference |D> and quantum borrow qubit |B>.

From the multi-valued DNA-quantum half subtractor circuit in Figure 6.17, it is found that 11 ANDs, 9 ORs, and 8 NOTs circuits are used to provide the output in quantum qubit. In this circuit, nine ANDs, five ORs, and eight NOTs DNA operations are performed with DNA sequence and the rest operations are performed with ternary quantum qubits. The output of all DNA operations goes through NMR at room temperature. It produces quantum qubits as an output and it can be used as an input in all quantum circuits.

Now, it can be divided into multiple pipelines to find the required time of DNA-quantum multi-valued half subtractor operation, where just five of them are presented below. Besides these five pipelines of DNA-quantum multi-valued operations, some of other basic quantum and DNA multi-valued operations are performed in parallel. Five pipelines are as follows:

1. DNA (NOT, OR, NOT, AND, OR) and Quantum (OR, OR);

2. DNA (NOT, NOT, OR, NOT, AND, OR) and Quantum (OR, OR);

3. DNA (NOT, NOT, OR, NOT, AND, OR) and Quantum (OR, AND, OR);

4. DNA (NOT, OR, NOT, AND, OR)and Quantum (OR, AND); and

5. DNA (NOT, NOT, OR, NOT, AND, OR) and Quantum (OR, AND).

Among all pipelines from input to output, the third pipeline is the largest pipeline for providing an output of the multi-valued DNA-quantum half subtractor which is used for measuring the total required time. Other DNA and quantum basic multi-valued operations are performed within this time in parallel.

Any DNA basic operation (i.e., AND, OR, NOT, and XOR) needs more or less 2 hours to perform its operation. In addition, it needs 6 hours for preparing any DNA basic operation and 2 hours for fluorescence detection which are fixed for any multi-valued basic operation.

So, the required time for DNA operations is (6×2) hrs
= 12 hrs.

The total time required for DNA operations is the summation of initial preparation time, fluorescence detection time and DNA operation time.

Therefore, the required time for DNA operations is

= (Basic operation preparation time + DNA operation time + Fluorescence detection time)

= $(6 + 12 + 2)$= 20 hours (approximately).

The DNA-quantum circuit for the multi-valued half subtractor is shown in Figure 6.17.

AND and OR quantum multi-valued operations need 60 μs and 80 μs, respectively.

So, the required time for quantum operation in multi-valued half subtractor is $(OR + AND + OR)\,\mu s$;

where the required time for basic multi-valued quantum AND operation is 60 μs;

the required time for basic multi-valued quantum OR operation is 80 μs; and

the required time for quantum operations is $(60 + 2 \times 80)\,\mu s$

= 220 μs.

Thus, the required time is the summation of quantum operations and DNA operations to find the expected output of a DNA-quantum multi-valued half subtractor circuit.

So, the total required time for DNA-quantum multi-valued half subtractor is

= (The required time for DNA operations + The required time for quantum operations)

= (220 microseconds + 20 hours)

= 20 hours (approximately).

Wave nature can be solved by a quantum computer. Quantum computers could aid in the development of improved climate models, allowing us to gain a better understanding of how humans affect the ecosystem. These models form the foundation for the projections of future warming, and they assist us in determining what steps need to be taken now to avoid calamities.

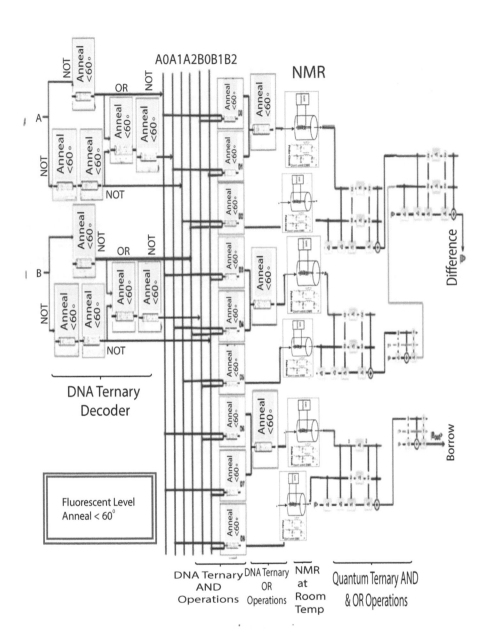

FIGURE 6.17
DNA-Quantum Circuit for Multi-Valued Half Subtractor

6.8 Summary

The required time for operations is important for any computation. It is found that, in quantum computing operation, the required time for performance depends on the number of qubits of the circuit. On the other hand, in DNA computing, the total required time for an operation depends on the number of used operations. This chapter describes the speed calculation for the performance measurement of multi-valued quantum, multi-valued DNA, multi-valued quantum-DNA and multi-valued DNA-quantum operation. The time of computation in quantum computing and the time in DNA computing are calculated in this chapter. All necessary techniques and procedures are shown here with appropriate figures.

Bibliography

[1] Wu, J., Shi, J., & Li, T. (2019). A novel image encryption approach based on a hyperchaotic system, pixel-level filtering with variable kernels, and DNA-level diffusion. Entropy, 22(1), 5.

[2] Troiani, F., Hohenester, U., & Molinari, E. (2001). Quantum–information processing in semiconductor Quantum dots. Physica Status Solidi (b), 224(3), 849-853.

[3] Levitin, L. B., Toffoli, T., & Walton, Z. (2002). Operation time of quantum gates. arXiv preprint quant-ph/0210076.

[4] Isailovic, N., Patel, Y., Whitney, M., & Kubiatowicz, J. (2006, June). Interconnection networks for scalable quantum computers. In 33rd International Symposium on Computer Architecture (ISCA'06) (pp. 366-377). IEEE.

[5] Adleman, L. M. (1994). Molecular computation of solutions to combinatorial problems. Science, 266(5187), 1021-1024.

[6] Zheng, X., Yang, J., Zhou, C., Zhang, C., Zhang, Q., & Wei, X. (2019). Allosteric DNAzyme-based DNA logic circuit: Operations and dynamic analysis. Nucleic Acids Research, 47(3), 1097-1109.

[7] SantaLucia Jr, J. (1998). A unified view of polymer, dumbbell, and oligonucleotide DNA nearest-neighbor thermodynamics. Proceedings of the National Academy of Sciences, 95(4), 1460-1465.

7

Heat Transfer

7.1 Introduction

Combinatorics is a type of calculation that traditional computers have trouble handling and it is one such challenge. These calculations involve arranging elements in a way that achieves a particular objective. As the number of things increases, so does the number of possible combinations. To find the ideal arrangement, today's digital computers must loop through each permutation until an outcome is found, and then determine which is the most effective at achieving the goal. In many circumstances, this will demand a large number of calculations. Combinatorics calculations prove problematic in a variety of sectors, ranging from banking to pharmaceuticals. Quantum computers come into play here. Quantum computing reduces the cost of calculating difficult combinatorial problems in the same way as classical computers reduced the cost of arithmetic. Multiple-valued quantum computing can bring more benefits to us.

Multi-Valued quantum computing is a field of study that focuses on the creation of computer-based technologies based on quantum-theoretical principles. On the quantum (atomic and subatomic) level, quantum theory describes the nature and behavior of energy and matter. To execute certain computational tasks, quantum computing employs a combination of qubits. All of this is done at a far higher rate than their traditional computing equipment. Quantum computers represent a significant advancement in computing capability, with enormous performance benefits for specific use cases. The ability of qubits to be in several states at the same time gives the quantum computer a lot of computing capability. They can accomplish jobs with a mix of $| 1 >$, $| 0 >$, and both $| 1 >$ and $| 0 >$ at the same time. Multiple-Valued Logic (MVL) is the non-binary-valued system, in which more than two levels of information content are available, i.e., L>2. In modern technologies, dual-level binary logic circuits have normally been used. So, multi-valued quantum computing can be defined as an area of computing that is focused on the development of computer technology based on the principles of quantum theory. In addition, quantum computing is much faster than classical bit-wise computing.

Multi-Valued quantum computing can be used in different sectors such as chemical and biological engineering, cybersecurity, artificial intelligence, financial service, and complex manufacturing. The discovery and manipulation of molecules are a part of chemical and biological engineering. Chemicals are made with subatomic particles and thus it involves quantum mechanics. For almost a thousand years,

DOI: 10.1201/9781003381938-7   119 combinatorics has been at the heart of cryptography. But using quantum computing cracking any encryption can be easier than other technology. Artificial intelligence frequently involves the combinatorics processing of very large amounts of data to make better predictions and decisions, and quantum computing is associated with this. Large amounts of data are the main domain of financial systems and manufacturing. Quantum computing has the potential to improve the speed of a critical set of financial calculations within a large amount of data. In recent years, billions of dollars have been invested in quantum computing because of its potential to tackle large-scale combinatorics problems faster and cheaper.

Every single cell that makes up a living organism contains information for various functions that are required for the cell's existence. Nucleic acids are molecules that contain genetic information in each cell. Deoxyribonucleic acid is the most stable type of nucleic acid (DNA). Each DNA strand forms helical structures, which are lengthy polymers made up of millions of nucleotides bonded together. One of four nitrogen bases, a five-carbon sugar, and a phosphate group make up these nucleotides. The genetic information is encoded by the nitrogen bases A (Adenine), T (Thymine), G (Guanine), and C (Cytosine), while the others offer structural stability. T with A and C with G are the base-pairing rules that connect the strands. The order in which these nucleotides are arranged is essential since it determines how different genes function.

So, DNA computing is a kind of natural computing that uses the molecular characteristics of DNA to conduct logical and arithmetic operations instead of typical carbon/silicon chips. The four-character genetic alphabets (A-adenine, G-guanine, C-cytosine, and T-thymine) are used in DNA computing instead of the binary alphabet (1 and 0) utilized by standard computers. This enables massively parallel computation, making it possible to answer difficult mathematical equations or problems in a fraction of the time. As a result, computation is far more efficient with a large volume of self-replicating DNA than with a standard computer, which would require a lot more hardware. Information or data will now be kept in the form of the bases A, T, G, and C, rather than binary digits. The capacity to generate short DNA sequences artificially allows these sequences to be used as inputs for algorithms. This is possible due to the ability to create small DNA molecules with any arbitrary sequence. The input of any DNA operation can be represented by DNA molecules with specific sequences. The instructions are carried out by laboratory operations on the molecules, and the result is defined as some property of the final set of molecules. DNA computing promises significant and meaningful linkages between computers and life systems, as well as massively parallel computations. DNA computing can carry out millions of operations at the same time.

Multi-Valued DNA computing is introduced based on the usage of DNA and molecular biology hardware instead of the typical silicon-based technology. Molecular computers could take advantage of DNA's physical properties to store information and perform calculations. These include extremely dense information storage, enormous parallelism, and extraordinary energy efficiency. DNA computing has an arresting performance in breaking Data Encryption Standards, Satisfiability problem solving, and traveling salesman problems.

It is already found that the quantum computation qubits produce large amounts of heat. On the other hand, DNA computation needs energy or heat to perform its expected operations. To process one or multiple DNA sequences, it is needed to perform multiple steps using different levels of heat or temperature. The main objective is to provide heat from multi-valued quantum logical gate operations to multi-valued DNA logical gate operations using a heat conduction system. Heat transfer is important in protecting the environment by reducing emissions and pollutants. In this chapter, a heat transferring procedure for the Quantum-DNA computation system will be shown.

7.2 Heat Transfer in Multiple-Valued Quantum-DNA Circuits

Multi-Valued quantum computation, according to multi-valued quantum computing, is far more advantageous than classical computation systems. Quantum computers are also more powerful in terms of computation than supercomputers. They process data 1000 times faster than normal computers and supercomputers. Quantum computers can perform computations in a fraction of a second that would take a traditional computer 1000 years to finish. Parallelism based on linear superposition of quantum states is a benefit of multi-valued quantum computing. It has been demonstrated that multi-valued quantum algorithms can solve some problems in fewer steps than conventional classical algorithms.

The use of multi-valued (more than two DNA strands) DNA strands to compute, on the other hand, has resulted in high parallel computation, which compensates for the chip's slow processing. When compared to traditional storage systems, DNA requires just about 1 bit per cubic nanometer of memory space. The chemical interactions in DNA provide energy to make or repair new strands, therefore there is essentially no power use. As a result, a multi-valued Quantum-DNA computation system can be created to find a super-fast computation system with a lot of memory. The advantages of multi-valued quantum computing and multi-valued DNA computing can be combined in this multi-valued Quantum-DNA computation system. In a multi-valued Quantum-DNA computing system, input is received as a qubit, which is then converted into DNA sequences by NMR relaxation after passing through a specific number of quantum operational gates.

7.2.1 Heat Transfer in Multi-Valued Quantum-DNA Full Subtractor (Difference)

A Multi-Valued Half Subtractor is a type of subtractor, an electronic circuit that performs the subtractions of ternary numbers. The half subtractor can subtract two single ternary digits and provide the output and a borrow value. It has two inputs, called A and B, and two outputs D (difference) and B (borrow). Quantum-DNA multi-valued

Half Subtractor will subtract two qubits from the quantum ternary digit |0>, |1>, and |2> and produce two outputs as difference |D> and borrow |B> in DNA sequence.

7.2.1.1   Design Procedure

To design a multi-valued Quantum-DNA Full Subtractor for heat transfer, it is needed to use Quantum and DNA operational gates so that it is easy to operate the input qubit for their corresponding outputs. The Quantum operational gates will be used for receiving the input qubits and the DNA operational gates will be used to produce the final output against the corresponding set of inputs. Each time, the Quantum-DNA Full Subtractor will receive three qubits as input. After operating in a certain number of Quantum operational gates, the qubit will be turned into a corresponding DNA sequence by using an NMR relaxation room temperature probe. By using a room temperature probe and corresponding components of NMR relaxation, the excited qubit turns into a ground state and produces a DNA sequence. Then the DNA sequence is processed through DNA operational gates and outputs are received. The multi-valued Quantum-DNA circuit for the Multi-Valued Full Subtractor is shown in Figure 7.1.

From the quantum operational gate, connect the junction to metal for transferring heat. The island is metal in a heat conduction circuit. There are two resistors and a photon bath working as a heat conductance at a distance of 1 meter. Heat is transferred by junction into the DNA circuit. Here, Figure 7.1 presents a multi-valued quantum-DNA Full Subtractor using quantum, DNA operational gates, and heat conduction nanotubes.

From the Figure 7.1, it is found that the multi-valued Quantum-DNA Full Subtractor consists of nineteen quantum operational gates and eight DNA operational gates. Here, ten AND, six OR, and three Decoder operational gates are used as Quantum operational gates and a further six AND and two OR DNA operational gates are used.

7.2.1.2   Working Procedure

Here, DNA sequence CCAGTC are used for Quantum qubit | 2 >, ACCTAG for Quantum qubit | 1 > and DNA sequence TGGATC for Quantum qubit | 0 >.

The heat conduction circuit receives heat from the Quantum operational gate. Firstly, the NIS junctions produce heat flows between the normal-metal islands and the superconducting leads. Secondly, the electrons in the normal metal exchange heat with the phonon bath. Then, the islands exchange heat between each other by photons traveling in the transmission line. Finally, the model takes into account the geometrical properties of the samples as well as properties specific to the measurement setup.

Here in this circuit heat passes through the junction into metal then the electron is heated and this electron transfers the heat to photon bath on the coplanar waveguide channel and then it goes to one 1-meter distance.

Using NMR relaxation quantum qubits turned into DNA sequences. Further, these DNA sequences and supplied heat operate the DNA operation of the Multiplier. The data conversion process will be discussed in the next chapter in detail.

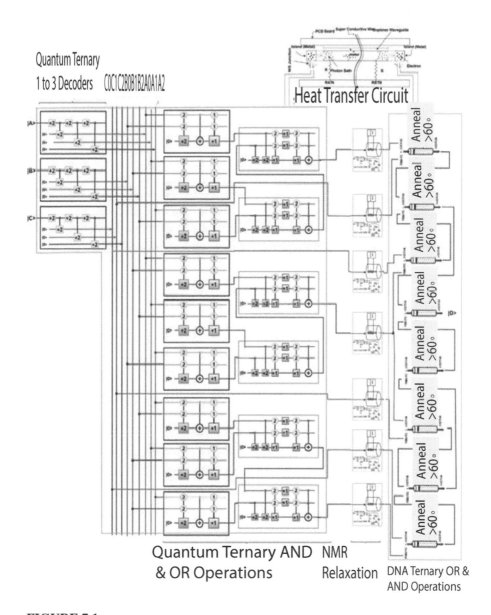

FIGURE 7.1
Multi-Valued Quantum-DNA Full Subtractor (Difference) Using Nanotubes for Heat Transfer

7.2.2   Heat Transfer in Multi-Valued Quantum-DNA Half Adder

A multi-valued Half Adder is a type of adder, an electronic circuit that performs the addition of ternary numbers. The half adder can add two single ternary digits and provide the output plus a carry value. It has two inputs, called A and B, and two outputs S (sum) and C (carry). Multi-Valued Quantum-DNA Half Adder will add two bits from the quantum ternary digit |0>, |1>, and |2> and produce two outputs as quantum sum |S> and quantum carry bit |C> in a DNA sequence. The multi-valued Quantum-DNA circuit for the multi-valued Half Adder is shown in Figure 7.2.

7.2.2.1   Design Procedure

To design a multi-valued Quantum-DNA Half Adder for heat transfer it is needed to use Quantum and DNA operational gates so that to operate the input qubit for their corresponding outputs. The Quantum operational gates will be used for receiving the input qubits and the DNA operational gates will be used to produce the final output against the corresponding set of inputs. Each time, the Quantum-DNA Half Adder will receive two qubits as input. After operating in a certain number of Quantum operational gates, the qubit will be turned into a corresponding DNA sequence by using an NMR relaxation room temperature probe. By using a room temperature probe and corresponding components of NMR relaxation, the excited qubit turns into a ground state and produces a DNA sequence. Then the DNA sequence is processed through DNA operational gates and outputs are received.

From the quantum operational gate, it needs to connect the junction to metal for transferring heat. The island is metal in a heat conduction circuit. There are two resistors and a photon bath working as a heat conductance at a distance of 1 meter. Heat is transferred by junction into the DNA circuit. Here, Figure 8 describes a multi-valued Quantum-DNA Full Subtractor using Quantum, DNA operational gates, and heat conduction nanotubes.

From the Figure 7.2, it is found that the multi-valued Quantum-DNA Full Subtractor consists of seven OR, eleven AND, and two Decoder operational gates. In addition, seventeen quantum operational gates and three DNA operational gates are required to find the expected output from the multi-valued quantum-DNA half adder.

7.2.2.2   Working Procedure

Here, DNA sequence CCAGTC for Quantum qubit | 2 >, ACCTAG for Quantum qubit | 1 > and DNA sequence TGGATC for Quantum qubit | 0 >.

The heat conduction circuit receives heat from the Quantum operational gate. Firstly, the NIS junctions produce heat flows between the normal-metal islands and the superconducting leads. Secondly, the electrons in the normal metal exchange heat with the phonon bath. Then, the islands exchange heat between each other by photons traveling in the transmission line. Finally, the model takes into account the geometrical properties of the samples as well as properties specific to the measurement setup.

Here in this circuit heat passes through the junction into metal then the electron is heated and this electron transfers the heat to photon bath on the coplanar waveguide

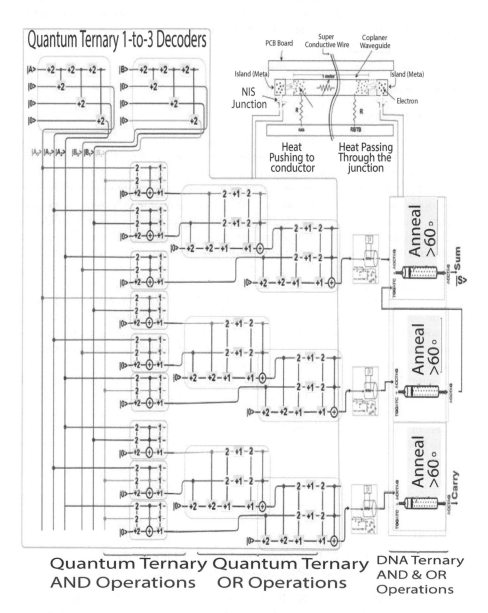

FIGURE 7.2
Multi-Valued Quantum-DNA Half Adder (Nanotubes for Heat Transfer)

channel and then it goes to one 1-meter distance. Using NMR relaxation quantum qubits turned into DNA sequences. Further, these DNA sequences and supplied heat operate the DNA operation of the Multiplier. The working procedure of a Multi-Valued Full Subtractor is also described in the next chapter.

7.3   Summary

According to the theory of thermodynamics, huge amounts of produced heat from a quantum circuit can deteriorate its performance. On the other hand, in the DNA operational circuit, to complete each step, a certain amount of heat is needed. If two of these findings are merged, it can be provided a novel procedure to provide heat in DNA operational circuits by taking away the produced heat in Quantum operational circuits. The main objective of this chapter is to transfer produced heat from Quantum operational circuit to DNA operational circuits by heat conduction circuits in Quantum-DNA operational circuit and multi-valued Quantum-DNA operational circuits. Heat transfer is important in protecting the environment by reducing emissions and pollutants. The heat conduction circuit receives heat from the quantum operational gate. The NIS junctions produce heat flows between the normal-metal islands and the superconducting leads. Then, the electrons in the normal metal exchange heat with the phonon bath. Next, the islands exchange heat between each other by photons traveling in the transmission line. Finally, the model takes into account the geometrical properties of the samples as well as properties specific to the measurement setup. Heat passes through the junction into metal then electrons are heated and this electron transfers the heat to photon bath on the coplanar wave guide channel, and then it goes to one 1-meter distance. This chapter has presented the way and process of multi-valued quantum-DNA computing where quantum circuits transfer heat to DNA circuits.

Bibliography

[1] Diósi, L. (2011). A Short Course in Quantum Information Theory: An Approach from Theoretical Physics (Vol. 827). Springer.

[2] Whitney, M., Isailovic, N., Patel, Y., & Kubiatowicz, J. (2007, May). Automated generation of layout and control for quantum circuits. In Proceedings of the 4th International Conference on Computing Frontiers (pp. 83-94).

[3] Raghavan, B. S., & Bhaaskaran, V. K. (2017, March). Design of novel multiple valued logic (mvl) circuits. In 2017 International Conference on Nextgen Electronic Technologies: Silicon to Software (ICNETS2) (pp. 371-378). IEEE.

[4] Partanen, M., Tan, K. Y., Govenius, J., Lake, R. E., Mäkelä, M. K., Tanttu, T., & Möttönen, M. (2016). Quantum-limited heat conduction over macroscopic distances. Nature Physics, 12(5), 460-464.

[5] Tagore, S., Bhattacharya, S., Islam, M., & Islam, M. L. (2010). DNA computation: application and perspectives. Journal of Proteomics and Bioinformatics, 3(07).

[6] Meschke, M., Guichard, W., & Pekola, J. P. (2006). Single-mode heat conduction by photons. Nature, 444(7116), 187-190.

[7] Dubi, Y., & Di Ventra, M. (2011). Colloquium: Heat flow and thermoelectricity in atomic and molecular junctions. Reviews of Modern Physics, 83(1), 131.

8

Data Conversion

8.1 Introduction

Classical computers have troubles with combination and permutation-related problems. These mathematical problems involve arranging elements in a way that achieves a particular objective. As the number of particles increases, so does the number of possible combinations also increase. To find the ideal arrangement, today's digital computers must use a loop through each permutation until an outcome is found, and then determine which is the most effective at achieving the goal. In many circumstances, this will demand a large number of calculations. Combinatorics calculations prove problematic in a variety of sectors, ranging from banking to pharmaceuticals. Quantum computers come into play here. Multiple-valued quantum computing reduces the cost of calculating difficult combinatorial problems in the same way as classical computers reduced the cost of arithmetic.

Multi-Valued quantum computing can be defined as a field of study that focuses on the creation of computer-based technologies grounded on quantum-theoretical principles. On the quantum (atomic and subatomic) level, quantum theory describes the nature and behavior of energy and matter. To execute certain computational tasks, quantum computing employs a combination of qubits. In multi-valued, it will be qutrits. All of this is done at a far higher rate than their traditional computing equipment. Quantum computers represent a significant advancement in computing capability, with enormous performance benefits for specific use cases. The ability of qutrits to be in several states at the same time gives the quantum computer a lot of computing capability.

Quantum computers are frequently referred to as a revolutionary advancement that is revolutionizing the world. It is extremely quick and efficient. They can conduct calculations that today's supercomputers would take decades or perhaps millennia to complete. Quantum supremacy is another term used by specialists to describe this occurrence. Quantum computing can be used in different sectors such as chemical and biological engineering, cybersecurity, artificial intelligence, financial service, and complex manufacturing. The discovery and manipulation of molecules are a part of chemical and biological engineering. Chemicals are made with subatomic particles and thus it involves quantum mechanics. For almost a thousand years, combinatorics has been at the heart of cryptography. But using quantum computing cracking any encryption can be easier than other technology. Artificial intelligence frequently

DOI: 10.1201/9781003381938-8

involves the combinatorics processing of very large amounts of data in order to make better predictions and decisions, and quantum computing is associated with this. Large amounts of data are the main domain of financial systems and manufacturing. Quantum computing has the potential to improve the speed of a critical set of financial calculations within a large amount of data.

Multi-Valued DNA computing is a kind of natural computing that uses the molecular characteristics of DNA to conduct logical and arithmetic operations instead of typical carbon/silicon chips. The four-character genetic alphabet is used in DNA computing instead of the binary alphabet (1 and 0) utilized by standard computers. This enables massively parallel computation, making it possible to answer difficult mathematical equations or problems in a fraction of the time. As a result, computation is far more efficient with a large volume of self-replicating DNA than with a standard computer, which would require a lot more hardware. Information or data will now be kept in the form of the bases A, T, G, and C, rather than binary digits. The capacity to generate short DNA sequences artificially allows these sequences to be used as inputs for algorithms. This is possible due to the ability to create small DNA molecules with any arbitrary sequence. The input of any DNA operation can be represented by DNA molecules with specific sequences. The instructions are carried out by laboratory operations on the molecules, and the result is defined as some property of the final set of molecules. DNA computing promises significant and meaningful linkages between computers and life systems, as well as massively parallel computations. DNA computing can actually carry out millions of operations at the same time.

The performance rate of DNA strands can be increased exponentially by doing millions of operations at the same time. DNA computer's massively parallel processing capabilities have the ability to speed up big, but otherwise solvable, polynomial-time problems with few operations. A mixture of 1,018 strands of DNA, for example, might run at 10,000 times the speed of today's fastest supercomputers. Traditional storage mediums, such as videotapes, require 10^{12} cubic nanometers of the area to store a single bit of data, but DNA molecules only require one cubic nanometer. To put it another way, a single cubic centimeter of DNA can store more data than a trillion CDs. This is due to the fact that DNA molecules have a storage density of around 18 Mbits per inch, whereas today's computer hard drives can only store about 1/100,000 of this data in the same amount of space.

This chapter will describe how to convert data from multi-valued quantum qubits to multi-valued DNA sequences and how to convert multi-valued DNA sequences to multi-valued quantum qubits.

8.2   Data Conversion in Multiple-Valued Quantum-DNA Circuits

The fastest computation system can be defined as a Quantum computation system, which works with qubits $| 1 >, | 0 >$, and both $| 1 >$ and $| 0 >$ at the same time. On the other hand, In place of standard silicon-based computer technology, DNA computing uses biological components such as DNA, biochemistry, and molecular biology. When applied to issues that can be separated into independent, non-sequential tasks, the DNA computer has demonstrable benefits over conventional computers. The reason for this is because DNA strands can store a lot of data and do numerous operations at the same time, allowing them to solve decomposable issues considerably faster. Again, Quantum computer calculations are especially promising for analyzing or simulating extremely complicated processes involving large volumes of data.

So, for finding a super faster computation system with huge memory can develop a multi-valued Quantum-DNA computation system. This multi-valued Quantum-DNA computation system can merge all advantages of multi-valued quantum computing and multi-valued DNA computing. It will be able to compute parallel operations at super-fast speed.

In a multi-valued Quantum-DNA circuit, inputs will be operated in quantum operation and provide output in DNA sequence. In this case, NMR relaxation or trap ion can be used for converting quantum qubit to a DNA sequence. In the next section, NMR relaxation will be discussed with an example.

8.2.1 NMR Relaxation at Room Temperature

Nuclear magnetic resonance (Figure 8.1), or NMR as it is abbreviated by scientists, is a phenomenon which occurs when the nuclei of certain atoms are immersed in a static magnetic field and exposed to a second oscillating magnetic field. Some nuclei experience this phenomenon, and others do not, dependent upon whether they possess a property called spin. Most of the matter with NMR is composed of molecules. Molecules are composed of atoms.

Nuclear magnetic resonance (NMR) is the study of molecules by recording the interaction of radiofrequency (Rf) electromagnetic radiations with the nuclei of molecules placed in a strong magnetic field.

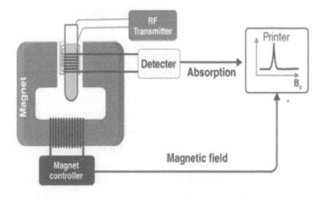

FIGURE 8.1
Nuclear Magnetic Resonance (NMR)

8.2.1.1   NMR Instrumentation

This instrument consists of nine major parts. They are discussed below:

1. **Sample holder** – It is a glass tube which is 8.5 cm long and 0.3 cm in diameter.

2. **Magnetic coils** – Magnetic coil generates magnetic field whenever current flows through it

3. **Permanent magnet** – It helps in providing a homogenous magnetic field at 60 – 100 MHZ

4. **Sweep generator** – Modifies the strength of the magnetic field which is already applied.

5. **Radiofrequency transmitter** – It produces a powerful but short pulse of the radio waves.

6. **Radiofrequency** – It helps in detecting receiver radio frequencies.

7. **RF detector** – It helps in determining unabsorbed radio frequencies.

8. **Recorder** – It records the NMR signals which are received by the RF detector.

9. **Readout system** – A computer that records the data.

8.2.1.2   NMR Working Techniques

1. Resonant Frequency

It refers to the energy of the absorption, and the intensity of the signal that is proportional to the strength of the magnetic field. NMR active nuclei absorb electromagnetic radiation at a frequency characteristic of the isotope when placed in a magnetic field.

2. Acquisition of Spectra

Upon excitation of the sample with a radiofrequency pulse, a nuclear magnetic resonance response is obtained. It is a very weak signal and requires sensitive radio receivers to pick up.

8.2.1.3   Multiple-Valued Quantum-DNA Multiplexer with NMR Relaxation

For multi-valued quantum-DNA 3-to-1 multiplexer, one 1-to-3 quantum decoder is needed to get the output of selection line S. Additionally, three DNA AND and one DNA OR operation is required as shown in Figure 8.2. To match the speed between Quantum and DNA operation a quantum cache memory is used which will be discussed in the next chapter.

Figure 6.8 shows the construction of multi-valued quantum-DNA 3-to-1 multiplexer circuit. In this circuit, all the inputs |I0>, |I1> and |I2> connect to three different DNA AND gates after performing the data conversion by NMR relaxation (convert qubits to DNA sequences) via quantum cache memory in which another input comes from the selection line |S>,

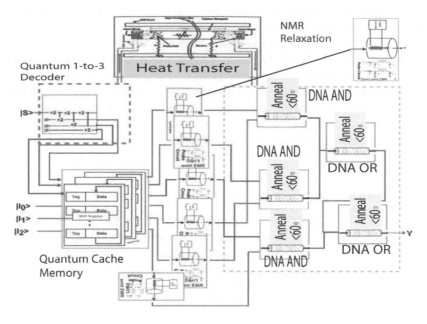

FIGURE 8.2
Circuit Architecture of Multi-Valued Quantum-DNA 3-to-1 Multiplexer

To select one particular input, qubit |0>, |1> or |2> is provided through |S> based on which one of the three output lines of the quantum 1-to-3 decoder gets activated and passes qubit |2> through it. Other output lines of the decoder provide qubit |0>. Thus only one DNA AND operation is open to take the input to the output. Finally, the DNA OR operation passes the input that comes via the DNA AND operation to the output of the multiplexer.

8.3 Data Conversion in Multiple-Valued DNA-Quantum Circuits

In place of standard silicon-based computer technology, DNA computing uses biological components such as DNA, biochemistry, and molecular biology. When applied to issues that can be separated into independent, non-sequential tasks, the DNA computer has demonstrable benefits over conventional computers. The reason for this is because DNA strands can store a lot of data and do numerous operations at the same time, allowing them to solve decomposable issues considerably faster. On the other hand, the fastest computation system can be defined as a Quantum computation system, which works with qubits | 1 >, | 0 >, and both | 1 > and | 0 > at the same time.

In addition, Quantum computer calculations are especially promising for analyzing or simulating extremely complicated processes involving large volumes of data.

So, for finding a super faster computation system with huge memory, it is possible to develop a multi-valued DNA-Quantum computation system. This multi-valued DNA-Quantum computation system can merge all advantages of multi-valued quantum computing and multi-valued DNA computing. It will be able to compute parallel operations at a super-fast speed too like multi-valued quantum-DNA computing.

In a multi-valued DNA-Quantum circuit, the inputs will be DNA sequences and provide output in qubits. In this case, it is needed to use NMR or quadrupole trap ion for converting DNA sequences to qubits.

8.3.1 Quadrupole Ion Trap

The Quandrupole Ion Trap (QIT) is an extraordinary device that functions both as an ion store in which gaseous ions can be confined for a period of time and as a mass spectrometer of large mass range, variable mass resolution, and high sensitivity. As a storage device, the QIT confines gaseous ions, either positively charged or negatively charged, in the absence of solvent. The confining capacity of the QIT arises from the formation of a trapping potential well when appropriate potentials are applied to the electrodes of the ion trap.

That the basic theory of operation of quadrupole devices was enunciated almost 100 years before the QIT and the related QMF were invented by Paul and Steinwedel is a shining example of the inherent value of sound basic research. The pio neering work of the inventors was recognized by the award of the 1989 Nobel Prize in Physics to Wolfgang Paul, together with Norman Ramsay and Hans Dehmelt.

8.3.1.1 The Structure of the QIT

The QIT mass spectrometer consists essentially of three shaped electrodes that are shown in open array in Figure 8.3. Two of the three electrodes are virtually identical and, while having hyperboloidal geometry, resemble small inverted saucers; these saucers are the so-called end-cap electrodes and each has one or more holes in the center. One end-cap electrode contains the "entrance" aperture through which electrons and/or ions can be gated periodically while the other is the "exit" electrode through which ions pass to a detector. The third "ring" electrode has an internal hyper boloidal surface: in some early designs of ion trap systems, a beam of electrons was gated through a hole in this electrode rather than an end-cap electrode.

The ring electrode is positioned symmetrically between two end-cap electrodes, as shown in Figures 8.4; Figure 8.4 shows a photograph of an ion trap cut in half along the axis of cylindrical symmetry while Figure 8.5 is a cross section of an ideal ion trap showing the asymptotes and the dimensions r_0 and z_0, where r_0 is the radius of the ring electrode in the central horizontal plane and $2z_0$ is the separation of the two end-cap electrodes measured along the axis of the ion trap.

The electrodes in Figure 8.4 and Figure 8.5 are truncated for practical purposes, but in theory they extend to infinity and meet the asymptotes shown in the figure. The

FIGURE 8.3
Three Electrodes of QIT Shown in Open Array

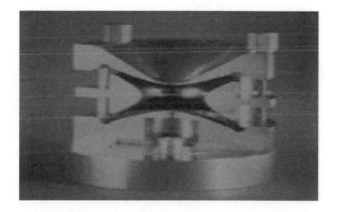

FIGURE 8.4
Quadrupole Ion Trap: Photograph of Ion Trap cut in half along Axis of Cylindrical Symmetry

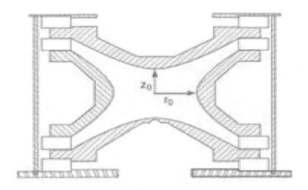

FIGURE 8.5
Quadrupole Ion Trap: Schematic Diagram of Three-Dimensional Ideal Ion Trap
Showing Asymptotes and Dimensions r_0 and z_0

asymptotes arise from the hyperboloidal geometries of the three electrodes. The geometries of the electrodes are defined so as to produce an ideal quadrupole potential
distribution that, in turn, will produce the necessary trapping field for the confinement
of ions.

8.3.1.2   Electrode Surfaces

It is known that for the cylindrically symmetric QIT, the values of λ, σ, and γ given
can satisfy the Laplace condition. Then

$$\varphi_{x,\,y,\,z} = A(x^2 + y^2 - 2z^2) + C \tag{8.1}$$

To proceed, it needs to convert Equation 8.1 into cylindrical polar coordinates
employing the standard transformations $x = r \cos\theta$, $y = r \sin\theta$, $z = z$. Thus Equation
8.1 becomes

$$\varphi_{x,\,y,\,z} = A(x^2 + y^2 - 2z^2) + C \tag{8.2}$$

It should be noted that in making this transformation the angular orientation of
the x–y coordinate plane is lost. The effect of this is that when the equations–of motion
of the ions are developed in a manner analogous to that presented earlier for the QMF
there is an implicit assumption that each ion possesses zero angular velocity around
the z axis.

The procedure adopted by Knight can be followed and write the equations for the
electrode surfaces in generalized forms as

$$\frac{r^2}{r_0^2} - \frac{z^2}{a^2} = 1 \, (ring\,electrode) \tag{8.3}$$

and

$$\frac{r^2}{b^2} - \frac{z^2}{z_0^2} = -1 (end - cap electrode). \tag{8.4}$$

It should be remembered that when plotted out these equations represent cross sections through the electrodes which, of course, possess cylindrical symmetry around the z axis. In terms of their respective geometric forms, the ring electrode is a single-sheet hyperboloid and the pair of end-cap electrodes comprises a double-sheet hyperboloid. As with the QMF, the conditions are noted as $r = \pm r_0$ when $z = 0$ and $z = \pm z_0$ when $r = 0$. From Figure 8.4b, it is seen thst $2r_0$ is the innermost diameter of the ring electrode and $2z_0$ is the closest distance between the innermost surfaces of the end-cap electrodes; a and b are geometric quantities which will be evaluated shortly.

Following standard mathematical procedures, it can be deduced that the slopes of the asymptotes of the ring electrode are

$$m = \pm \frac{a}{r_0} \tag{8.5}$$

and those of the end-cap electrodes are

$$m' = \pm \frac{z_0}{b}. \tag{8.6}$$

In order to establish a quadrupolar field the ring and the end-cap electrodes must share common asymp- totes, so that $m = m\prime$, and from Equation 8.5 and equation 8.6,

$$a^2 b^2 = r_0^2 z_0^2. \tag{8.7}$$

Hence substituting for a^2 in Equation 8.3 for the ring electrode,

$$\frac{r^2}{r_0^2} - \frac{z^2 b^2}{r_0^2 z_0^2} = 1 \tag{8.8}$$

thus

$$r^2 = r_0^2 + \frac{z^2 b^2}{z_0^2}. \tag{8.9}$$

Since the value of the potential given must be a constant across the electrode surfaces, it is possible to establish conditions under which $\varphi_{r,z}$ is independent of r and z by the following procedure, in which r^2 is replaced in the term $r^2 - 2z^2$ to obtain, for the ring electrode,

$$r^2 - 2z^2 = r_0^2 + \frac{z^2 b^2}{z_0^2} - 2z^2$$

$$= r_0^2 + \frac{z^2 \left(b^2 - 2z_0^2\right)}{z_0^2}. \tag{8.10}$$

Similarly for the end-cap electrodes, from Equation 8.4,

$$r^2 = \frac{z^2 b^2}{z_0^2} - b^2 \tag{8.11}$$

$$r^2 - 2z^2 = \frac{z^2 \left(b^2 - 2z_0^2\right)}{z_0^2} - b^2. \tag{8.12}$$

Thus $\varphi_{r,z}$ becomes constant when $b^2 - 2z^2{}_0 = 0$

$$\therefore b^2 = 2z_0^2 \tag{8.13}$$

and therefore from Equation 8.7,

$$a^2 = \frac{1}{2} r_0^2. \tag{8.14}$$

Hence the equations for the electrode surfaces now become

$$\frac{r^2}{r_0^2} - \frac{2z^2}{r_0^2} = 1 \; (Ring\,electrode) \tag{8.15}$$

and

$$\frac{r^2}{2z_0^2} - \frac{z^2}{z_0^2} = -1 \; (End-cap\,electrode) \tag{8.16}$$

Also from Equation 8.5 and equation 8.6, equating the gradients of the asymptotes, it is obtained as

$$m = \pm \frac{a}{r_0} = \pm \frac{r_0}{\sqrt{2} \, r_0} = m' = \pm \frac{z_0}{b} = \pm \frac{z_0}{\sqrt{2} \, z_0} \tag{8.17}$$

$$= \pm \frac{1}{\sqrt{2}} \tag{8.18}$$

This relationship corresponds to the asymptotes having an angle of 35.264° with respect to the radial plane of the ion trap.

8.3.1.3 Multiple-Valued DNA-Quantum Multiplexer

For multi-valued DNA-Quantum 3-to-1 Multiplexer, one 1-to-3 DNA decoder is needed to get the output of selection line S. Additionally, three Quantum AND and one Quantum OR operation is required as shown in Figure 8.6. To match the speed between DNA and Quantum operation a DNA cache Memory is used which will be discussed in the next chapter. Figure 8.6 shows the construction of multi-valued DNA-Quantum 3-to-1 multiplexer circuit.

In this circuit, all the inputs I0, I1, and I2 connect to three different DNA AND gates after performing data conversion by trap ion, via DNA cache memory in which another input comes from the Selection line S.

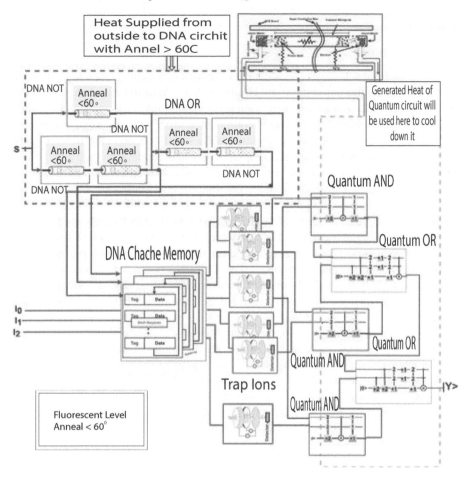

FIGURE 8.6
Circuit Architecture of Multi-Valued DNA-Quantum 3-to-1 Multiplexer

To select one particular input, molecular sequence **ACCTAG, CAAGCT** or **TG-GATC** is provided through S to represent 1, 2, or 3, respectively. Based on this output of the selection line one of the three output lines of the DNA 1-to-3 Decoder gets activated and passes molecular sequence **TGGATC** through it. Other output lines of the decoder provide molecular sequence **ACCTAG**. Thus, only one Quantum AND operation is open after the trap ion to take the input to the output. Finally, the Quantum OR operation passes the input that comes via the Quantum AND operation to the output of the multiplexer.

8.4 Summary

Quantum molecular biology needs data conversions in a multi-valued logic system. Two different computing systems are combined here, so to transfer data, it needs to convert data. Multi-Valued quantum computer calculations are especially promising for analyzing or simulating extremely complicated processes with large volumes of data. Natural science fields, in particular, see the significant potential here, in addition to digital marketing. Quantum computers may help researchers gain a better and more detailed understanding of how specific particles, components, and processes interact in live cells. However, there are possible medical applications. Most importantly, experts believe quantum computers will advance artificial intelligence (AI) significantly. These could then safely and reliably take over tasks such as data evaluation or forecasting in the future. Furthermore, DNA computing is a new technology that uses DNA molecules to create computers that are quicker than the most powerful human-built computers on the market. The DNA computers of the future will be able to work in a massively parallel fashion, completing several calculations at the same time. In practically every sector, DNA computers have made significant development.

Multi-Valued DNA-Quantum computing systems, which will merge all the advantages of both multi-valued DNA computing and multi-valued quantum computing. In DNA-Quantum computing, input data is in quantum qubit and the output is in DNA sequence. It uses NMR or quadrupole trap ion to convert DNA sequences into quantum qubits. Again, in multi-valued Quantum-DNA computing, input data is in DNA sequences and provided output is in quantum qubits. Quantum-DNA uses NMR relaxation or trap ion to convert quantum qubits into DNA sequences.

Bibliography

[1] Takahashi, Y., & Tani, S. (2021). Power of uninitialized qubits in shallow quantum circuits. Theoretical Computer Science, 851, 129-153.

[2] Jones, J. A. (2001). Quantum computing and nuclear magnetic resonance. PhysChemComm, 4(11), 49-56.

[3] Wei, Q., Kais, S., Friedrich, B., & Herschbach, D. (2011). Entanglement of polar molecules in pendular states. The Journal of Chemical Physics, 134(12), 124107.

[4] Anders, J., Oi, D. K., Kashefi, E., Browne, D. E., & Andersson, E. (2010). Ancilla-driven universal quantum computation. Physical Review A, 82(2), 020301.

[5] Zhang, C., Ge, L., Zhuang, Y., Shen, Z., Zhong, Z., Zhang, Z., & You, X. (2019). DNA computing for combinational logic. Science China Information Sciences, 62(6), 1-16.

[6] March, R. E., & Todd, J. F. (2005). Quadrupole Ion Trap Mass Spectrometry. John Wiley & Sons.

9

Data Management

9.1 Introduction

Quantum computing is a branch of computing that focuses on improving computer technology using quantum theory's principles and postulates (which emphasizes the behavior of energy and material on the atomic and subatomic levels). Quantum computing uses quantum qubits rather than binary bits. Multi-Valued quantum computing uses qutrits. It supplies subatomic particles with the one-of-a-kind ability to exist in several states (i.e., a 1 and a 0 at the same time). DNA Computing is another advanced computing approach that uses living molecules rather than regular silicon chips to execute computations. The four-character genetic alphabet (A [adenine], G [guanine], C [cytosine], and T [thymine]) is used in DNA computing instead of the binary alphabet (1 and 0) utilized by standard computers. Small DNA molecules of any arbitrary sequence may be manufactured to order any data, which can solve memory limitation problems for any computing.

Both quantum and DNA computing approaches can be used to accomplish operations at a very high speed and efficiency. There is a critical issue that occurs when merging both computing systems for a single activity. According to IBM, quantum computing is so quick that the system completed a theoretically defined computation in 200 seconds that would take the world's most powerful supercomputer 10,000 years to complete. Quantum computers are 158 million times quicker than the world's fastest supercomputer as a result of this. As a result, quantum operations produce instantaneous results. However, DNA procedures take a long time to prepare.

The quantum computer's output qubits cannot be instantly inserted into the DNA system. When working with the Quantum-DNA system, a temporary storage device or system is needed where the qubits may be held for a very short time. So, a Quantum Cache memory is needed to operate a Quantum and DNA computing circuit for its operation, which can store data for a short amount of time and can pass it to another. DNA cache memory is needed while working with DNA-quantum computing.

Cache memory is a supplementary memory system that temporarily stores frequently used instructions and data for quicker processing. It is an extremely fast memory type that holds frequently requested data and instructions so that they are immediately available for further processing. In most microprocessors, Static Random-Access Memory (SRAM) is used as cache memory as SRAM is very high speed. Due

DOI: 10.1201/9781003381938-9

to this high speed, cache memory is used to store data temporarily, which will be designed for the multi-valued quantum computer. Anyone can design a cache memory using the D flip-flops which will create a one-bit SRAM along with *Read/Write* and *select* bit as input.

9.2   Data Management in Quantum-DNA Circuits

According to quantum computing, quantum computation is faster than classical computation systems. Quantum computers are also more powerful than supercomputers in terms of computing. They are thousands of times faster than regular computers and supercomputers at processing data. Quantum computers can execute calculations that would take a regular computer 1000 years to complete in a matter of seconds. On the other hand, the use of DNA strands to compute has led to high parallel computation that makes up for the slow processing of the chip. So, for finding a super faster computation system with huge memory, a Quantum-DNA computation system can be developed. This Quantum-DNA computation system can merge all advantages of quantum computing and DNA computing.

In a multi-valued quantum-DNA computing system, input will be received as a qubit and after performing in a certain number of quantum operational gates these qubits will be turned into DNA sequences by NMR relaxation which is discussed in the previous chapter. But before entering into the NMR relaxation, the speedy qubits will enter into a multi-valued cache memory for a short period of time.

In the multiple-valued logic system, the same thing will be done. The general organization of the Quantum-DNA cross-platform for the multiple-valued logic system is depicted in Figure 9.1. Every ternary operation in the Quantum-DNA system will contain some basic parts.

1. Multiple-valued Quantum System,
2. Multiple-valued Quantum Cache Memory,
3. Data Conversion Unit, and
4. Multiple-valued DNA System.

Among them, the multiple-valued quantum system, multiple-valued DNA system, and the data conversion unit have beebn discussed and those will remain the same.

So, one thing that needs to consider now is the quantum ternary cache memory to store the qutrits. But that cannot store the qutrit data unless it is made in a way to allow, store and retrieve the qutrits also. So, need to construct a quantum cache memory for the ternary system to store the qutrits and to retrieve the qutrits from the cache memory for further process.

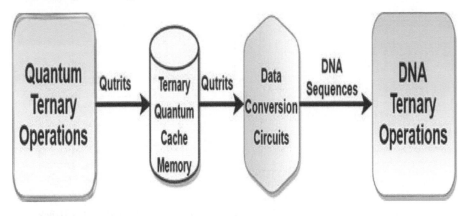

FIGURE 9.1
The General Organization of Multiple-Valued Quantum-DNA Computing to Perform Ternary Logical Operations

9.2.1 Construction of an Intermediary System to Control Quantum-DNA Data Flow

Quantum computing is so fast that the DNA system cannot receive the quantum system's outputs spontaneously. That is why a cache memory is used where the data will be stored temporarily. Now, for the ternary system, a ternary quantum cache memory needs to be developed.

In the quantum ternary system, the number of qutrits as the base is three which are |0>, |1>, and |2>. Quantum cache memory can only store |0>, and |1>. That is why the design of the quantum cache memory should be in the ternary system.

9.2.1.1 Design Procedure of the Quantum Cache Memory in Ternary System

To design ternary quantum cache memory, it is needed to design quantum D flip-flops and quantum cache memory cells (one-bit cache memory) with qutrits.

The circuit diagram for the ternary quantum D flip-flop is shown in Figure 9.2.

To design a quantum ternary D flip-flop, five quantum ternary NAND operations are needed along with one clock pulse. Here, |D> is the quantum ternary bits or qutrits that are |0>, |1>, and |2>. When CLK pulse is 1 then the quantum ternary D flip-flop transfers the |D> input to output |Q> and if the CLK pulse is 0, then the input remains unchanged.

9.2.1.2 Designing of Quantum Ternary One-bit Cache Memory

A one-qutrit quantum ternary cache memory is designed using the previously designed quantum ternary D flip-flops. The circuit diagram of the quantum one-qutrit cache memory is shown in Figure 9.3.

The circuit of one-qutrit quantum ternary cache memory contains six quantum ternary NAND and three quantum ternary AND operations. Again, |R/$\overline{W}$ > = |1>

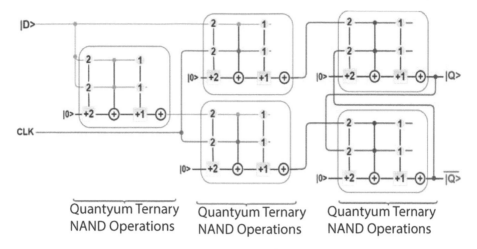

Quantyum Ternary   Quantyum Ternary   Quantyum Ternary
NAND Operations   NAND Operations   NAND Operations

FIGURE 9.2
Circuit Diagram of Quantum Ternary D Flip-Flops

indicates the READ operation and $|R/\overline{W}> = |0>$ indicates the WRITE operation. And here $|S>$ indicates the select qutrit where $|S> = |1>$ means the memory is selected.

Using these quantum ternary cache memory cells it is possible to construct a 9×2 quantum ternary cache memory. A 2×9 quantum ternary decoder and eighteen one-qutrit quantum ternary cache memory cells will be needed to construct a 9×2 quantum ternary cache memory.

9.2.1.3   Designing of 9×2 Quantum Ternary Cache Memory

Figure 9.4 shows the circuit architecture of 9×2 quantum ternary cache memory that contains eighteen quantum one-qutrit cache memory cells, providing 2-qutrit output and nine locations.

The quantum ternary 9×2 cache memory contains a quantum ternary 2-to-9 decoder which decodes the input qubits into nine output qutrits. Here, **R** indicates the quantum ternary one-qutrit cache memory cell. The architecture of the quantum 9×2 cache memory includes a 2×9 quantum ternary decoder, eighteen quantum cache memory cells, and quantum-OR operations, where inputs are two qutrits and $|R/\overline{W}>$ signal. Besides, DI inputs are actual inputs (Data inputs) that will be written to the memory or read from the memory. The output from the decoder will set the locations for two qutrit inputs to be stored.

9.2.1.4   Working Procedure of the Quantum Ternary Cache Memory

Figure 9.4 shows a 9×2 quantum ternary cache memory . It includes eighteen quantum cache memory cells providing 2-qutrit output and nine locations. The 9×2 quantum ternary cache memory also contains a 2-to-9 quantum ternary decoder and eighteen quantum cache memory cells implemented with quantum ternary D flip-flops

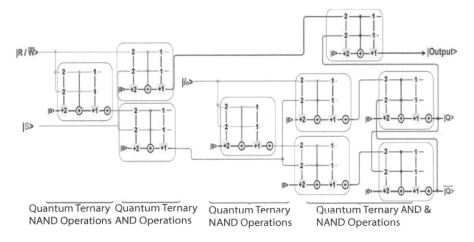

Quantum Ternary Quantum Ternary   Quantum Ternary   Quantum Ternary AND &
NAND Operations AND Operations   NAND Operations   NAND Operations

FIGURE 9.3
The Circuit Diagram of One-Qutrit Quantum Ternary Cache Memory

and quantum gates. The nine locations (00, 01, 02, 10, 11, 12, 20, 21, and 22) in the cache memory are addressed by 2 qutrits (A_1, A_0) with the help of the quantum ternary 2-to-9 decoder. In order to read from location 00, the address $A_1 A_0 = 00$ and $|R/\overline{W}> = |1>$. The decoder selects $|0>$, high. $|R/\overline{W}> = |1>$ will apply 0 at the clock inputs of the two quantum ternary cache memory cells of the top row and will apply 1 at the inputs of the output quantum AND operations, thus transferring the outputs of the two quantum D flip-flops to the inputs of the two quantum-OR operations. The other inputs of the quantum-OR operations will be 0. Thus, the outputs of the two quantum cache memory cells of the top row will be transferred to DO_1, and DO_0, performing a READ operation. On the other hand, consider a WRITE operation: The 2-qutrit data to be written is presented at $|DI_1>|DI_0>$. Suppose $|A_1 A_0> = |02>$. The top row is selected ($0_2 = 1$). Input bits at $|DI_1>$ and $|DI_0>$ will respectively be applied at the inputs of the D flip-flops of the third row from the top. Because $|R/\overline{W}> = |0>$, the clock inputs of both the quantum D flip-flops of the third row are $|1>$; thus, the D inputs are transferred to the outputs of the flip-flops. Therefore, data at DI_1 DI_0 will be written into the Quantum Ternary Cache Memory.

9.2.2 Multiple-Valued Quantum-DNA Half Adder

Figure 9.5 shows the circuit diagram of the quantum-DNA ternary half-adder operation. From the Figure 9.5, it is seen that the quantum system does some portion of the operation. Then quantum system's outputs are stored in the quantum ternary cache memory. From the ternary quantum cache memory, the qutrits are passing through the NMR relaxation process. The NMR relaxation converts the qutrits into the

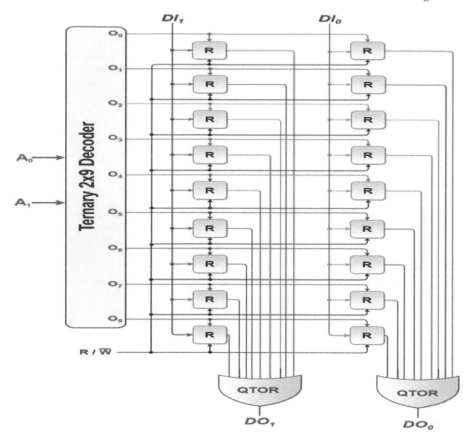

FIGURE 9.4
Circuit Architecture of Quantum Ternary 9×2 Cache Memory

equivalent DNA base sequence (i.e. from |0>, |1>, and |2> to ACCTAG, CAAGCT, and TGGATC, respectively). Note that, the NMR relaxation processes here are conducted at zero Kelvin temperature. Then the converted DNA base sequences are entered into the DNA system as inputs. And the DNA system does the rest of the operations and produces the final outputs. Notice that the excessive heat produced by the quantum system is transferred to a cold storage by a quantum heat conductance circuit. And the required heat to the DNA system can be transferred from the quantum system as well shown in the Figure 9.5. The detailed design and working principle of quantum-DNA ternary half adder and other arithmetic circuits will be described in the next part (part 3) of the book. The Figure 9.5 is depicted to show the quantum ternary cache memory in the multi-valued quantum-DNA circuit.

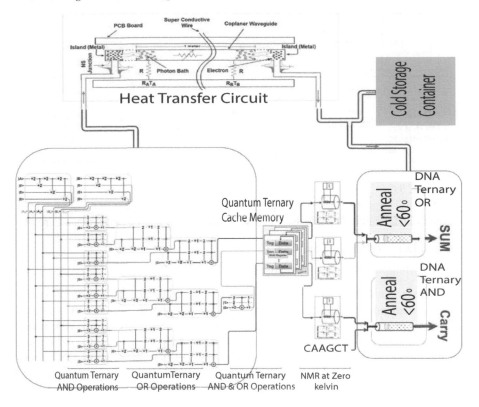

FIGURE 9.5
The Circuit Diagram of the Quantum-DNA Ternary Half-Adder Operation

9.3 Data Management in DNA-Quantum Circuits

Quantum computers are also more powerful in terms of computation than super-computers. They process data thousands of times faster than normal computers and supercomputers. As a result, a DNA-Quantum computation system can be created to find a super-fast computation system with a lot of memory. This DNA-Quantum computation device combines the benefits of both quantum and DNA computing.

In a DNA-Quantum computing system, input will be received in DNA sequences and after performing in a certain number of DNA operational gates these DNA sequences will be turned into quantum qubits by the NMR process.

In the multiple-valued logic system, the general organization of the DNA-Quantum cross-platform for the multiple-valued logic system is depicted in Figure 9.6.

Every ternary operation in the DNA-Quantum system will contain

1. Multiple-valued DNA System,
2. Multiple-valued DNA Cache Memory,

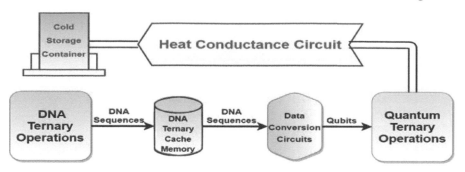

FIGURE 9.6
The General Organization of Multiple-Valued DNA-Quantum Computing with Data
Conversion Circuits, DNA Ternary Cache Memory, and Heat Conductance Circuit

3. Data Conversion Unit, and

4. Multiple-valued Quantum System.

Among them, the multiple-valued quantum system is learned, multiple-valued
DNA system, and the data conversion unit which will remain the same. That means
the NMR process is used, and the trap ion to convert the DNA information (DNA
strands) into the equivalent quantum ternary data (qutrits). Another thing to consider
now is – the DNA ternary cache memory to store the DNA information. DNA cache
memory for the ternary system is needed to store the ternary DNA sequences and
to retrieve the ternary sequences from the DNA ternary cache memory for further
process.

9.3.1 Construction of an Intermediary System to Control DNA-Quantum Data Flow

Quantum computing is so fast that collaboration is difficult for the DNA system.
That is why it is needed to use a cache memory where the DNA data will be stored
temporarily. And while the conversion process the DNA sequence will be retrieved
from the DNA cache memory. Now, for the ternary system, a ternary DNA cache
memory is needed to develop.

In the DNA ternary system, the number of DNA sequences as the base is three
which are represented as ACCTAG, CAAGCT, and TGGATC (for ternary values 0,
1, and 2, respectively). DNA cache memory could only store TGGATC (for 0), and
ACCTAG (for 1). That is why need to design the DNA cache memory in the ternary
system to store the ternary data.

9.3.1.1 Design Procedure of the Ternary One-bit DNA Ternary Cache Memory

One-bit DNA ternary cache memory can store one DNA sequence in the ternary
system. A design is done for the two-valued data also. Figure 9.7 shows the one-bit
DNA ternary cache memory cell.

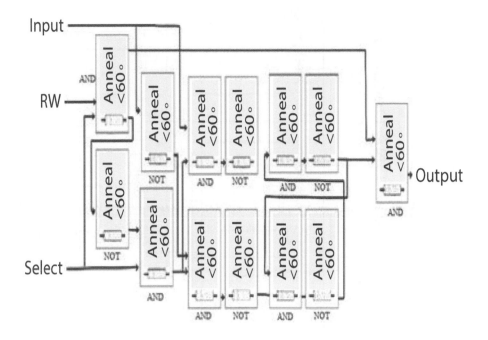

FIGURE 9.7
Circuit Diagram of One-Bit DNA Ternary Cache Memory

It is considered as the *DNA ternary cache memory cell* because each cell can store one bit of ternary DNA data, and using several DNA cache memory cells can store more DNA information into the DNA ternary cache memory. The DNA cache memory cell has three inputs and one output. First $|R/\overline{W}>$ and select line go through the DNA ternary AND operation. Inverted $|R/\overline{W}>$ and select line go through another DNA ternary AND operation. This AND operation's output and an input bit will go to the DNA ternary D flip-flop. The output of the D flip-flop ternary AND with the first DNA ternary AND operation to produce output (store or retrieve data based on the $|R/\overline{W}>$ mode).

Using these DNA ternary cache memory cells it is possible to construct a 9×2 DNA ternary cache memory. A 2-to-9 DNA ternary decoder will be needed and eighteen one-bit DNA ternary cache memory cells to construct a 9×2 DNA ternary cache memory.

9.3.1.2 Constructing of 9×2 DNA Ternary Cache Memory

Figure 9.8 shows the general organization of a 9×2 DNA ternary cache memory that contains eighteen DNA one-bit cache memory cells, providing 2-bit output and

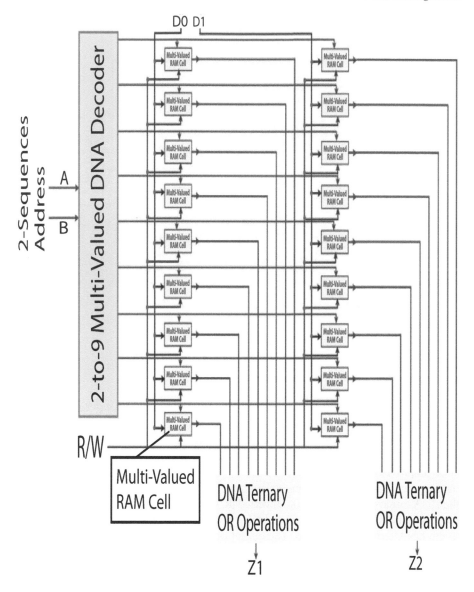

FIGURE 9.8
The General Architecture of DNA Ternary 9×2 Cache Memory

nine locations. The DNA ternary 9×2 cache memory contains a DNA ternary 2-to-9 decoder which decodes the input qubits into nine output bits.

Figure 9.9 depics the circuit architecture of DNA ternary 9×2 cache memory. From the Figure 9.9, it is seen that a 2-to-9 decoder, eighteen DNA ternary one-bit

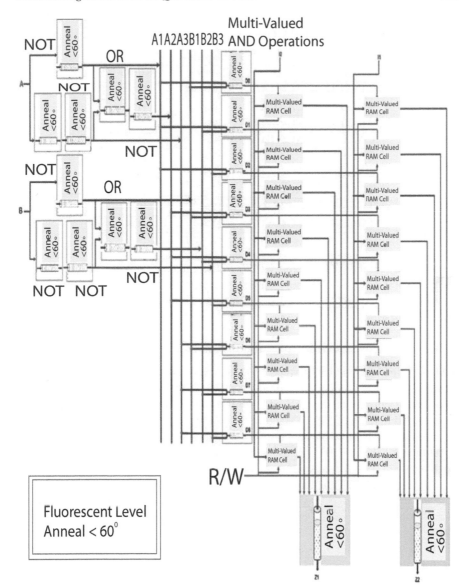

FIGURE 9.9
Circuit Architecture of DNA Ternary 9×2 Cache Memory

cache memory cells, and two DNA ternary OR operations are needed to construct the circuit of the DNA ternary 9×2 cache memory.

Each DNA ternary cache memory cell has three inputs and one output. Three inputs are the select, $|R/\overline{W}>$ and one input bit. A 2-to-9 decoder's outputs are used to select input locations for the cache memory cells. A DNA ternary OR operation

TABLE 9.1

Multi-Valued DNA-Quantum Half Adder Truth Table

Inputs		Outputs	
A	**B**	**|Carry>**	**|Sum>**
ACCTAG	ACCTAG	|0>	|0>
ACCTAG	CAAGCT	|0>	|1>
ACCTAG	TGGATC	|0>	|2>
CAAGCT	ACCTAG	|0>	|1>
CAAGCT	CAAGCT	|0>	|2>
CAAGCT	TGGATC	|1>	|0>
TGGATC	ACCTAG	|0>	|2>
TGGATC	CAAGCT	|1>	|0>
TGGATC	TGGATC	|1>	|1>

is used to OR all the cache memory cell's outputs that are connected to the D_0 line and perform Z_0 as cache memory output. Another OR operation is used to OR all the cache memory cell's outputs that are connected to the D_1 line and perform Z_1 as cache memory output.

9.3.1.3   Multiple-Valued DNA-Quantum Half Adder

A Multi-Valued half adder is a type of adder, an electronic circuit that performs the addition of ternary numbers. A half adder in which DNA sequences act as the input however the outputs of this half adder will produce in Quantum qubits. Table 9.1 shows the truth table of DNA-Quantum Half adder.

From the truth table, the equations for the sum and carry are found as –

Sum = $A^0.B^2 + A^1.B^1 + A^2.B^0 + 1. (A^0.B^1 + A^1.B^0 + A^2.B^2)$

Carry = $1. (A^1.B^2 + A^2.B^1 + A^2.B^2)$

To design a DNA-Quantum ternary half adder, DNA decoders are needed to decode the input sequence into the corresponding bits. Except for the two DNA decoders at the input level, other operations will be conducted using Quantum operations.

As shown in Figure 9.10, at first, multi-valued DNA decoder produces 3 outputs line for each input value. If 1-to-3 decoders are not used, it will not be possible to access all combinations of these two inputs. Decoder for input A produces A0, A1, A2, for input B produce B0, B1, B2. Only one output line is true (**TGGATC**) and the other remains false (**ACCTAG**) for an input. These nine DNA sequence outputs of the DNA decoder are temporarily stored in DNA cache memory. Then they convert into Quantum qubits with the help of trap ions. After the conversion, they act as the inputs of the remaining Quantum operations. The Quantum part will produce the final output. In the Figure 9.10, ternary DNA cache memory is used to store information.

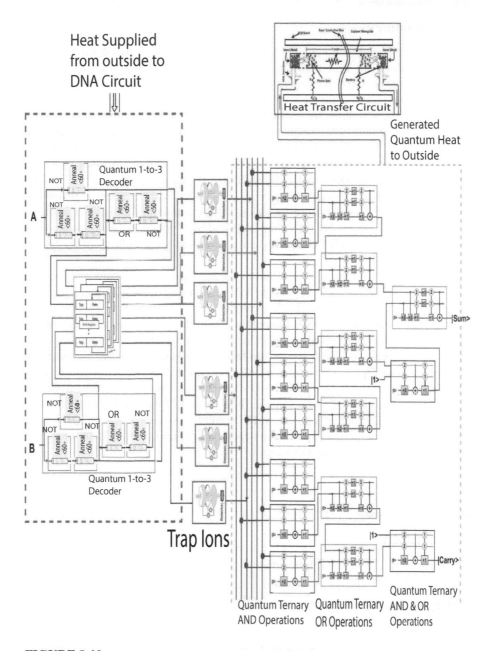

FIGURE 9.10
Circuit Architecture of Multi-Valued Quantum Half Adder

9.4 Summary

Multi-Valued quantum computing is a speedy technology based on quantum-theoretical principles, which is the behavior of energy and matter of a qurit. Combination of qutrits is used to perform any specific task in quantum computing. Quantum computers represent a significant advancement in computing capability, with enormous performance benefits for specific use cases. The ability of bits to be in several states at the same time gives the quantum computer a lot of computing capability and it is much faster than classical computing. Furthermore, multi-valued DNA computing uses biological molecules to do computations. The input of any DNA operation can be represented by DNA molecules with specific sequences. The instructions are carried out by laboratory operations on the molecules, and the result is defined as some property of the final set of molecules. DNA computing promises significant and meaningful linkages between computers and life systems, as well as massively parallel computations. DNA computing can actually carry out millions of operations at the same time.

Multi-Valued computing in quantum molecular biology that means multi-valued DNA-Quantum computing and Quantum-DNA computing systems will merge all the advantages of both multi-valued DNA computing and multi-valued quantum computing. Both quantum and DNA computing approaches can be used to accomplish operations at a very high speed and efficiency. Quantum operations produce instantaneous results. However, DNA procedures take a long time to prepare. The quantum computer's output qubits cannot be instantly inserted into the DNA system. As a result, when working with the Quantum-DNA system, a temporary storage device or system is needed where the qubits may be held for a very short time. So, a Quantum Cache memory operates in a Quantum-DNA and DNA cache memory operates on DNA-Quantum computing circuit for its operation, which can store data for a short amount of time and can pass it to another.

Bibliography

[1] Steane, A. (1998). Quantum computing. Reports on Progress in Physics, 61(2), 117.

[2] Hirvensalo, M. (2003). Quantum computing. Springer Science & Business Media.

[3] National Academies of Sciences, Engineering, and Medicine. (2019). Quantum Computing: Progress and Prospects.

[4] Mavroeidis, V., Vishi, K., Zych, M. D., & Jøsang, A. (2018). The impact of quantum computing on present cryptography. arXiv preprint arXiv:1804.00200.

[5] Chatterjee, G., Dalchau, N., Muscat, R. A., Phillips, A., & Seelig, G. (2017). A spatially localized architecture for fast and modular DNA computing. Nature Nanotechnology, 12(9), 920-927.

[6] Tao, J., Zhang, R., & Zhu, Y. (2020). DNA Computing Based Genetic Algorithm. Springer Singapore.

Part III

Multiple-Valued Arithmetic Circuits in Quantum Molecular Biology

Overview

Both multi-valued quantum and multi-valued DNA computing approaches can be used to accomplish operations at a very high speed and efficiency. The combination of these two form multi-valued quantum-DNA computing and multi-valued DNA-quantum computing can also be called quantum molecular biology where the advantages of both can be merged. Arithmetic circuits are the base of all computation in the computer system. The arithmetic Logic Unit is one of the main components of any computer. All logical operations are performed in the arithmetic logic unit. In both quantum and DNA computers, the arithmetic logic unit exists. So, this part will discuss about multi-valued arithmetic circuits in quantum and DNA computing systems and the combinations of both systems. Now the diagram, general organization, operational architecture, circuit diagram, and the algorithm will be demonstrated for the multiple-valued system here. The ternary number system is used as the multiple-valued logic system. Ternary arithmetic operations in Quantum computing and DNA computing will be discussed here.

The basic gates of ternary quantum and DNA systems are also discussed. The architecture and working procedures, the general architecture of some of the fundamental operations in the ternary logic system are described also. All these operations in two-valued logic system are already introduced. Now the multi-valued logic system will be introduced here. The readers are recommended to read Part I of this book first. Then read Part II sequentially. To understand the arithmetic operations. First, understand the basic quantum ternary gates and quantum ternary basic operations, then each operation is explained with proper examples and Figures. Follow the Table carefully which will help to understand the operations better. Finally, try to implement the operations with suitable simulation software with implementing the given algorithms for the operations.

This part is divided into six chapters. Chapters 10 and 11 will describe all arithmetic operations in multi-valued quantum computing. Chapters 12 and 13 will discuss about all arithmetic operations in multi-valued DNA computing. Chapters 14 and 15 are about arithmetic operations in quantum molecular biology which means multi-valued quantum-DNA and DNA-quantum computing, respectively.

10

Multiple-Valued Logic Operations in Quantum Computing

10.1 Introduction

Multi-Valued quantum computing is a challenging field for researchers. Two-valued quantum computing is not so old, but multi-valued quantum computing is a new topic to discuss. The benefits of working with other number systems over the binary number system are known. It is possible to work with more data with less effort in a multiple-valued system. Possibilities in which there are no middle options between true and false are familiar to computer scientists, computer engineers, applied mathematicians, and physicists. The soft logic of probability is familiar to statisticians, while the logic of uncertainty is familiar to physicists. When attempting to determine whether the status of a computer system is go, wait, or stop, the lack of better options is inconvenient and critical. These intermediate choices are the focus of multiple-valued logic. The difficulty in checking, correcting, or modifying overcomplicated flowcharts created by computer programmers is a severe disadvantage.

Designing with multiple-valued logic has obtained a lot of attention in the previous three decades. Multiple-valued logic (MVL) emerged as a separate study in the early 1920s, thanks to a Polish philosopher named Lukasiewicz, his goal was to add a third value to the binary system. The Lukasiewicz system is the result of this investigation. Emil Post, an American mathematician, invented multiple-valued algebra, sometimes known as post algebra, in response to this technique.

In this chapter, the quantum world with multiple-valued logic systems will be explored. All operations will be covered that learned before in the binary quantum logic system. The basic operations of quantum computing in the *ternary logic system* will be implemented.

10.2 Quantum Ternary Logic

The qubits are known to all and have two states - $|0>$, and $|1>$. The information unit in a three-valued quantum system (ternary quantum system) is termed as *qutrit*. The ternary quantum system represents one type of three-dimensional quantum system

DOI: 10.1201/9781003381938-10

with the basis states |0>, |1>, and |2>. These basis states are called qutrit states and can be represented by 3×1 vectors: $|0> = \begin{bmatrix} 1 \\ 0 \\ 0 \end{bmatrix}$, $|1> = \begin{bmatrix} 0 \\ 1 \\ 0 \end{bmatrix}$, and $|2> = \begin{bmatrix} 0 \\ 0 \\ 1 \end{bmatrix}$

In a ternary quantum system, a qutrit can be defined as a linear superposition of the above-mentioned basis states with the following equation:

$\psi = \alpha \, |0> + \beta \, |1> + \gamma \, |2>$

Where α, β and γ are the complex quantities to represent the probability amplitudes of the basis states and ψ is the wave function.'

10.2.1 Why Ternary Logic in Quantum Computing?

A ternary computer (sometimes known as a trinary computer) is a computer that does computations using ternary logic (three possible values) rather than binary logic (two possible values). In ternary quantum computing the superposition will be formed in the range of |0> to |2>. Ternary computing has many basic benefits over binary computing, some are given below.

1. Higher data throughput.

2. Access to additional instructions.

3. Back-compatibility with legacy binary codes.

4. Preventing malware and viruses.

5. Providing more security.

However, the usual aim of ternary computers has not met with overwhelming success thus far, because they are not as efficient as binary computers in computing binary codes, which are widely used and appear to be nearly ubiquitous. Besides, constructing the ternary operational circuits is much more difficult than the binary operational circuit.

10.3 Quantum Fundamental Gates in Multi-Valued Logic

In the two-valued quantum computing, several quantum fundamental gates for reversible logic are learned which include - the Pauli gate, Pauli-X gate, Hadamard gate, Toffoli gate, Fredkin gate, Deutsch gate, and Swap gate. All of them will not work in the ternary quantum system. The fundamental quantum gates which will work in ternary quantum computing are

1. Quantum Ternary Shift Gates

2. Quantum Ternary Toffoli Gates

3. Quantum Ternary C2 NOT Gate

TABLE 10.1

Operations of 1-Qutrit Permutation Gates/Ternary Shift Gates

A	Z(0)=A	Z(+1)=A+1	Z(+2)=A+2	Z (12)=2A	Z (1)=2A+1	Z (02)=2A+2
0	0	1	2	1	1	2
1	1	2	0	0	0	1
2	2	0	1	2	2	0

These three forms of quantum ternary gates are discussed in the next section.

10.3.1 Quantum Ternary Shift Gates

Six ternary permutation matrices are widely used which are termed Quantum Ternary Shift Gates .

1. $Z(+0)$ is the primary state. Its columns correspond to 0, 1, and 2, respectively.
2. Transform $Z(+1)$ shifts the qutrit states by 1.
3. Transform $Z(+2)$ shifts the qutrit states by 2.
4. Transform $Z(12)$ swaps (permutes) the qutrit states $|1>$ and $|2>$.
5. Transform $Z(01)$ swaps the qutrit states $|0>$ and $|1>$.
6. Transform $Z(02)$ swaps the qutrit states $|0>$ and $|2>$.

Those transformations are depicted below.

(1-Qutrit Ternary Permutation Transformations)

$$Z_3(+0) = \begin{bmatrix} 1 & 0 & 0 \\ 0 & 1 & 0 \\ 0 & 0 & 1 \end{bmatrix} Z_3(+1) = \begin{bmatrix} 0 & 0 & 1 \\ 1 & 0 & 0 \\ 0 & 1 & 0 \end{bmatrix} Z_3(+2) = \begin{bmatrix} 0 & 1 & 0 \\ 0 & 0 & 1 \\ 1 & 0 & 0 \end{bmatrix}$$

$$Z_3(12) = \begin{bmatrix} 1 & 0 & 0 \\ 0 & 0 & 1 \\ 0 & 1 & 0 \end{bmatrix} Z_3(01) = \begin{bmatrix} 0 & 1 & 0 \\ 1 & 0 & 0 \\ 0 & 0 & 1 \end{bmatrix} Z_3(02) = \begin{bmatrix} 0 & 0 & 1 \\ 0 & 1 & 0 \\ 1 & 0 & 0 \end{bmatrix}$$

And the above table (Table 10.1) shows the result of the operations.

From Table 10.1 and the matrix transformations, it is easy to understand the operations in the ternary shift gates. Figure 10.1 shows the general diagram of the ternary shift gates.

Figure 10.1 (a) shows the $Z(+1)$ operations, where the given input will be shifted to the next one (i.e. $|0>$ will become $|1>$, $|1>$ will become $|2>$, and $|2>$ will become to $|0>$). Figure 10.1 (b) shows the $Z(+2)$ operations, where the given input will be

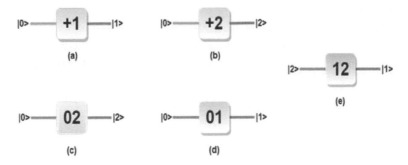

FIGURE 10.1
The General Architecture of the Ternary Permutation Gates

shifted by +2 (i.e. |0> will become |2>, |1> will become |0>, and |2> will become to |1>). Figure 10.1 (c) shows the Z (02) operations, where if the given input is |0> it will be swapped by |2> and vice versa. Note that input |1> will not affect this gate. That means input |1> will pass as |1> itself (i.e. no swapping). Figure 10.1 (d) shows the Z (01) operations, in the same way, if the given input is |0> it will be swapped by |1> and vice versa. And, now input |2> will not affect this gate. That means input |2> will pass as |2> itself (i.e. no swapping). And Figure 10.1 (e) shows the Z (12) operations, where if the given input is |1> it will be swapped by |2> and vice versa. Eventually, input |0> will not affect this gate. That means input |0> will pass as |0> itself (i.e. no swapping).

10.3.1.1 Quantum Ternary Toffoli Gates

The Ternary Toffoli gate is another quantum ternary gate. Its inputs are A, B, and C, where A and B are the controlling inputs and C is the controlled input. Its outputs are

P = A
Q = B
R = Z transforms of C ; if A = X_1 and B = X_2
R = C ; otherwise

The symbol of the generalized 3-qutrit Ternary Permutation/Shift operations is shown in Figure 10.2.

As shown in Figure 10.2, the outputs of the ternary Toffoli gate are P, Q, and R. The outputs P and Q are equal to A and B, and the output R is equal to the Z transform of C if the inputs A and B are equal to X1 and X2, where Z = {+1, +2, 01, 02, 12}. Otherwise, the output R is equal to the input C.

Figure 10.3 shows the Toffoli gate for one controlled input. Here, the gate will open only if the controlled qutrit is |2>.

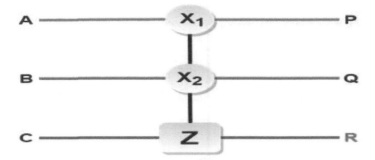

FIGURE 10.2
Symbol of Generalized 3-Qutrit Ternary Permutation Operations

10.3.2 Quantum Ternary C^2 NOT Gate

In ternary logic, several multi-qutrit control operations are possible. A 3-qutrit C^2NOT which is used for realizing the simplification rules for ternary minterms as

C^2 NOT (A, B, C) = NOT(C) ; if A != B and A,B != 0
C^2 NOT (A, B, C) = C ; otherwise

where A, B are control inputs and C is the target input. Here, the final NOT (C) will provide the (+1) operation of the given input.

Figure 10.4 shows the symbol of the ternary C^2 NOT gate.

There might be confusion in this gate because it is called C^2 NOT gate but isn't doing the actual NOT operation. This gate will only activate if the input A = |1> and B = |2> or input A = |2> and B = |1> (See the logical expression). And if the control inputs are fulfilled the condition and input C is |0>, it will perform a Z (+1) operation and produce output |1>, and for input C as |1> and |2> it will produce |2> and |0>, respectively.

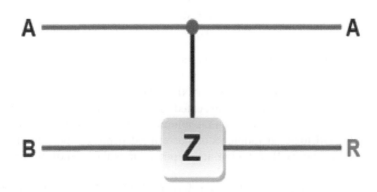

FIGURE 10.3
Ternary 1-Bit Controlled Operation

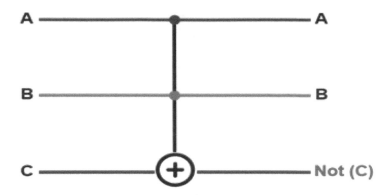

FIGURE 10.4
The General Structure Ternary 3-Qutrit C^2 NOT Gate

10.4   Quantum Multi-Valued Basic Logic Operations

Quantum ternary logic functions are those functions that have significance if a third value is acquainted with the quantum binary logic. Quantum computing in the ternary logic system is quite interesting. Here, |0>, |1>, and |2> denote the ternary levels for basic logic gates to represent false, undefined, and true, respectively. The basic operations of quantum ternary logic can be defined as follows:

$$y_{OR} = \max(x, y)$$
$$y_{NOR} = \overline{\max(x, y)}$$
$$y_{AND} = \min(x, y)$$
$$y_{NAND} = \overline{\min(x, y)}$$
$$y_{XOR} = \text{sum}(x, y)$$
$$y_{XNOR} = \overline{\text{sum}(x, y)}$$

Where, $x, y = \{0, 1, 2\}$

The truth table of those operations is shown in Table 10.2. Assume every operation is in the quantum system. For convenience, classical names are used.

The truth Table of Quantum Ternary AND, NAND, OR, NOR, XOR, and XNOR logic operations for ternary logic is presented in the above Table [6]. For AND logic operation its output value depends on the minimum value of its inputs. Similarly, in the case of the OR logic operation, its output value depends on the maximum value of its inputs. For the XOR operation, its output value is the sum of the value of its inputs.

Therefore, the outputs of Quantum Ternary NAND, NOR, and XNOR logic operations become the inverted of quantum ternary AND, OR, and XNOR logic operations.

TABLE 10.2

Truth Table for Ternary AND, NAND, OR, NOR, XOR, XNOR

Input 1	Input 2	AND	NAND	OR	NOR	XOR	XNOR
0	0	0	2	0	2	0	2
0	1	0	2	1	1	1	1
0	2	0	2	2	0	2	0
1	0	0	2	1	1	1	1
1	1	1	1	1	1	2	0
1	2	1	1	2	0	0	2
2	0	0	2	2	0	2	0
2	1	1	1	2	0	0	2
2	2	2	0	2	0	1	1

10.4.1 Ternary Quantum-AND Operation

Quantum ternary AND operation is defined as $Y_{QAND} = \min(X, Y)$, where X and Y are the input from $\{|0>, |1>, |2>\}$. The truth table of quantum ternary AND operation is given in Table 10.3.

10.4.1.1 The Architecture of Ternary Quantum-AND Operation

The circuit diagram of the Quantum Ternary AND operation is shown in Figure 10.5, both (a) and (b) show the circuit diagram of the Quantum Ternary AND operation. Any of those can be used.

The Quantum Ternary AND operation require the shifting operations that include Z (+1) and Z (+2) shifting. Where one Z (+2) operation is controlled by both inputs and will open only if the value of both inputs is $|2>$. And following a C^2 NOT gate is used which will work only if the inputs A and B met conditions of A! =B and A, B! $= |0>$. And the following Z (+1) operation is controlled by both inputs and will open only if the value of both inputs is $|1>$ (Figure 10.5 a). This will give the expected outputs that were shown in the truth table. Figure 10.5 (b) is the equivalent of 10.5 (a) but requires more quantum ternary gates.

TABLE 10.3

Truth Table of Quantum Ternary AND
Operations

	$|0>$	$|1>$	$|2>$
$|0>$	$|0>$	$|0>$	$|0>$
$|1>$	$|0>$	$|1>$	$|1>$
$|2>$	$|0>$	$|1>$	$|2>$

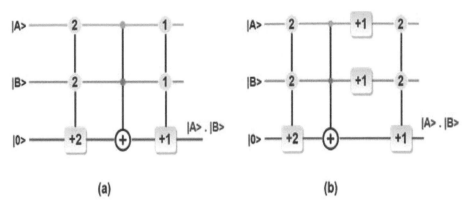

FIGURE 10.5
Circuit Diagram of Quantum Ternary AND Operation

10.4.1.2 Working Principles of Ternary Quantum-AND Operation

From Figure 10.5 (a), it is observed that the 1st Z (+2) and last Z (+1) operations are controlled by the input A and B. And the Z (+2) and Z (+1) gates will open only if the |A> and |B> are |2> and |1>, respectively. And the next quantum C^2NOT gate will open only if the A, B! = 0 && A! = B. The operations at each quantum ternary gate are shown in Table 10.4.

Table 10.4 shows the outputs from each gate of the Quantum Ternary AND operation shown in Figure 10.5 (a) for every input pattern that can be possible for two inputs in the ternary system. For simplicity, in the Table, the Kendal notation of qutrits is not used.

1. For input |A> = |1>, and |B> = |2>.

    (a) The Z (+2) will not active because control qutrits both are not equal to|2>. Therefore, it will pass the input unchanged which is an ancilla qutrit |0>.

TABLE 10.4
Quantum Ternary AND Operations for the Circuit Shown in Figure 10.5 (a)

A	B	Z(+2)	C^2 NOT	OutputZ(+1)
0	0	0	0	0
0	1	0	0	0
0	2	0	0	0
1	0	0	0	0
1	1	0	0	1
1	2	0	1	1
2	0	0	0	0
2	1	0	1	1
2	2	2	2	2

TABLE 10.5
Truth Table of Quantum Ternary NAND Operations

	|0>	|1>	|2>
|0>	|2>	|2>	|2>
|1>	|2>	|1>	|1>
|2>	|2>	|1>	|0>

(b) The next operation is the quantum ternary C^2NOT gate. It will be activated because the conditions are fulfilled (A! =B and A, B! = \|0>). Therefore it will increment the input by +1. So the output from this gate is \|1>.

(c) The last gate is Z (+1), which will also not be active because control qutrits both are not equal to \|1>. Therefore, it will pass the input unchanged which is \|1> (from step b). So, obtained output as \|1> for the given input.

2. For input \|A> = \|1>, and \|B> = \|0>,

(a) The Z (+2) will not active because control qutrits both are not equal to \|2>. Therefore, it will pass the input unchanged which is an ancilla qutrit \|0>.

(b) The next operation is the quantum ternary C^2NOT gate. It will not be activated because the conditions are not fulfilled (A! =B and A, B! = \|0>). So the output from this gate is \|0> as well.

(c) The last gate is Z (+1), which will also not be active because control qutrits both are not equal to \|1>. Therefore, it will pass the input unchanged which is \|0> (from step b). So, the output is obtained as \|0> for the given input.

10.4.2 Ternary Quantum-NAND Operation

As it is known how to design the quantum ternary AND operation , it is understandable that only a NOT operation is required to get the result of quantum NAND operation. Quantum NAND operation is defined as $Y_{QNAND} = \overline{(\min(X , Y))}$ where X and Y are the input from {\|0>, \|1>, \|2>}. Table 10.5 shows the truth table of Quantum Ternary NAND Operations.

From the truth table, it can be understood that this operation will not affect the quantum ternary AND output value \|1>. If the output value \|0>, it will convert it to the \|2> and vice versa.

10.4.2.1 The Architecture of Ternary Quantum-NAND Operation

The quantum ternary NAND operation is nothing but the $\overline{AND}$ operation. Therefore, the only need is to invert the output of the quantum ternary AND operation outputs. The circuit diagram of Quantum Ternary NAND operation is shown in Figure 10.6.

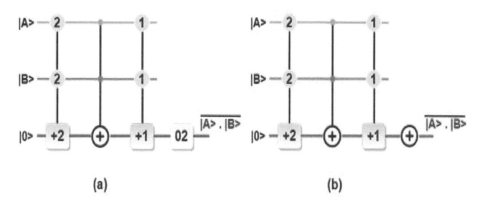

FIGURE 10.6
Circuit Diagram of Quantum Ternary NAND Operation

The output can be inverted in two ways. Figures 10.6 (a) and 10.6 (b) both can perform quantum ternary NAND operations. Because Z(02) swapping is the same as quantum NOT operation.

10.4.2.2 Working Principles of Ternary Quantum-NAND Operation

As the circuit is shown, it needs to perform a quantum ternary NOT operation after the quantum ternary AND operation. So it is not necessary to again write the working procedure of quantum ternary AND followed by quantum ternary NOT operation. Note that, quantum ternary NOT converts the input |0> to |2> and input |2> to |0>. And input |1> remains the same (see Table 10.2). It is known that operation Z (02) also does the same. That is why Z (02) is used in Figure 10.6 (a). The working procedure of the ternary NAND operation is shown in Table 10.6.

TABLE 10.6
Quantum Ternary NAND Operations for the Circuit Shown in Figure 10.6a

A	B	Z(+2)	C^2 NOT	Z(+1)	Output Z(02)
0	0	0	0	0	2
0	1	0	0	0	2
0	2	0	0	0	2
1	0	0	0	0	2
1	1	0	0	1	1
1	2	0	1	1	1
2	0	0	0	0	2
2	1	0	1	1	1
2	2	2	2	2	0

TABLE 10.7
Truth Table of Quantum Ternary OR
Operations

	$\lvert 0 \rangle$	$\lvert 1 \rangle$	$\lvert 2 \rangle$
$\lvert 0 \rangle$	$\lvert 0 \rangle$	$\lvert 1 \rangle$	$\lvert 2 \rangle$
$\lvert 1 \rangle$	$\lvert 1 \rangle$	$\lvert 1 \rangle$	$\lvert 2 \rangle$
$\lvert 2 \rangle$	$\lvert 2 \rangle$	$\lvert 2 \rangle$	$\lvert 2 \rangle$

10.4.3 Ternary Quantum-OR Operation

Quantum OR operation is defined as $Y_{QOR} = \max (X, Y)$, where X and Y are the input from $\{\lvert 0 \rangle, \lvert 1 \rangle, \lvert 2 \rangle\}$. The operations in the quantum ternary OR operation are shown in Table 10.7.

10.4.3.1 The Architecture of Ternary Quantum-OR Operation

The Quantum Ternary OR operation requires the shifting operations that include $Z(+1)$ and $Z(+2)$ shifting. Where two $Z(+1)$ operation is not controlled, and two $Z(+1)$ operation is controlled by two input. And two $Z(+2)$ operation is controlled by respective one input(input $\lvert B \rangle$ for the first $Z(+2)$, and input $\lvert A \rangle$ for the 2nd $Z(+2)$). And, a two-input controlled C^2 NOT gate is also required to get the expected result that was shown above in the truth table. The circuit diagram of The Quantum Ternary OR operation is shown in Figure 10.7.

10.4.3.2 Working Principles of Ternary Quantum-OR Operation

From Figure 10.7, it is found that the 1st Z $(+2)$ and 2nd Z $(+2)$ operations are controlled by the input $\lvert B \rangle$ and $\lvert A \rangle$, respectively which will open only if the input is $\lvert 2 \rangle$. The following Z $(+1)$ is controlled by both $\lvert A \rangle$ and $\lvert B \rangle$ input that will only open if both inputs are $\lvert 2 \rangle$. The next two Z $(+1)$ are not controlled. And the following Z $(+1)$

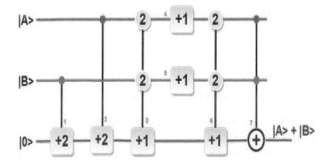

FIGURE 10.7
The Circuit Diagram of Quantum Ternary OR Operation

is again controlled by two inputs and will open only if get |2> from both inputs. And the final C^2NOT gate will open only if the A, B! = 0 && A! = B.

To understand the operation let's take an example.

Suppose, |A> = |1> and |B> = |2>,

1. The 1st Z (+2) gate will get input |0> and the controlled input is |B> = |2>. Therefore, this gate will be activated. And generate the output |2> (see shifting operations). This will work as an input to the next Z (+2) gate.

2. The next Z (+2) gate will get input |2> (from step a) and the controlled input is |A> = |1>. Therefore, this gate will not be activated. And generate the output |2> (without changing). This will work as an input to the next Z (+1) gate.

3. The next operation is a 2-input controlled Z (+1) gate operation which will act only if both the control inputs are |2>. So, it will not be activated. Therefore, the input to the gate will pass as the same. As a result, it generates |2> (from step b). This output will deliver to the next controlled Z(+1) operation.

4. In this step, two Z(+1) operations will be performed. They are not controlled. Therefore, they will increase the value of |A> and |B> by +1. So, |A> will be |2> and |B> will be |0>.

5. The next operation is again a 2-input controlled Z(+1) gate which gets input |2>(from step c) and will be activated only if the input values of |A> and |B> both are |2>. Therefore, it will also not be activated and it delivers |2> to the last operation.

6. The last operation is the ternary C^2NOT operation. It will also not be activated because now the value of |A> and |B> is |2> and |0> (from step d), respectively. Therefore, they do not fulfill the condition of the ternary C^2NOT gate. Thus, the final output will be |2> (the input to the C^2NOT gate.

The operations for all the input combinations is shown in Table 10.8.

10.4.4 Ternary Quantum-NOR Operation

Quantum Ternary NOR operation is defined as $Y_{QNOR} = \overline{(max(X , Y))}$ where X and Y are the input from {|0>, |1>, |2>}.

The operations in the quantum ternary NOR operation are shown in Table 10.9.

10.4.4.1 The Architecture of Ternary Quantum-NOR Operation

The Quantum Ternary NOR NOR operation is nothing but the $\overline{OR}$ operation. Therefore, it is needed to invert the output of the quantum ternary OR operation's outputs.

TABLE 10.8

Quantum Ternary OR Operations for the Circuit Shown in Figure 10.7

A	B	1 Z(+2)	2 Z(+2)	3 Z(+1)	4 Z(+1)	5 Z(+1)	6 Z(+1)	Output C²NOT
0	0	0	0	0	1	1	0	0
0	1	0	0	0	1	2	0	1
0	2	2	2	2	1	0	2	2
1	0	0	0	0	2	1	0	1
1	1	0	0	0	2	2	1	1
1	2	2	2	2	2	0	2	2
2	0	0	2	2	0	1	2	2
2	1	0	2	2	0	2	2	2
2	2	2	1	2	2	0	0	2

Therefore the circuit diagram will be the same as the quantum ternary OR operation with an extra NOT operation at the end. The circuit diagram of Quantum Ternary NOR operation is shown in Figure 10.8.

In Figure 10.8, both (a) and (b) will perform the quantum ternary NOR operations.

10.4.4.2 Working Principles of Ternary Quantum-NOR Operation

The working procedure of the quantum ternary OR operation is discussed already. In the quantum ternary NOR operation, all that have to do is to invert the output of the OR operation. The following table (Table 10.10) shows the operations of each gate of the circuit shown in Figure 10.8 (a).

10.4.5 Ternary Quantum XOR Operation

Quantum Ternary XOR operation is defined as Y_{QXOR} = sum (X, Y), where X and Y are the input from {|0>, |1>, |2>}. The operations in the quantum ternary XOR operation are shown in Table 10.11.

TABLE 10.9

Truth Table of Quantum Ternary NOR Operations

	|0>	|1>	|2>
|0>	|2>	|1>	|0>
|1>	|1>	|1>	|0>
|2>	|0>	|0>	|0>

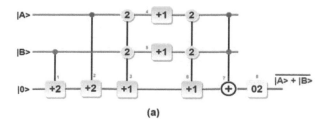

(a)

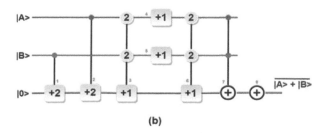

(b)

FIGURE 10.8
The Circuit Diagram of Quantum Ternary NOR Operation

TABLE 10.10
Quantum Ternary NOR Operations for the Circuit Shown in Figure 10.8a

A	B	1 Z(+2)	2 Z(+2)	3 Z(+1)	4 Z(+1)	5 Z(+1)	6 Z(+1)	7 C²NOT	Output Z (02)
0	0	0	0	0	1	1	0	0	2
0	1	0	0	0	1	2	0	1	1
0	2	2	2	2	1	0	2	2	0
1	0	0	0	0	2	1	0	1	1
1	1	0	0	0	2	2	1	1	1
1	2	2	2	2	2	0	2	2	0
2	0	0	2	2	0	1	2	2	0
2	1	0	2	2	0	2	2	2	0
2	2	2	1	2	2	0	0	2	0

TABLE 10.11
Truth Table of Quantum Ternary XOR Operation

	|0>	|1>	|2>
|0>	|0>	|1>	|2>
|1>	|1>	|2>	|0>
|2>	|2>	|0>	|1>

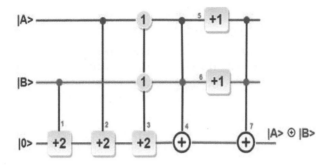

FIGURE 10.9
The Circuit Diagram of Quantum Ternary XOR Operation

10.4.5.1 The Architecture of Ternary Quantum XOR Operation

The circuit diagram of the Quantum Ternary XOR operation is shown in Figure 10.9. The Quantum Ternary XOR operation requires the shifting operations that include Z (+1) and Z (+2) shifting.

Where two Z (+1) operation is not controlled, and two Z (+2) operation is controlled by respective one input. And one Z (+2) operation is controlled by both inputs. And, two two-input controlled C^2 NOT gate is also required to get the expected result that was shown in the above truth table (Table 10.11).

10.4.5.2 Working Principles of Ternary Quantum XOR Operation

From Figure 10.9, the 1st Z (+2) and 2nd Z (+2) operations are controlled by the input B and A, respectively which will open only if the input is |2>. The following Z (+2) is controlled by both A and B input that will only open if both inputs are |1>. The next two Z (+1) are not controlled. And the following C^2 NOT is controlled by two input and will open only if [|2>, |1>] or [|1>, |2>] input combination is obtained from both input, which was mentioned earlier (i.e. gate will open only if at A, B! = 0 && A! = B). The last C^2 NOT will also open only if the input combination fulfills the condition of C^2 NOT.

Let's take some examples to observe the working principle of the designed circuit for the quantum ternary XOR operation shown in Figure 10.9.

1. Suppose, the inputs |A> = |2> and |B> = |1>,

    (a) The 1st Z(+2) is controlled by the value of |B> and it will activate only if the control qutrit is |2>. Here |B> = |1>, therefore the gate will not be activated. And it will pass its input value |0> unchanged. So, the next Z(+2) will get input |0> as well.

    (b) The 1st Z(+2) is controlled by the value of |A> and it will activate only if the control qutrit is |2>. Here |A> = |2>, therefore the gate will be activated. And it will pass its input value |0> by shifting the

Z(+2) operation. So, it will generate |2> as output which will be the input value of the next Z(+2) operation.

(c) The 3rd Z (+2) is controlled by the value of both |A> and |B>, and it will activate only if both of the control qutrits are |1>. Here |A> = |2> and |B> = |1>, therefore the gate will not be activated. And it will pass its input value |2> (from step b) unchanged.

(d) The next operation is the C^2NOT gate which gets the input |2> (from step c). It will open because the input values fulfill its conditions. Therefore it will produce the output |0> (C^2 NOT increases by Z(+1)). This value will work as the input to the last C^2NOT gate.

(e) Now both the Z(+1) operations will be held. They are not controlled; therefore, they will increase the value of |A> and |B> by +1. So, |A> = |0>, and |B> = |2>.

(f) The last C2NOT gate gets input |0> (from step d). But this gate will not open because the controlled value of |A> and |B> do not fulfill the conditions of the C^2NOT gate. Therefore, the final output will be |0>.

2. Suppose, the inputs |A> = |0> and |B> = |2>,

(a) The 1st Z(+2) is controlled by the value of |B> and it will activate only if the control qutrit is |2>. Here |B> = |2>, therefore the gate will be activated. And it will pass its input value |0> by shifting +2, which is |2>. So, the next Z(+2) will get input |2>.

(b) The 1st Z(+2) is controlled by the value of |A> and it will activate only if the control qutrit is |2>. Here |A> = |0>, therefore the gate will not be activated. And it will pass its input value |2> unchanged. So, it will generate |2> as output which will be the input value of the next Z(+2) operation.

(c) The 3rd Z(+2) is controlled by the value of both |A> and |B>, and it will activate only if both of the control qutrits are |1>. Here |A> = |0> and |B> = |2>, therefore the gate will not be activated. And it will pass its input value |2> (from step b) unchanged.

(d) The next operation is the C^2NOT gate which gets the input |2> (from step c). It will not open because the input values do not fulfill its conditions. Therefore, it will produce the output |2> (no change will occur). This value will work as the input to the last C^2NOT gate.

(e) Now both the Z(+1) operations will be held. They are not controlled; therefore, they will increase the value of |A> and |B> by +1. So, |A> = |1>, and |B> = |0>.

(f) The last C2NOT gate gets input |2> (from step d). But this gate will not open because the controlled value of |A> and |B> do not fulfill the conditions of the C^2NOT gate. Consequently, the final output will be |2>.

TABLE 10.12

Quantum Ternary XOR Operations for the Circuit Shown in Figure 10.9

A	B	1 Z (+2)	2 Z (+2)	3 Z (+2)	4 C²NOT	5 Z (+1)	6 Z (+1)	Output C²NOT
0	0	0	0	0	0	1	1	0
0	1	0	0	0	0	1	2	1
0	2	2	2	2	2	1	0	2
1	0	0	0	0	0	2	1	1
1	1	0	0	2	2	2	2	2
1	2	2	2	2	0	2	0	0
2	0	0	2	2	2	0	1	2
2	1	0	2	2	0	0	2	0
2	2	2	1	1	1	0	0	1

Therefore, it is found that the designed circuit generates the expected outputs of the quantum ternary XOR operations which were shown in the truth table of the quantum ternary XOR operation.

The following table (Table 10.12) shows the operations of each gate of the circuit shown in Figure 10.9.

10.4.6 Ternary Quantum XNOR Operation

Quantum Ternary XNOR operation is defined as $Y_{QXNOR} = \overline{sum(X, Y)}$, where X and Y are the input from {|0>, |1>, |2>}. The operations in the quantum ternary XOR operations are shown in Table 10.13.

10.4.6.1 The Architecture of Ternary Quantum XNOR Operation

The Quantum Ternary XNOR operation is nothing but the ternary $\overline{QXOR}$ operation. Therefore, it is needed to invert the output of the quantum ternary XOR operation's outputs. In order to do that, an inverter is needed to add to the QXOR's

TABLE 10.13

Truth Table of Quantum Ternary XNOR Operations

	|0>	|1>	|2>
|0>	|2>	|1>	|0>
|1>	|1>	|0>	|2>
|2>	|0>	|2>	|1>

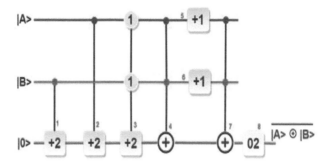

FIGURE 10.10
The Circuit Diagram of Quantum Ternary XNOR Operation

circuit. The circuit diagram of the quantum ternary XNOR operation is shown in Figure 10.10.

Z (O2) shift operation is added which will invert the output value of the quantum ternary XNOR operations. Quantum NOT operation can be added instead of Z (02) as shown earlier.

10.4.6.2 Working Principles of Ternary Quantum-XNOR Operation

Again, the working procedure is the same as the quantum ternary XOR operation with an extra Z(02) operation. The table below (Table 10.14) shows the operation at each gate of the quantum ternary XNOR operations.

TABLE 10.14
Quantum Ternary XNOR Operations for the Circuit Shown in Figure 10.10

A	B	1 Z(+2)	2 Z(+2)	3 Z(+2)	4 C²NOT	5 Z(+1)	6 Z(+1)	7 C²NOT	Output Z(02)
0	0	0	0	0	0	1	1	0	2
0	1	0	0	0	0	1	2	1	1
0	2	2	2	2	2	1	0	2	0
1	0	0	0	0	0	2	1	1	1
1	1	0	0	2	2	2	2	2	0
1	2	2	2	2	0	2	0	0	2
2	0	0	2	2	2	0	1	2	0
2	1	0	2	2	0	0	2	0	2
2	2	2	1	1	1	0	0	1	1

10.5 Summary

This chapter has focused on multi-valued quantum computing with some basic operations. The ternary quantum system represents one type of three-dimensional quantum system with the basis states |0>, |1>, and |2>. These basis states are called qutrit states. A ternary computer is a computer that does computations using ternary logic (three possible values) rather than binary logic (two possible values). The fundamental quantum gates which will work in ternary quantum computing are Quantum Ternary Shift Gates, Quantum Ternary Toffoli Gates, and Quantum Ternary C2 NOT Gate where quantum ternary shift gates are working along with the Toffoli gate. This chapter has presented basic multi-valued quantum operations with their architectures and working principles. Necessary figures are also shown in their description for better understanding.

Bibliography

[1] Marella, S. T., & Parisa, H. S. K. Introduction to Quantum Computing.

[2] Miller, M. D., & Thornton, M. A. (2008). MVL concepts and algebra. In Multiple Valued Logic: Concepts and Representations (pp. 21-42). Springer, Cham.

[3] Haghparast, M., Wille, R., & Monfared, A. T. (2017). Towards quantum reversible ternary coded decimal adder. Quantum Information Processing, 16(11), 1-25.

[4] Mandal, S. B., Chakrabarti, A., & Sur-Kolay, S. (2011, May). Synthesis techniques for ternary quantum logic. In 2011 41st IEEE International Symposium on Multiple-Valued Logic (pp. 218-223). IEEE.

[5] Hallworth, R. P., & Heath, F. G. (1962). Semiconductor circuits for ternary logic. Proceedings of the IEE-Part C: Monographs, 109(15), 219-225.

[6] Di, Y., Zhang, J., & Wei, H. (2008). Cartan decomposition of a two-qutrit gate. Science in China Series G: Physics, Mechanics and Astronomy, 51(11), 1668-1676.

11

Multiple-Valued Quantum Arithmetic Operations

11.1 Introduction

Quantum computing is a new and challenging field in research and arithmetic circuits in multi-valued quantum computing are a recent attraction for all. It is a speedy technology in modern science. Electrical circuits with more than two distinct levels of signal are now achievable because to major advances in integrated circuit technology, which have sparked a lot of interest in them in recent years. Multiple-valued logic circuits (MVLCs) are a type of circuit that offer several potential opportunities for the improvement of present VLSI circuit designs. Multiple-valued logic displays to us the phenomena that have never been witnessed in binary logic since the only two values available are the null and unit elements of Boolean algebra, which have extremely precise features. When these phenomena are reflected on a two-valued scale, fresh and deeper knowledge of the matter has gained.

Multiple-valued logic has a wide range of applications in addition to providing a better understanding of binary difficulties. They can be divided into two categories. To handle binary issues more quickly, the first group use a multiple-valued logic domain. For example, converting a multiple-output Boolean function to a single-output multiple-valued function by considering its output part as a single multiple-valued variable is a well-known method of representing it. Berkeley's verification and synthesis tool VIS, for example, uses this approach. The second group focuses on electrical circuits that use more than two discrete levels of signals, such as multiple-valued memory, arithmetic circuits, and Field Programmable Gate Arrays. Multiple-valued logic (MVL) circuits propose a number of potential improvements to current VLSI circuit designs. For example, serious difficulties with limitations on the number of connections an integrated circuit has with the outside world (pinout problem) as well as the number of connections within the circuit encountered in some VLSI circuit synthesis could be greatly reduced if signals in the circuit could assume four or more states instead of only two. Furthermore, employing multiple-valued number representation in many applications has a strong mathematical appeal. Residue and redundant number systems, for example, allow for the reduction or elimination of ripple-through carries in regular binary addition and subtraction, resulting in faster arithmetic.

DOI: 10.1201/9781003381938-11

On the other hand, designing and implementing a multiple-valued logic circuit is a substantial challenge. Because the circuits are more complicated to build than binary logic circuits. When it comes to quantum computing, things get a little trickier. In this chapter, the following topics will be discussed.

1. Multiple-Valued Quantum Half Subtractor

2. Multiple-Valued Quantum Full Subtractor

3. Multiple-Valued Quantum N-qutrit Parallel Adder

4. Multiple-Valued Quantum Carry Skip Adder

5. Multiple-Valued Quantum Carry Lookahead Adder

6. Multiple-Valued Quantum Multiplier

7. Multiple-Valued Quantum Divider

8. Multiple-Valued Quantum Comparator

These topics will be discussed in the ternary logic system. Multi-Valued can be ternary, quaternary and so on.

11.2   Multiple-Valued Quantum Half-Adder

A multi-valued half adder for a ternary logic system is a type of adder where an electronic circuit performs the addition of ternary numbers. The half adder is able to add two single ternary digits and provide the output plus a carry value. It has two inputs, called A and B, and two outputs called S (sum) and C (carry).

Quantum Ternary Half Adder will add two qutrits from the quantum ternary digit |0>, |1>, and |2> and produce two outputs as quantum sum |S> and quantum carry qutrit |C>.

The Table 11.1 shows the outputs of a quantum ternary half adder.

From the Table 11.1, the logical expressions for the ternary sum and carry are found as

   Sum = $A^0.B^2 + A^1.B^1 + A^2.B^0 + 1. (A0.B1 + A^1.B^0 + A^2.B^{2)}$

   Carry = $1. (A1.B2 + A^2.B^1 + A^2.B^{2)}$

So, using these equations the circuit diagram can be constructed for the quantum ternary half-adder operation.

11.2.1   The Architecture of Quantum Ternary Half-Adder Operation

Decoders are required to create a ternary half adder so that the input qutrit can be decoded into the appropriate qutrit. Two decoders will be used to decode two qutrits |A> and |B> and they will produce 3-qutrit outputs. The architecture of the Quantum Ternary half-adder is shown in Figure 11.1.

TABLE 11.1

The Truth Table of Quantum Ternary Half-Adder Operation

A	B	Carry	Sum
$|0>$	$|0>$	$|0>$	$|0>$
$|0>$	$|1>$	$|0>$	$|1>$
$|0>$	$|2>$	$|0>$	$|2>$
$|1>$	$|0>$	$|0>$	$|1>$
$|1>$	$|1>$	$|0>$	$|2>$
$|1>$	$|2>$	$|1>$	$|0>$
$|2>$	$|0>$	$|0>$	$|2>$
$|2>$	$|1>$	$|1>$	$|0>$
$|2>$	$|2>$	$|1>$	$|1>$

From the Figure 11.1, it is found that two quantum ternary 1-to-3 decoders, eleven quantum ternary AND operations, and seven quantum ternary OR operations are required to build the quantum ternary half-adder.

The quantum ternary AND and OR operations are nothing but the products-of-sum (which was shown in the logical expression of the quantum ternary half-adder) that produce the expected output that was mentioned in the truth table 11.1.

The quantum ternary AND and the quantum ternary OR operations are already discussed in the previous chapter. The architecture may seem very complex and difficult to understand for the first time.

11.2.2 Working Principles of Quantum Ternary Half-Adder Operation

Since the architecture of the quantum ternary half-adder is designed completely following the logical expression derived from the truth table (Table 11.1). The logical expression can be used to demonstrate the working methods. If the expressions provide absolute outputs, the architecture will produce the expected results as well.

The working procedure of the quantum ternary half adder is given below for each pattern of the input set of qutrits.

1. For input A,B = 0,

    $A_0= 2$, $A_1=0$, $A_2=0$

    $B_0= 2$, $B_1=0$, $B_2=0$

    Sum = $A_2.B_0 + A_1.B_1 + A_0.B_2 + 1.(A_1.B_0 + A_0.B_1 + A_2.B_2)$

    $= 0.2 + 0.0 + 2.0 + 1.(0.2 + 2.0 + 0.0)$

    $= 0 + 1.0$

    $= 0$

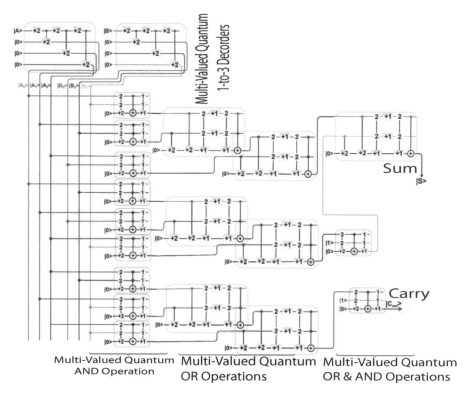

FIGURE 11.1

The Circuit Architecture of Quantum Ternary Half-Adder Operation

$\textbf{Carry} = 1. \ (A_2.B_1 + A_2.B_2 + A_1.B_2)$

$= 1. \ (0.0 + 0.0 + 0.0)$

$\qquad\qquad\qquad\qquad = 1.0$

$\qquad\qquad\qquad\qquad = 0$

2. For input A, B = 0, 1

$A_0 = 2, \ A_1 = 0, \ A_2 = 0$

$B_0 = 0, \ B_1 = 2, \ B_2 = 0$

$\textbf{Sum} = A_2.B_0 + A_1.B_1 + A_0.B_2 + 1. \ (A_1.B_0 + A_0.B_1 + A_2.B_2)$

$= 0.0 + 0.2 + 2.0 + 1.(0.2 + 2.2 + 0.0)$

$= 0 + 1.2$

$= 1$

$\textbf{Carry} = 1. \ (A_2.B_1 + A_2.B_2 + A_1.B_2)$

$= 1. \ (0.2 + 0.0 + 0.0)$

$= 1.0$

$= 0$

3. For input A, B = 0, 2

    $A_0 = 2, A_1 = 0, A_2 = 0$

    $B_0 = 0, B_1 = 0, B_2 = 2$

    Sum $= A_2.B_0 + A_1.B_1 + A_0.B_2 + 1. (A_1.B_0 + A_0.B_1 + A_2.B_2)$

    $= 0.0 + 0.0 + 2.2 + 1.(0.2 + 2.0 + 0.0)$

    $= 2 + 1.0$

    $= 2$

    Carry $= 1. (A_2.B_1 + A_2.B_2 + A_1.B_2)$

    $= 1. (0.0 + 0.2 + 0.2)$

    $= 1.0$

    $= 0$

4. For input A, B = 1, 0

    $A_0 = 0, A_1 - 2, A_2 = 0$

    $B_0 = 2, B_1 - 0, B_2 = 0$

    Sum $= A_2.B_0 + A_1.B_1 + A_0.B_2 + 1. (A_1.B_0 + A_0.B_1 + A_2.B_2)$

    $= 0.2 + 2.0 + 0.0 + 1.(2.2 + 0.0 + 0.0)$

    $= 0 + 1.2$

    $= 1$

    Carry $= 1. (A_2.B_1 + A_2.B_2 + A_1.B_2)$

    $= 1. (0.0 + 0.0 + 2.0)$

    $= 1.0$

    $= 0$

5. For input A, B = 1, 1

    $A_0 = 0, A_1 = 2, A_2 = 0$

    $B_0 = 0, B_1 = 2, B_2 = 0$

    Sum $= A_2.B_0 + A_1.B_1 + A_0.B_2 + 1. (A_1.B_0 + A_0.B_1 + A_2.B_2)$

    $= 0.0 + 2.2 + 0.0 + 1.(2.0 + 0.2 + 0.0)$

    $= 2 + 1.0$

    $= 2$

    Carry $= 1. (A_2.B_1 + A_2.B_2 + A_1.B_2)$

    $= 1. (0.2 + 0.0 + 2.0)$

    $= 1.0$

    $= 0$

6. For input A, B = 1, 2 A_0= 0, A_1=2, A_2=0

   B_0= 0, B_1=0, B_2=2

   Sum = $A_2.B_0 + A_1.B_1 + A_0.B_2 + 1. (A_1.B_0 + A_0.B_1 + A_2.B_2)$

   = 0.0 + 2.0 +0.2 +1.(2.0 + 0.0 + 0.2)

   = 0 +1.0

   = 0

   Carry = $1. (A_2.B_1 + A_2.B_2 + A_1.B_2)$

   = 1. (0.0 + 0.2 + 2.2)

   = 1.2

   = 1

7. For input A, B = 2, 0

   A_0= 0, A_1=0, A_2=2

   B_0= 2, B_1=0, B_2=0

   Sum = $A_2.B_0 + A_1.B_1 + A_0.B_2 + 1. (A_1.B_0 + A_0.B_1 + A_2.B_2)$

   = 2.2 + 0.0 +0.0 +1.(0.2 + 0.0 + 2.0)

   = 2 +1.0

   = 2

   Carry = $1. (A_2.B_1 + A_2.B_2 + A_1.B_2)$

   = 1. (2.0 + 2.0 + 0.0)

   = 1.0

   = 0

8. For input A, B = 2, 1

   A_0= 0, A_1=0, A_2=2

   B_0= 0, B_1=2, B_2=0

   Sum = $A_2.B_0 + A_1.B_1 + A_0.B_2 + 1. (A_1.B_0 + A_0.B_1 + A_2.B_2)$

   = 2.0 + 0.2 +0.0 +1.(0.0 + 0.2 + 2.0)

   = 0 +1.0

   = 0

   Carry = $1. (A_2.B_1 + A_2.B_2 + A_1.B_2)$

   = 1. (2.2 + 2.0 + 0.0)

   = 1.2

   = 1

9. For input A, B = 2, 2

$A_0 = 0, A_1 = 0, A_2 = 2$

$B_0 = 0, B_1 = 0, B_2 = 2$

Sum $= A_2.B_0 + A_1.B_1 + A_0.B_2 + 1. (A_1.B_0 + A_0.B_1 + A_2.B_2)$

$= 2.0 + 0.0 + 0.2 + 1.(0.0 + 0.0 + 2.2)$

$= 0 + 1.2$

$= 1$

Carry $= 1. (A_2.B_1 + A_2.B_2 + A_1.B_2)$

$= 1. (2.0 + 2.2 + 0.2)$

$= 1.2$

$= 1$

To keep the calculations simple, the input values were not expressed in quantum notation.

The developed architecture for the quantum ternary half-adder is found to perfectly perform the half-adder computation for the qutrits.

11.3   Multiple-Valued Quantum Full-Adder

In this section, it will be discussed how to construct a quantum full-adder in ternary logic. In the quantum ternary full adder, there are a total of three inputs $|A>$, $|B>$, and $|C>$. Here the value of A and B can be $|0>$, $|1>$ or $|2>$. But for $|C>$ in which acts as a carry input in the full-adder, can be either $|0>$ or $|1>$. The reason is that the carry qutrit can never be $|2>$. Thus a total number of eighteen combinations of the input values are obtained. Table 11.2 shows the operations in a quantum ternary full-adder.

Again, from the truth table, it is possible to derive the logical expression to find the sum and carry outputs.

The logical expression for the sum and carry out can be derived as follows:

Sum = A2.B0.C0 +A1.B0.C1 + A1.B1.C0+ A0.B1.C1 +A0.B2.C0 +A2.B2.C1+1 (A1.B0.C0+A0.B0.C1 +A0.B1.C0+A2.B1.C1+A2.B2.C0+A1.B2.C1)

= (A2.B0+A1B1+A0B2).C0 + (A1.B0+A0.B1+A2.B2).C1+ 1((A1.B0+A0.B1+A2.B2).C0 + (A0B0+A2B1+A1B2).C1)

Carry = 1 (A0.B2.C0+A2.B1.C0+A2.B2.C0 +A1.B1.C1+B2.C1+A2.C1)

= 1. ((A0.B2+A2.B1+A2B2).C0 + (A1.B1+ B2+A2).C1)

Here, A0, A1, and A2 refer that the value of input $|A>$ might be $|0>$,$|1>$, and $|2>$, respectively. Similarly, B0, B1, and B2 refer that the value of b might be $|0>$, $|1>$, and $|2>$, respectively. Again the equations are nothing but the sum-of-products form.

TABLE 11.2

Truth Table of Quantum Ternary Full-Adder
Operation

A	B	C	Sum	C_{out}
0	0	0	0	0
0	0	1	1	0
0	1	0	1	0
0	1	1	2	0
0	2	0	2	0
0	2	1	0	1
1	0	0	1	0
1	0	1	2	0
1	1	0	2	0
1	1	1	0	1
1	2	0	0	1
1	2	1	1	1
2	0	0	2	0
2	0	1	0	1
2	1	0	0	1
2	1	1	1	1
2	2	0	1	1
2	2	1	2	1

11.3.1 The Circuit Architecture of Quantum Ternary Full-Adder Operation

Figure 11.2 shows the carry output generation part and Figure 11.3 shows the sum output generation part of the quantum ternary full-adder operation. Like the quantum ternary half-adder, it is possible to construct the circuit using the equations. Decoders are used to extract the ternary values from the inputs. Three decoders are utilized for three inputs.

The quantum ternary decoder produces three output lines for each input value. Decoder for input $|A>$ produces $|A0>$, $|A1>$, $|A2>$, for input $|B>$ produces $|B0>$, $|B1>$, $|B2>$ and for input $|C>$ produce $|C0>$, $|C1>$, $|C2>$. Only one output line is true (which gives $|2>$; see decoder section) and the other remains false (produce $|0>$) for an input. From the truth table (Table 11.2), the sum of the full adder produces six true (i.e. $|2>$) values and six $|1>$ values. For these six true values, it is possible to design six ternary quantum-AND operations (TAND) and each of the quantum ternary AND operation outputs will TOR (quantum ternary OR) to produce the true output when one of the outputs of the quantum ternary AND operation is true, also have nine $|1>$ values in sum. For these six $|1>$ values, it is possible to design six quantum ternary AND operations (TAND) and each of the TAND operation outputs will TOR to produce the true ($|2>$) output when one of the outputs of the TAND operation is true. But as the output $|1>$, so TAND is performed with the TOR operation output with $|1\rangle$. TAND operation produces the minimum value. So the output of the TAND operation will always $|1\rangle$. Finally, a TOR operation is used to generate the maximum output among $|2\rangle$, $|0\rangle$ and $|0\rangle$, $|1\rangle$.

From the truth table (Table 11.2), carry of the full adder produces nine $|1\rangle$ values. It is possible to design nine quantum ternary AND operations (TAND) and each of the TAND operation's output will TOR to produce the true ($|2\rangle$) output when one of the outputs of the TAND operation is true. But as it needs the output $|1\rangle$, so TAND on the TOR operation's output with $|1\rangle$. TAND operation produces the minimum value. So the output of the TAND operation will always $|1\rangle$ for one of the true inputs among nine TAND operations. Again for one $|2\rangle$ value, one TAND operation is used and the output of this TAND will perform TOR to produce the carry output.

Therefore, it is possible to construct the architecture using the equations for the sum and carry output. It will be divided into two parts because the circuit architecture will be much larger.

11.3.2 Working Principles of Quantum Ternary Full-Adder Operation

A multiple-valued quantum Full-Adder contains three quantum qutrits $|A\rangle$, $|B\rangle$ and $|C\rangle$ as input and two output quantum qutrits, $|Sum\rangle$ and $|C_{out}\rangle$. Here, the value of $|A\rangle$ and $|B\rangle$ can be any of the three qutrits from $|0\rangle$, $|1\rangle$ and $|2\rangle$ and for $|C\rangle$ it can be $|0\rangle$ and $|1\rangle$ (in no situation carry input can be $|2\rangle$). Here, it needs three 1-to-3 multiple-valued quantum decoders, seventeen multiple-valued quantum AND operations, eleven multiple-valued quantum OR operations to get the result of $|Sum\rangle$. Similarly, another seven multiple-valued quantum AND operations and five multiple-valued quantum OR operations are required to generate the output of $|C_{out}\rangle$.

Let's have a look for one input pattern. When the input of $|A\rangle$, $|B\rangle$ and $|C\rangle$ are $|2\rangle$, $|0\rangle$ and $|0\rangle$, then decoder active $|A2\rangle$, $|B2\rangle$, and $|C0\rangle$ bypassing the qubit $|2\rangle$ among them and $|0\rangle$ through other lines. As a result first multiple-valued quantum AND operation is executed and produces the output $|2\rangle$. This output pass through some multiple-valued quantum OR operation resulting in output $|2\rangle$ again. Then this output $|2\rangle$ performs a multiple-valued quantum AND operation with $|C0\rangle$ which value is also $|2\rangle$. The output then passes through two other multiple-valued quantum OR operations to produce the result of $|Sum\rangle$ which is $|2\rangle$. And for the output value of $|C_{out}\rangle$, a similar procedure is followed. But before generating the result of $|C_{out}\rangle$, one last multiple-valued quantum AND operation is executed with a fixed input qubit $|1\rangle$. And the value of $|C_{out}\rangle$ for this input combination is $|0\rangle$.

11.4 Multiple-Valued Quantum Half-Subtractor

A ternary half subtractor is a sort of subtractor, which is an electrical circuit that performs ternary number subtractions. The half subtractor is able to subtract two single ternary digits and provide the output plus a borrow value. It has two inputs,

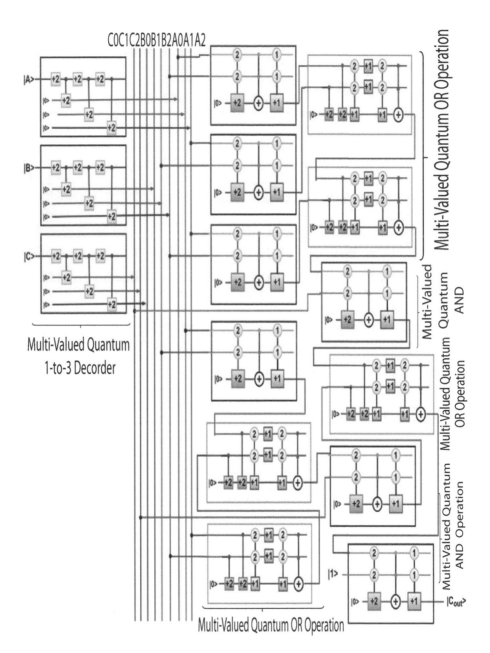

FIGURE 11.2
Circuit Architecture of Quantum Ternary Full-Adder Carry Generation

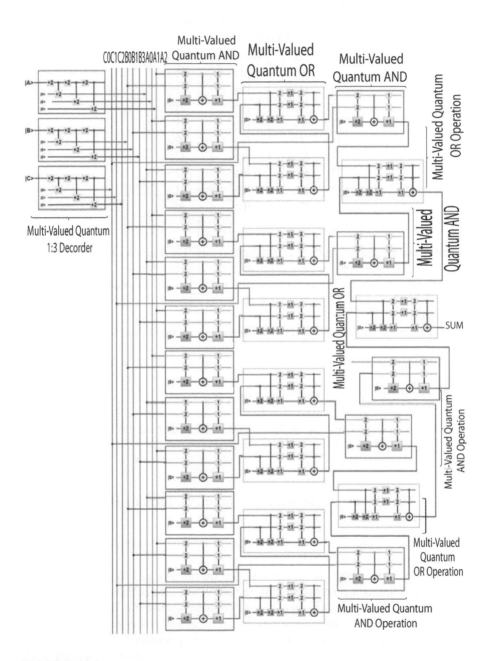

FIGURE 11.3
Circuit Architecture of Quantum Ternary Full-Adder Sum Output Generation

TABLE 11.3

Truth Table of Quantum Ternary Half-Subtractor Operation

A	B		Borrow	Difference
|0>	|0>		|0>	|0>
|0>	|1>		|1>	|2>
|0>	|2>		|1>	|1>
|1>	|0>		|0>	|1>
|1>	|1>		|0>	|0>
|1>	|2>		|1>	|2>
|2>	|0>		|0>	|2>
|2>	|1>		|0>	|1>
|2>	|2>		|0>	|0>

called A and B, and two outputs D (difference) and B_{out} (borrow). Quantum Ternary Half Subtractor will subtract two qutrits from the quantum ternary digit |0>, |1>, and |2> and produce two outputs as quantum difference |D> and quantum borrow bit |B>. The truth table below (Table 11.3) will show the outputs of a quantum ternary half subtractor.

From the truth table (Table 11.3), the equations for the difference and borrow are given as follows:

Difference $= A^0.B^1 + A^1.B^2 + A^2.B^0 + 1.(A^0.B^2 + A^1.B^0 + A^2.B^1)$

Borrow $= 1.(A^0.B^1 + A^0.B^2 + A^1.B^2)$

Here, A^0, A^1, and A^2 refer to the |0>, |1>, and |2> which are the value of input |A>. Similarly, B^0, B^1, and B^2 refer to the |0>, |1>, and |2> which are the value of input |B>.

11.4.1 The Circuit Architecture of Quantum Ternary Half-Subtractor Operation

Decoders are required to decode the input qutrit into the appropriate qutrits in order to create a quantum ternary half subtractor.

Two decoders are be used to decode two qutrits |A> and |B> that have to be subtracted. They will produce 3-qutrit output. Where one output qutrit will active (i.e. having value |2>) and the other two will remain inactive (i.e. having value |0>).

Figure 11.4 depicts the architecture of the quantum ternary half-subtractor operation. The Figure 11.4 shows that for the quantum ternary half subtractor, it is needed to use two quantum ternary 1-to-3 decoders, eleven quantum ternary AND operations, and seven quantum ternary OR operations.

The quantum ternary AND and OR operations are nothing but the products-of-sum that produces the expected output that was mentioned in the above truth table 11.3.

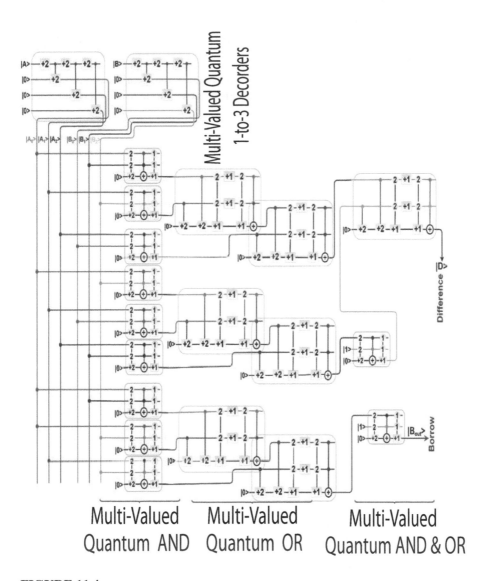

FIGURE 11.4

The Circuit Diagram for Quantum Ternary Half-Subtractor Operation

The quantum ternary half adder was constructed using the designed quantum ternary AND and OR operations, which use shifting operations and C^2 NOT operations.

11.4.2 Working Principles of Quantum Ternary Half-Subtractor Operation

The logical formulations generated for the quantum ternary half-subtractor are explained in this section. The working procedure of the quantum ternary half subtractor is given below for each pattern of input bits.

1. For input A, B = 0,

    $A_0 = 2$, $A_1 = 0$, $A_2 = 0$

    $B_0 = 2$, $B_1 = 0$, $B_2 = 0$

    Difference = $A_0.B_1 + A_1.B_2 + A_2.B_0 + 1.(A_0.B_2 + A_1.B_0 + A_2.B_1)$

    $= 2.0 + 0.0 + 0.2 + 1.(2.0 + 0.2 + 0.0)$

    $= 0 + 1.0$

    $= 0$

    Borrow = $1.(A_0.B_1 + A_0.B_2 + A_1.B_2)$

    $= 1.(2.0 + 2.0 + 0.0)$

    $= 1.0$

    $= 0$

2. For input A, B = 0, 1

    $A_0 = 2$, $A_1 = 0$, $A_2 = 0$

    $B_0 = 0$, $B_1 = 2$, $B_2 = 0$

    Difference = $A_0.B_1 + A_1.B_2 + A_2.B_0 + 1.(A_0.B_2 + A_1.B_0 + A_2.B_1)$

    $= 2.2 + 0.0 + 0.0 + 1.(2.0 + 0.0 + 0.2)$

    $= 2 + 1.0$

    $= 2$

    Borrow = $1.(A_0.B_1 + A_0.B_2 + A_1.B_2)$

    $= 1.(2.2 + 2.0 + 0.0)$

    $= 1.2$

    $= 1$

3. For input A, B = 0, 2

    $A_0 = 2$, $A_1 = 0$, $A_2 = 0$

    $B_0 = 0$, $B_1 = 0$, $B_2 = 2$

    Difference = $A_0.B_1 + A_1.B_2 + A_2.B_0 + 1.(A_0.B_2 + A_1.B_0 + A_2.B_1)$

$= 2.0 + 0.2 + 0.0 + 1. (2.2 + 0.0 + 0.0)$

$= 0 + 1.2$

$= 1$

Borrow $= 1. (A_0.B_1 + A_0.B_2 + A_1.B_2)$

$= 1. (2.0 + 2.2 + 0.2)$

$= 1.2$

$= 1$

4. For input A, B $= 1, 0$

   $A_0 = 0, A_1 = 2, A_2 = 0$

   $B_0 = 2, B_1 = 0, B_2 = 0$

   Difference $= A_0.B_1 + A_1.B_2 + A_2.B_0 + 1. (A_0.B_2 + A_1.B_0 + A_2.B_1)$

   $= 0.0 + 2.0 + 0.2 + 1.(0.2 + 2.2 + 0.0)$

   $= 0 + 1.2$

   $= 1$

   Borrow $= 1. (A_0.B_1 + A_0.B_2 + A_1.B_2)$

   $= 1. (0.0 + 0.0 + 2.0)$

   $= 1.0$

   $= 0$

5. For input A, B $= 1, 1$

   $A_0 = 0, A_1 = 2, A_2 = 0$

   $B_0 = 0, B_1 = 2, B_2 = 0$

   Difference $= A_0.B_1 + A_1.B_2 + A_2.B_0 + 1. (A_0.B_2 + A_1.B_0 + A_2.B_1)$

   $= 0.2 + 2.0 + 0.0 + 1.(0.0 + 2.0 + 0.2)$

   $= 0 + 1.0$

   $= 0$

   Borrow $= 1. (A_0.B_1 + A_0.B_2 + A_1.B_2)$

   $= 1. (0.2 + 0.0 + 2.0)$

   $= 1.0$

   $= 0$

6. For input A, B $= 1, 2$

   $A_0 = 0, A_1 = 2, A_2 = 0$

   $B_0 = 0, B_1 = 0, B_2 = 2$

   Difference $= A_0.B_1 + A_1.B_2 + A_2.B_0 + 1. (A_0.B_2 + A_1.B_0 + A_2.B_1)$

   $= 0.0 + 2.2 + 0.0 + 1.(0.2 + 2.0 + 0.0)$

$= 2 + 1.0$

$= 2$

Borrow $= 1. (A_0.B_1 + A_0.B_2 + A_1.B_2)$

$= 1. (0.0 + 0.2 + 2.2)$

$= 1.2$

$= 1$

7. For input A, B = 2, 0

$A_0= 0, A_1=0, A_2=2$

$B_0= 2, B_1=0, B_2=0$

Difference $= A_0.B_1 + A_1.B_2 + A_2.B_0 + 1. (A_0.B_2 + A_1.B_0 + A_2.B_1)$

$= 0.0 + 0.0 + 2.2 + 1.(0.0 + 0.2 + 2.0)$

$= 2 + 1.0$

$= 2$

Borrow $= 1. (A_0.B_1 + A_0.B_2 + A_1.B_2)$

$= 1. (0.0 + 0.0 + 0.0)$

$= 1.0$

$= 0$

8. For input A, B = 2, 1

$A_0= 0, A_1=0, A_2=2$

$B_0= 0, B_1=2, B_2=0$

Difference $= A_0.B_1 + A_1.B_2 + A_2.B_0 + 1. (A_0.B_2 + A_1.B_0 + A_2.B_1)$

$= 0.2 + 0.0 + 2.0 + 1.(0.0 + 0.0 + 2.2)$

$= 0 + 1.2$

$= 1$

Borrow $= 1. (A_0.B_1 + A_0.B_2 + A_1.B_2)$

$= 1. (0.0 + 0.0 + 0.0)$

$= 1.0$

$= 0$

9. For input A, B = 2, 2

$A_0= 0, A_1=0, A_2=2$

$B_0= 0, B_1=0, B_2=2$

Difference $= A_0.B_1 + A_1.B_2 + A_2.B_0 + 1. (A_0.B_2 + A_1.B_0 + A_2.B_1)$

$= 0.0 + 0.2 + 2.0 + 1.(0.2 + 0.0 + 2.0)$

$= 0 + 1.0$

$= 0$

Borrow $= 1. (A_0.B_1 + A_0.B_2 + A_1.B_2)$

$= 1. (0.0 + 0.2 + 0.2)$

$= 1.0$

$= 0$

The circuit reflects those two expressions for the difference and borrows outputs. Therefore, the results will be the same for the inputs to the circuit shown in Figure 11.4.

11.5 Multiple-Valued Quantum Full-Subtractor Operation

In this section, a quantum full-subtractor in ternary logic is constructed. In the quantum ternary full subtractor, there are a total of three inputs $|A>$, $|B>$, and $|C>$. Here the value of $|A>$ and $|B>$ can be $|0>$, $|1>$ or $|2>$. But for $|C>$ in which acts as a borrow input in the full-subtractor, can be either $|0>$ or $|1>$ (see the half-subtractor section; no borrow was generated as $|2>$). The reason is that the borrow qutrit can never be $|2>$. Thus a total number of eighteen combinations as of the input values. Table 11.4 shows the operations in a quantum ternary full-subtractor.

Again, from the truth table, the logical expressions are derived to find the difference and borrow outputs.

The logical expressions for the difference and borrow outputs are derived as follows:

Dfference $= A0.B1.C0 + A1.B2.C0 + A2.B0.C0 + A0.B0.C1 + A1.B1.C1 + A2.B2.C1$

$+ 1. (A0.B2.C0 + A1.B0.C0 + A2.B1.C0 + A0.B1.C1 + A1.B2.C1 + A2.B0.C1)$

$= (A0.B1 + A1.B2 + A2.B0).C0 + (A0.B0 + A1.B1 + A2.B2).C1$

$+ 1. ((A0.B2 + A1.B0 + A2.B1).C0 + (A0.B1 + A1.B2 + A2.B0).C1)$

Borrow$_{out}$ $= 1.(A0.B1.C0+A0.B2.C0 + A1.B2.C0 + A0.C1 +B2.C1+A1.B1.C1)$

$= 1.((A0.B1+A0.B2+A1.B2).C0 + (A0+B2+A1.B1).C1)$

Here, A0, A1, and A2 refer that the value of input $|A>$ might be $|0>$, $|1>$, and $|2>$, respectively. Similarly, B0, B1, and B2 refer that the value of $|b>$ might be $|0>$, $|1>$, and $|2>$, respectively. Again the equations are nothing but the sum-of-products form. Therefore, it is possible to construct the architecture using the equations for the difference and borrow output. Because the architecture will be significantly larger, it will be splited into two halves.

11.5.1 The Circuit Architecture of Quantum Ternary Full-Subtractor Operation

Figure 11.5 shows the difference output generation part and Figure 11.6 shows the borrow output generation part of the quantum ternary full-subtractor operation. For

TABLE 11.4

Truth Table of Quantum Ternary Full-Subtractor
Operation

A	B	C	Difference	Borrow
0	0	0	0	0
0	0	1	2	1
0	1	0	2	1
0	1	1	1	1
0	2	0	1	1
0	2	1	0	1
1	0	0	1	0
1	0	1	0	0
1	1	0	0	0
1	1	1	2	1
1	2	0	2	1
1	2	1	1	1
2	0	0	2	0
2	0	1	1	0
2	1	0	1	0
2	1	1	0	0
2	2	0	0	0
2	2	1	2	1

the difference output part, three multi-valued decoders, fourteen quantum ternary AND and nine quantum ternary OR operations are needed. On the other hand, for the borrow output part, three quantum multi-valued decoders, seven quantum ternary AND and five quantum ternary OR operations are needed.

Like the quantum ternary half-adder, full-adder, and half-subtractor, the circuit is using the equations. The input decoders are used to get the ternary values. For three inputs, three decoders are needed.

To get the result of the difference, the simplified equation is used which is generated from the truth table. In the truth table, there are a six of $|2\rangle$ value for the difference. So it is needed to pass six input combinations through quantum ternary AND operations. Finally, to get one final output from them, quantum ternary OR operation is used. The same procedure is repeated for truth value $|1\rangle$ in the difference equation. But, to confirm that from this part, never get the truth value $|2\rangle$, the output of the quantum ternary OR operation passes through a quantum ternary AND operation whose other input is $|1\rangle$. Finally, the output of the quantum ternary OR operation and quantum ternary AND operation pass as the input of the last quantum ternary OR operation to generate the desired result of $|D\rangle$ (difference). Figure 11.5 shows the procedure in detail.

In the case of the borrow output, the truth table gives nine $|1\rangle$ values (for input $|A\rangle$, $|B\rangle$, and $|C\rangle$). After using K-map, it is possible to simplify the equation of borrow output. The equation is divided into two parts, one for $|C_0\rangle$ and the other for

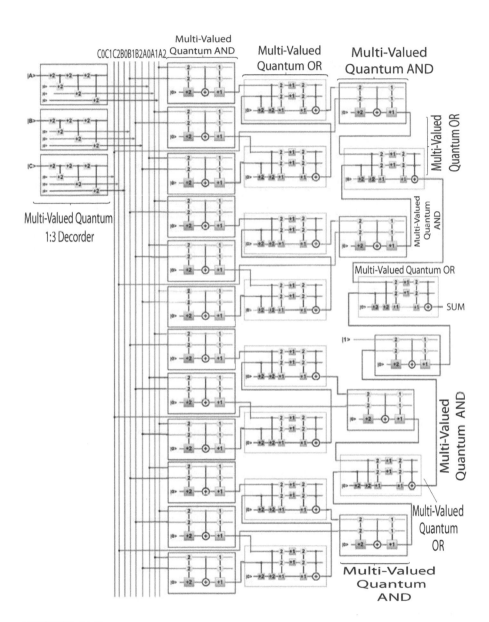

FIGURE 11.5
The Circuit Diagram for Quantum Ternary Full-Subtractor Difference Output Generation

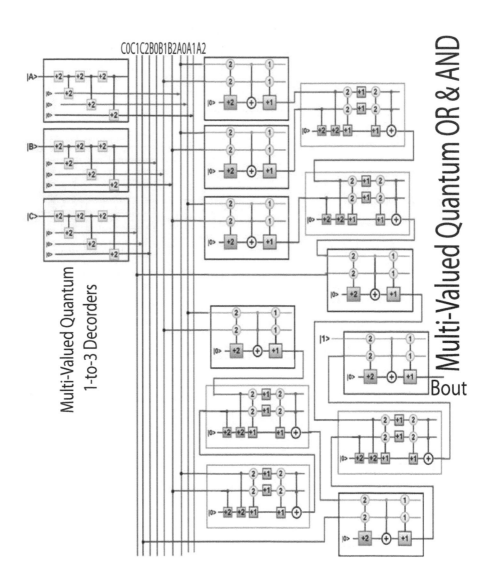

FIGURE 11.6
The Circuit Diagram for Quantum Ternary Full-Subtractor Borrow Output Generation

$|C_1>$. In the first part, three different pairs of input values of $|A>$ and $|B>$ go through three quantum ternary AND operations. The output of these three-quantum ternary AND operations sequentially executes quantum ternary OR operation among them and the output of the second quantum ternary OR operation then performs a quantum ternary AND operation with $|C_0>$ input. In the second part, after one quantum ternary AND and two quantum ternary OR operations, the output of these takes part in a quantum ternary AND operation with $|C_1>$ input. The output of both the parts then passes through a quantum ternary OR operation. At last, to ensure that the maximum output of this circuit is $|1>$, another quantum ternary AND operation is performed with input value $|1>$. Finally, the result of the borrow output is obtained.

11.5.2 Working Principles of Quantum Ternary Full-Subtractor Operation

In multiple-valued Quantum Full-Subtractor, three 1-to-3 multiple-valued quantum decoders are needed, fourteen multiple-valued quantum-AND operations, and nine multiple-valued quantum-OR operations to get the result of $|D>$ (difference output). Similarly, another seven multiple-valued quantum-AND operations and five multiple-valued quantum-OR operations are required to generate the output of $|B_{out}>$ (borrow output).

When the inputs of $|A>$, $|B>$ and $|C>$ are $|0>$, $|1>$ and $|0>$, then decoder actives $|A_0>$, $|B_1>$ and $|C_0>$ bypassing the qutrit $|2>$ among them and $|0>$ through the other lines. As a result, the first multiple-valued quantum-AND operation is executed and produces the output $|2>$. This output pass through some multiple-valued quantum-OR operation resulting in output $|2>$ again. Then this output $|2>$ performs a multiple-valued quantum AND operation with $|C_0>$ which value is also $|2>$. The output then passes through two other multiple-valued quantum-OR operations to produce the result of difference $|D>$ which is $|2>$. And for the borrow output $|B_{out}>$, a similar procedure is followed. But before generating the result of $|B_{out}>$, one last multiple-valued quantum-AND operation is executed with a fixed input qutrit $|1>$.

11.6 Multiple-Valued Quantum Parallel Adder

A single quantum ternary complete adder adds two single-qutrit values and one input carry qutrit. A quantum ternary Parallel Adder is a quantum circuit that finds the arithmetic sum of two ternary numbers that are more than one qutrit in length by operating on analogous pairs of qutrits in parallel. It consists of a chain of quantum ternary full-adders, with each quantum ternary full adder's output carry connected to the carry input of the next higher-order quantum ternary full-adder in the chain. An n-qutrit parallel adder requires n quantum ternary full-adders to operate. So two quantum ternary full-adders are required for a two-qutrit number, four quantum ternary full-adders are required for a four-qutrit value, and so on.

11.6.1 General Organizations of Multiple-Valued Quantum Parallel Adder

Figure 11.7 shows the general organizations of a quantum ternary 4-qutrit parallel adder. From the Figure, it is seen that to perform the ternary addition of two numbers of 4-qutrits, four quantum ternary full-adders are required connected in a manner where the carry output of the 1st ternary adder is the carry input of the 2nd ternary adder, and the the 2nd ternary adder's carry output is 3rd ternary adder's carry input, and the carry output of the 3rd ternary adder is the carry input of the 4th ternary adder.

And finally, the carry output of the 4th ternary adder will be the most significant qutrit of the ternary addition result.

Figure 11.8 shows the general organizations of a quantum ternary n-qutrit parallel adder. From Figure 11.8, it is understood that how n number of quantum ternary full-adders are connected to each other to perform the ternary addition process of two values of n-qutrits. In an n-qutrit quantum ternary parallel adder, it is possible to add any number of qutrits length values. To design a circuit for that operation with the help of the circuit shown in Figure 11.8.

The carry input qutrit value can be either $|0>$ or $|1>$ (see the full-adder section), where the output sum and carry qutrit can contain $|0>$, or $|1>$, or $|2>$.

The design procedure is exactly the same as it is done in the quantum n-qubit parallel adder. But the difference is in the logical operation- the ternary addition process. The ternary addition process is described already.

11.6.2 Circuit Architecture of Quantum Multiple-Valued Parallel Adder

It is necessary to connect a number of quantum ternary full-adders in order to create the quantum ternary parallel adder. The architecture of the quantum ternary full-adder was explained already.

The circuit diagram for sum output is shown in Figure 11.3, and the carry output is shown in Figure 11.2. The full-adders must be connected to make a quantum ternary parallel adder. The carry output will be the carry input qutrit to the next quantum ternary full-adder.

11.6.3 The Working Principles of Quantum Multiple-Valued Parallel Adder

In quantum multiple valued parallel adder, it is required to add two qutrits (the addend and the augend) which are only single-qutrit input. The procedures are the same, the only difference is the logical operation will be in the ternary system.

A 4-qutrit quantum ternary parallel adder operation's example can be shown as follows:

$|C_3>$ $|C_2>$ $|C_1>$ $|C_0>$

$|A>$: $|A_3>$ $|A_2>$ $|A_1>$ $|A_0>$ (Augend)

$|B>$: $|B_3>$ $|B_2>$ $|B_1>$ $|B_0>$ (Addend)

$|S>$: $|S_4>$ $|S_3>$ $|S_2>$ $|S_1>$ $|S_0>$

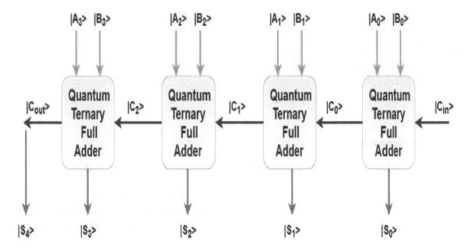

FIGURE 11.7
The General Organizations of Quantum Ternary 4-Qutrit Parallel Adder

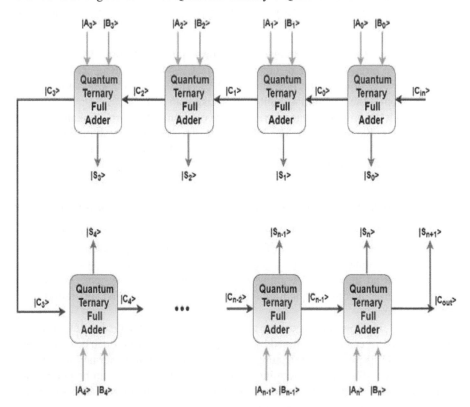

FIGURE 11.8
The General Organizations of Quantum Ternary N-Qutrit Parallel Adder

Let, $|A> = |1210>$, and $|B> = |0221>$
|0 1 1 0>
|A>: |1 2 1 0> (Augend)
|B>: |0 2 2 1> (Addend)
|S>: | 0 2 2 0 1>

11.7   Multiple-Valued Quantum Carry-Lookahead Adder

In the quantum ternary carry-lookahead adder, all the carry outputs are calculated without performing the *FullAdderCarryValue ()* operations – the similar way it is done earlier. That means the carry output will be calculated using combinational circuits. Remember that, no carry will be generated as $|2>$, therefore it will not be a problem to determine the carry output values in the ternary system.

11.7.1   General Organizations of Quantum Ternary Carry-Lookahead Adder

Figure 11.9 shows the general architecture of the 4-qutrit carry-lookahead adder. The combinational circuits consist of quantum ternary AND and OR operations which is shown several times in the previous chapters. This representation is the ternary representation of the quantum 4-qubit carry-lookahead adder.

Figure 11.10 shows the general architecture of the n-qutrit carry-lookahead adder where it is possible to perform the addition of n-qutrit inputs, and all operations will be executed at the same time after determining the value of $|C_1>$ which will reduce the total computational time.

11.7.2   The Architecture of Quantum Ternary Carry-Lookahead Adder

The value of $|C_i>$ is determined to perform the addition operation concurrently. Figure 11.11 shows the architecture of the quantum ternary 3-qutrit carry-lookahead adder where the quantum ternary 3-qutrit carry-lookahead adder is constructed using the most familiar universal operations in the quantum ternary system. That means the architecture contains the quantum ternary XOR operations, AND operations, and OR operations.

Figure 11.11 shows the procedure to calculate the $|P_i>$ and $|G_i>$ to determine the carry qutrits. $|P_i>$ can be determined by $|A_i>$ QXOR $|B_i>$ and $|G_i>$ can be determined by $|A_i>$ QAND $|B_i>$. The architecture does the same thing in the ternary system.

So, the value C_2 can be determined from $P_1C_1 + G_1$, and C_3 can be determined from $P_2C_2 + G_2$. Figure 11.11 shows the same thing in the circuit. The values are determined and connected accordingly.

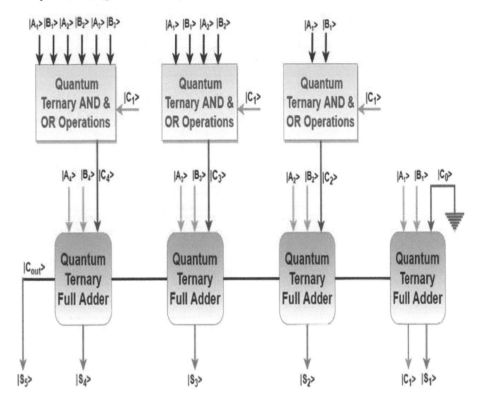

FIGURE 11.9
The General Organizations of Quantum Ternary 4-Qutrit Carry-Lookahead Adder

11.7.3 The Working Principles of Quantum Ternary Carry-Lookahead Adder

The working principles of the quantum ternary carry-lookahead adder are straightforward. The equations are devised and implemented in the circuit. As a result, the circuit should provide the desired outcome if the equations are right. However, as an example, it is possible to simulate the inputs to the circuit and observe how the circuit behaves.

Suppose, $|A> = |120>$ and $|B> = |101>$.

1. The 1st addition operation will be performed between $|A_1> = |0>$, and $|B1> = |1>$. And for 1st operation the value of $|C_0> = |0>$. The 1st ternary QXOR will produce $|1>$. And the 2nd ternary QXOR will produce also $|1>$. Therefore, the value of $|S_1>$ will be $|1>$. And the carry $|C_1>$ will generate using the ternary QAND and ternary QOR operations, which will be $|0>$. Thus the $|C_1> = |0>$ will be used to determine all the values of other $|C_i>$.

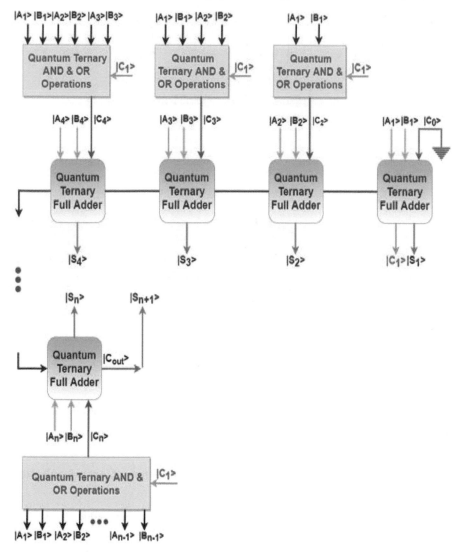

FIGURE 11.10
The General Organizations of Quantum Ternary n-Qutrit Carry-Lookahead Adder

2. The value of $|P_1\rangle$ will be $|1\rangle$, and the value of $|G_1\rangle$ will be $|0\rangle$. Therefore, the value of $|C_2\rangle = |0\rangle$ $(P_1C_1 + G_1)$.

3. Now the 2nd addition operation will be performed between $|A_2\rangle = |2\rangle$, and $|B_2\rangle = |0\rangle$. And $|C_2\rangle = |0\rangle$ is determined. Therefore, the value of $|S_2\rangle = |2\rangle$. Now the value of $|C_3\rangle$ is determined.

4. Here, the value of $|P_2\rangle$ will be $|2\rangle$, and the value of $|G_2\rangle$ will be $|0\rangle$ again. Therefore, the value of $|C_3\rangle$ will be $|0\rangle$.

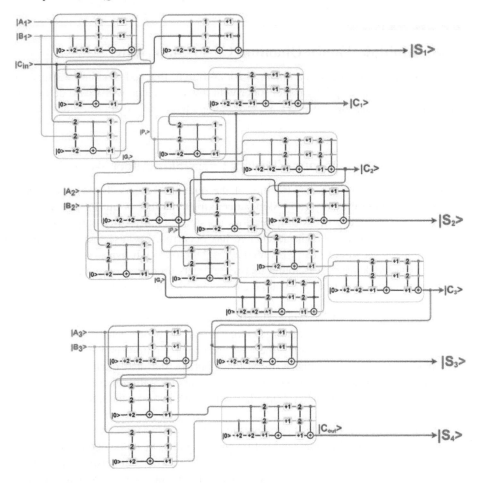

FIGURE 11.11
The Circuit Diagram of Quantum Ternary 3-Qutrit Carry-Lookahead Adder

5. Now the 3rd addition operation will be performed between $|A_3> = |1>$, and $|B_3> = |1>$. And $|C_3> = |0>$ is determined. Therefore, the value of $|S_3> = |2>$. And the value of $|C_4> = |0>$ will be the most significant qubit of the result.

6. Therefore, the value of $|S_4>$ will be $|0>$.

11.8 Multiple-Valued Quantum Carry-Skip Adder

Quantum carry-skip adder is a quantum adder system that reduces the propagation delay of a quantum n-qubit parallel adder with low effort. The worst-case delay can be decreased by merging several carry-skip adders to create a quantum block-carry-skip adder.

11.8.1 How Does Quantum Ternary Carry-Skip Adder Work?

Figure 11.12 shows the basic structure of the quantum ternary 4-qutrit carry-skip adder. The quantum ternary carry-skip adder skips a number of qutrits (often called block) if the addition process of qubit does not generate any carry output which is equivalent to $|1>$ (in the logic no change has occurred, because there will be no addition operation where a carry is found equivalent to $|2>$). It bypasses the 1st carry qutrit of the block when it is found that there exists no operation which generates carry output $|1>$. That is why it can be called the quantum ternary bypass adder as well.

To determine the block propagate BP using the value of $|P_i>$ in the quantum two-valued system. Here, BP stands for block propagate which determines whether a carry will generate as $|1>$ or not. If there will be no carry output as $|1>$, a multiplexer will bypass the 1st carry input $|C_0>$ as the input for the next quantum full-adder operation. And, if there will be generated a carry as $|1>$ within these four-quantum full-adder operations, this carry skip adder will work just like the quantum parallel adder.

In the quantum full-adder operation the carry qubit was generated as $|1>$ only when the input addend and augend both were $|1>$. Otherwise, the carry was always $|0>$. Now in the ternary system carry generates as $|1>$ while adding two qutrits either $|1>$ and $|2>$, or $|2>$ or $|1>$ are added. And for another input pattern carry can generate as $|1>$ is if both inputs are $|2>$. Therefore, avoiding these three input patterns, it is possible to find the block propagate when no carry will be generated as $|1>$.

So, to perform quantum carry-skip adder in the ternary system, $|2>$ is needed to be considered also as the block propagates along with the $|1>$. So, the quantum ternary carry-skip adder will bypass the value of $|C0>$ as the final carry output of the block if the value of BP is either $|1>$ or $|2>$. When the BP is $|0>$ then this carry-skip adder will work like the parallel adder. And one thing that can be noted, to optimize the result, a condition with the system is needed, that is, if the input values both are $|2>$ the block will not perform the carry-skip addition, rather it will perform the regular parallel addition as well. Because this input set will also generate a carry qutrit $|1>$ but the ternary QXOR will generate $|1>$ as well. Therefore, this input set will break the logic for the other input set if it is not set the condition by the algorithm.

Therefore, the block propagate BP in the quantum ternary system can be determined as−

$$BP_i = |P_0> . |P_1> . |P_2> . \ldots . |P_i>$$

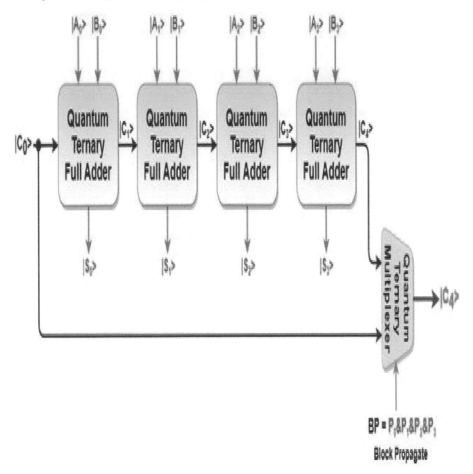

FIGURE 11.12
Basic Structure of a 4-Qutrit Quantum Ternary Carry-Skip Adder

Where, $|P_i\rangle = (A_i$ **QXOR** $B_i)$

All operations will be for ternary qutrits. And if the value of BP is $|1\rangle$ or $|2\rangle$, the multiplexer will bypass the carry input of the block to the next block.

11.8.2 The General Architecture of Quantum Ternary Carry-Skip Adder

Figure 11.13 shows the general organizations of a 4-qutrit quantum ternary carry-skip adder where the block propagate will generate first, if it is BP = $|1\rangle$ or $|2\rangle$ then the circuit will bypass the first carry qutrit to the next quantum ternary full-adder using a quantum ternary 3-to-1 multiplexer.

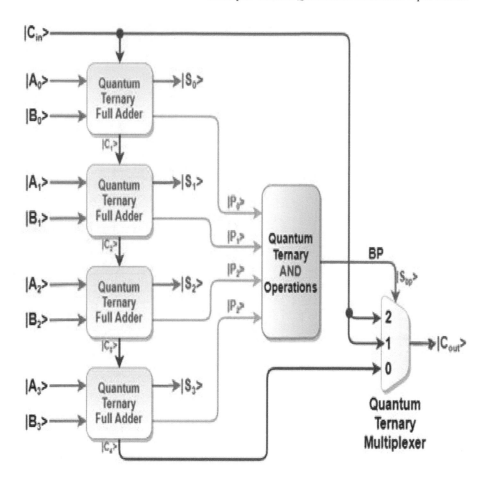

FIGURE 11.13
General Organizations of a 4-Qutrit Quantum Ternary Carry-Skip Adder

In Figure 11.13, if it is BP = |1> or |2> then the circuit will bypass the first carry qutrit to the next quantum ternary full-adder using a quantum ternary 3-to-1 multiplexer.

11.8.3   The Working Principles of Quantum Ternary Carry-Skip Adder

To make the thing easier, as always, the working procedure of the quantum ternary carry-skip adder through example will be understood.

Suppose during the addition of two inputs |A> and |B>, where |A> = |0120102101> and |B> = |2001020020>.

|A>: |0120102101>
|B>: |2001020020>

|S>: |2121122121>

In the ternary system for such cases, carry does not generate as |1>. Therefore determining carry for such cases will increase the computational time. Let's see how BP will tell if carry is generated or not.

For the above example, block propagate (BP) can be calculated as –

$$BP = \prod_{i=0}^{n-1} P_i$$

And, $|P_i> = |A_i>$ **QXOR** $|B_i>$, in ternary logic.

If BP = |1> or |2>, then the carry of the last adder operation will be the carry output of the 1st adder operation. This means no carry is been generated as |1>. Let's calculate BP for the above example.

BP = |1> . |2> . |1> . |2> . |2> . |1> . |1> . |2> . |1> . |2> = **|1>**

And suppose the inputs are $|A> = |20002020220>$, and $|B> = |02220202002>$, then the block propagate will be –

BP = |2> . |2> . |2> . |2> . |2> . |2> . |2> . |2> . |2> . |2> . |2> = **|2>**

Therefore, in both cases, no carry will be generated as |1>. Therefore, the last carry output will be |0> which was the 1st carry input of this operation.

Now let's take an example for the designed 4-qutrit quantum ternary carry-skip adder and the working principle is given below.

As the circuit is for 4-qutrit, suppose $|A> = |1012>$, and $|B> = |0200>$. So, $|A_0> = |2>$, $|A_1> = |1>$, $|A_2> = |0>$, and $|A_3> = |1>$. And $|B_0> = |0>$, $|B_1> = |0>$, $|B_2> = |2>$, and $|B_3> = |0>$.

1. 1st quantum full-adder will receive inputs $|A_0> = |2>$, $|B_0> = |0>$, and $|C_0> = |0>$. Remember the carry will determine only if the value of BP is not |1> or |2>. Therefore, the sum will be |2>, and the value of $|P_0> = |2>$.

2. For 2nd quantum full-adder operation, input qutrits are $|A_1> = |1>$, $|B_1> = |0>$. Therefore, $|P_1> = |1>$.

3. For 3rd quantum full-adder operation, input qutrits are $|A_2> = |0>$, $|B_2> = |2>$. Therefore, $|P_2> = |2>$.

4. For 4th quantum full-adder operation, input qutrits are $|A_3> = |1>$, $|B_3> = |0>$. Therefore, $|P_3> = |1>$.

All the values of P_i will be found, so the block propagate will be,

BP = $|P_0>$. $|P_1>$. $|P_2>$. $|P_3>$

= |2> . |1> . |2> . |1>

= |1>

Here, BP = |1> is found therefore, no carry was generated as |1>. And that is why the sum of the operations will be the values of all $|P_i>$'s for each A_i and B_i.

And the value of BP will activate the quantum multiplexer. And consequently, the carry output for $|C_4>$ will be the value of $|C_0>$ which is |0> for this example.

This circuit can be tested for input qutrits that generate carry qutrit as |1> in any of the quantum full adder operations.

A condition is given that if both the input qutrits are |2> then the carry-skip adder will perform as like the quantum ternary parallel adder. Because, if both inputs are |2>, the quantum ternary XOR will generate |1>, which is correct for the carry-skip condition, but at the same time, they produce a carry as |1> as well. That is why this specific input pattern is taken under an extra condition.

11.9   Multiple-Valued Quantum Multiplier

A quantum multiplier is a quantum circuit which multiplies two qubit numbers. There are different types of quantum multiplier circuits. It is seen how the quantum multiplier works. The architecture in quantum logic is constructed already. Everything that is done was for the two-valued system. But now the multiple-valued system is introduced. Therefore, it is needed to design the quantum multiplier in the ternary system.

11.9.1   How Does the Quantum Ternary Multiplier Work?

The quantum ternary multiplier will take two input qutrit and produce a product qutrit and a carry qutrit. The next multiplication will consider the carry to determine the product as well. Figure 11.14 shows the general working block of a quantum ternary multiplier.

To establish the quantum ternary multiplier, it is needed to start from the core of the ternary multiplication. That means need to understand how 1-qutrit multiplication can be formed. Table 11.5 shows the rules of quantum ternary multiplication.

From the Table 11.5, the logical expressions for the product and the carry are obtained and given as follows:

Product = $A^1.B^2 + A^2.B^1 + 1.(A^1.B^1 + A^2.B^2)$
Carry = $1. (A^2.B^2)$

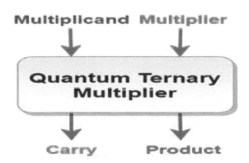

FIGURE 11.14
The General Working Block of a Quantum Ternary Multiplier

TABLE 11.5

Truth Table for 1-Qutrit Quantum Ternary Multiplication

MultiplicandA	MultiplierB	ProductP	CarryC	ResultR
$|0>$	$|0>$	$|0>$	$|0>$	$|00>$
$|0>$	$|1>$	$|0>$	$|0>$	$|00>$
$|0>$	$|2>$	$|0>$	$|0>$	$|00>$
$|1>$	$|0>$	$|0>$	$|0>$	$|00>$
$|1>$	$|1>$	$|1>$	$|0>$	$|01>$
$|1>$	$|2>$	$|2>$	$|0>$	$|02>$
$|2>$	$|0>$	$|0>$	$|0>$	$|00>$
$|2>$	$|1>$	$|2>$	$|0>$	$|02>$
$|2>$	$|2>$	$|1>$	$|1>$	$|11>$

Here, A^1 means the value of A is $|1>$, similarly A^2 means that the value of A is $|2>$. The same thing goes for the others as well.

From the equation, it is possible to generate the 1-qutrit quantum ternary multiplier.

11.9.2 The Architecture of the Quantum Ternary 2 × 2 Multiplier

From the logical expression, a ternary 1-qutrit multiplier is generated which is shown in Figure 11.15. This 1-qutrit multiplier can perform the multiplication operation as shown in the truth table of the multiplication rules. From Figure 11.15, it is seen that two decoders decode the given input qutrit for the multiplicand and multiplier. They produce the corresponding outputs and get the product and the carry output by using the outputs of the decoders as the input of the quantum ternary AND and OR operations. The circuit reflects the equations for the product and carry.

So, this was just to perform 1-qutrit multiplication which will produce only one qutrit as the product, and just for one case it will produce a carry as $|1>$.

What will be the result if the multiplicand $|A> = |A_1A_0>$, and $|B> = |B_1B_0>$?

Figure 10.16 shows the architecture of the quantum 2x2 qutrit ternary multiplier where the multiplicand and the multiplier are in 2-qutrit in size.

In the above Figure two inputs, the multiplicand $|A> = |A_1A_0>$, and multiplier $|B> = |B_1B_0>$ are performing multiplication operation and generate the product of $|A>$ and $|B>$ as $|p> = |P_4P_3P_2P_1P_0>$. From the above Figure, to perform the multiplication operation of the inputs of 2-qutrit size, need four 1-qutrit quantum ternary multipliers. These multipliers will perform the 1-qutrit multiplication operation. But to get the final product outputs it is needed to connect five half adders (which will add two qutrit values) and two full adders (which will also add a carry input qutrit along with two input qutrits) accordingly to Figure 11.16. The circuit diagram for the quantum ternary half-adder and quantum ternary full-adder was discussed already. So, connect every component and construct the circuit to perform the quantum ternary 2x2 multiplication operation.

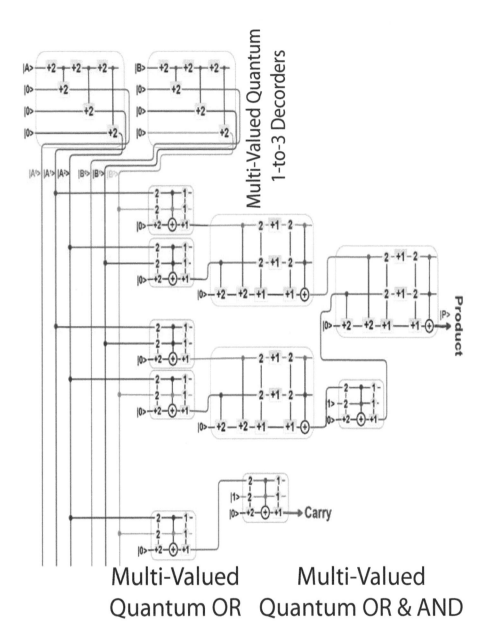

FIGURE 11.15
Circuit Diagram of the Quantum Ternary 1-Qutrit Multiplier

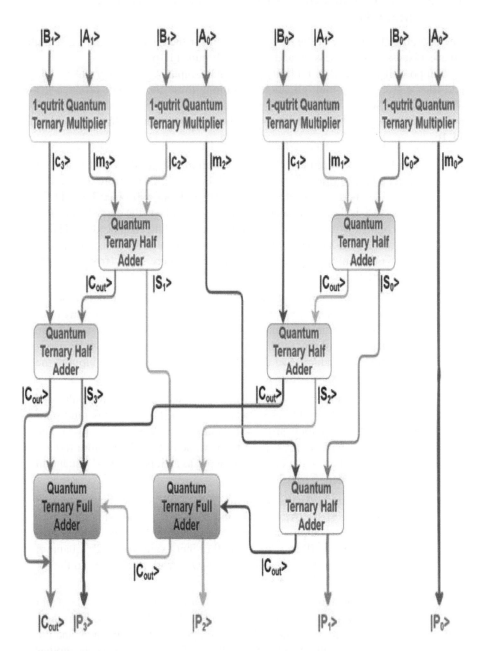

FIGURE 11.16
The Architecture of the Quantum Ternary 2 × 2 Multiplier

11.9.3   The Working Procedure of the Quantum Ternary 2 × 2 Multiplier

The operational behavior of the designed circuit for the quantum ternary 2x2 multiplier is discussed in this session and shown in Figure 11.16.

Suppose, it is needed to multiply two ternary input values, where $|A> = |02>$, and $|B> = |10>$. After multiplication operations these inputs will produce $|P> = |P_4 P_3 P_2 P_1 P_0>$. So, the value of $|A_0> = |2>$, and $|A_1> = |0>$. And the value of $|B_0> = |0>$, and $|B_1> = |1>$. The operations will be described from right to left of Figure 11.16.

1. The 1st quantum ternary 1-qutrit multiplier will get the input values of $|A_0>$ and $|B_0>$. Here, $|A_0> = |2>$, and $|B_0> = |0>$. Therefore, this multiplier will generate two outputs $|m_0> = |0>$ and $|c_0> = |0>$ (see Table 11.5). Here, $|m_0>$ is the multiplication product, and $|c_0>$ is the carry value. The value of $|m_0>$ will be the least significant qutrit of the final product. Therefore, $|P_0> = |0>$ is found, and the value of $|c_0>$ will work as an input to the 1st quantum ternary half-adder shown in Figure 11.16 (from the right side of the Figure).

2. The 2nd quantum ternary 1-qutrit multiplier will get the input values of $|A_1>$ and $|B_0>$. Here, $|A_1> = |0>$, and $|B_0> = |0>$. Therefore, this multiplier will generate two outputs $|m_1> = |0>$ and $|c_1> = |0>$ (see Table 11.5). The value of $|m_1>$ will work as the 2nd input value to the 1st quantum ternary half-adder. And the value of $|c_1>$ will work as an input to the 3rd quantum ternary half-adder.

3. The 3rd quantum ternary 1-qutrit multiplier will get the input values of $|A_0>$ and $|B_1>$. Here, $|A_0> = |2>$, and $|B_1> = |1>$. And, this multiplier will generate two outputs $|m_2> = |2>$ and $|c_2> = |0>$ (see Table 11.5). The value of $|m_2>$ will be work as an input value to the 5th quantum ternary half-adder. And the value of $|c_2>$ will work as an input to the 2nd quantum ternary half-adder.

4. The 4th quantum ternary 1-qutrit multiplier will get the input values of $|A_1>$ and $|B_1>$. Here, $|A_1> = |0>$, and $|B_1> = |1>$. And, this multiplier will generate two outputs $|m_3> = |0>$ and $|c_3> = |0>$ (see Table 11.5). The value of $|m_3>$ will be work as another input value to the 2nd quantum ternary half-adder. And the value of $|c_3>$ will work as an input to the 4th quantum ternary half-adder.

5. Now, the 1st quantum ternary half-adder will get inputs $|c_0> = |0>$ (from step 1), and $|m_1> = |0>$ (from step 2). So, this half-adder will produce two outputs. Here, from the 1st half adder, $|S_0> = |0>$ is obtained, and the carry output will also be $|0>$. The value of $|S_0>$ will be work as another input of the 5th quantum ternary half-adder. And the carry value from this half-adder will work as another input value of the 3rd quantum ternary half-adder (according to Figure 11.16).

6. The 2nd quantum ternary half-adder will get inputs $|c_2> = |0>$ (from step 3), and $|m_3> = |0>$ (from step 4). So, this half-adder will produce two outputs, $|S_1> = |0>$, and the carry output will also be $|0>$ (see the truth table of the ternary half-adder). The value of $|S_1>$ will be work as an input of the 1st quantum ternary full-adder. And the carry value from this half-adder will work as another input value of the 4th quantum ternary half-adder (according to Figure 11.16).

7. The 3rd quantum ternary half-adder will get inputs $|c_1> = |0>$ (from step 2), and also $|0>$ from the carry output of the 1st half-adder (from step 5). So, this half-adder will produce two outputs, $|S_2> = |0>$, and the carry output will also be $|0>$ (see the truth table of the ternary half-adder). The value of $|S_2>$ will work as an input of the 1st quantum ternary full-adder. And the carry value from this half-adder will work as an input value of the 2nd quantum ternary full-adder (according to Figure 11.16).

8. The 4th quantum ternary half-adder will get inputs $|c_3> = |0>$ (from step 4), and also $|0>$ from the carry output of the 2nd half-adder (from step 6). So, this half-adder will produce two outputs, $|S_3> = |0>$, and the carry output will also be $|0>$ (see the truth table of ternary half-adder). The value of $|S_3>$ will work as an input of the 2nd quantum ternary full-adder. And the carry value from this half-adder produces the most significant qutrit of the final product value.

9. The 5th quantum ternary half-adder will get inputs $|S_0> = |0>$ (from step 5), and also $|m_2> = |2>$ from step 3. So, this half-adder will produce the value of $|P_1> = |2>$. And a carry out value $|0>$ which will work as an carry input to the 1st quantum full-adder.

10. Now, the 1st quantum ternary full-adder operations can be performed. It gets three inputs as, $|S_1> = |0>$, $|S_2> = |0>$, and $|0>$ (from step 9). Therefore the sum of the addition will be $|0>$ which is the value of $|P_2>$. And the carry will also be $|0>$ which will be the carry input to the 2nd quantum ternary full-adder.

11. The last operation is the 2nd quantum ternary full-adder. It gets the three inputs as - $|S_3> = |0>$, $|0>$ from step 7, and also $|0>$ from step 10. Therefore it will also produce sum $|0>$ and carry output as $|0>$. The sum output will be the value of $|P_3>$ and the carry output will be the value of $|P_4>$. And therefore, both the value as $|0>$ has obtained.

So, finally, all the production values are obtained.

Here, the input values are $|A> = |02>$, and $|B> = |10>$. And the product values are obtained as- $|P_0> = |0>$, $|P_1> = |2>$, $|P_2> = |0>$, $|P_3> = |0>$, and $|P_4> = |0>$. Therefore, the multiplication result for the two input values is $|P> = |00020>$. That is entirely accurate.

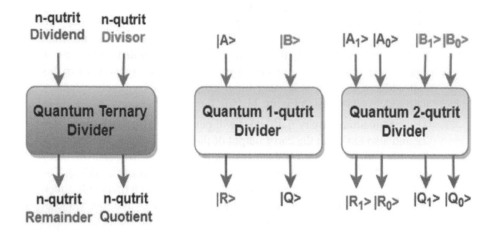

FIGURE 11.17
The General Working Block of a Quantum Ternary Divider; (left) n-Qutrit Quantum
Ternary Divider, (middle) 1-Qutrit Quantum Ternary Divider, and (right) 2-Qutrit
Quantum Ternary Divider

11.10 Multiple-Valued Quantum Divider

One of the most challenging mathematical procedures is the division of ternary numbers. Although the division operation can be performed logically using the subtraction operations iteratively, which is most suitable for the software program. In this section, an attempt is made to construct the ternary divider circuit for 1-qutrit and 2-qutrit division operations.

11.10.1 How Does the Quantum Ternary Divider Work?

The logic of ternary division operation is as same as it was in the binary division operation. The input and the output of the quantum ternary divider are shown in Figure 11.17.

The following table (Table 11.6) shows the 1-qutrit ternary division operation for the quantum ternary divider. To keep it simple the values are not in quantum notation.

Table 11.6 shows the operational truth table for the ternary 1-qutrit division operations, where A, B, Q, and R are the ternary values of dividend, divisor, quotient, and the remainder, respectively. From the operational Table, it is possible to derive the logical expression to construct the operational circuit. The logical expressions for the 1-qutrit ternary divider are –

$$Q = A^2.B^1 + 1.(A^1.B^1 + A^2.B^2)$$

(11.1)

TABLE 11.6

Truth Table for 1-Qutrit Quantum Ternary Division Operation

A	B	Q	R
0	0	-	-
0	1	**0**	**0**
0	2	**0**	**0**
1	0	-	-
1	1	**1**	**0**
1	2	**0**	**1**
2	0	-	-
2	1	**2**	**0**
2	2	**1**	**0**

$$R = 1.(A^1 . B^2) \qquad (11.2)$$

It takes a long time to find the expression for the 2-qutrit divider. It can be derived by considering all the possible inputs and outputs for 2-qutrit values and then mapping them to derive the logical expression to construct the operational circuit.

For the 2-qutrit values of divisor and dividend, there will be 81 possible combinations for the inputs (Table 11.7). To make it more understandable it is possible to segment the total input combinations into three equal parts. Each part contains 27 combinations. It can be named Table A, Table B, and Table C. It is possible to derive expressions separately for each portion of the segmented input-output mapping.

Observe the following input-output mapping very scrupulously and try to derive logical expressions for each Table.

$$Q_{1A} = NOT\ APPLICABLE$$

$$Q_{0A} = A_1^0 A_0^2 B_1^0 B_0^1 + 1.(A_1^0 A_0^1 B_1^0 B_0^1 + A_1^0 A_0^2 B_1^0 B_0^2) = A_1^0 B_1^0 . [\ A_0^2 B_0^1 + 1.(A_0^1 B_0^1 + A_0^2 B_0^2)\ ]$$

$$R_{1A} = NOT\ APPLICABLE$$

$$R_{0A} = A_1^0 A_0^2 B_1^1 B_0^0 + A_1^0 A_0^2 B_1^1 B_0^1 + A_1^0 A_0^2 B_1^1 B_0^2 + A_1^0 A_0^2 B_1^2 B_0^0 + A_1^0 A_0^2 B_1^2 B_0^1 + A_1^0 A_0^2 B_1^2 B_0^2$$

$$+A_1^0 A_0^1 B_1^0 B_0^2 + A_1^0 A_0^0 B_1^1 B_0^0 + A_1^0 A_0^1 B_1^1 B_0^0 + A_1^0 A_0^1 B_1^1 B_0^2 + A_1^0 A_0^1 B_1^2 B_0^0 + A_1^0 A_0^1 B_1^2 B_0^1 + A_1^0 A_0^1 B_1^2 B_0^2$$

TABLE 11.7

Truth Table for 2-Qutrit Quantum Ternary Division Operation

A

A_1	A_0	B_1	B_0	Q_1	Q_0	R_1	R_0
0	0	0	0	-	-	-	-
0	0	0	1	**0**	**0**	**0**	**0**
0	0	0	2	**0**	**0**	**0**	**0**
0	0	1	0	-	-	-	-
0	0	1	1	**0**	**0**	**0**	**0**
0	0	1	2	**0**	**0**	**0**	**0**
0	0	2	0	-	-	-	-
0	0	2	1	**0**	**0**	**0**	**0**
0	0	2	2	**0**	**0**	**0**	**0**
0	1	0	0	-	-	-	-
0	1	0	1	**0**	**1**	**0**	**0**
0	1	0	2	**0**	**0**	**0**	**1**
0	1	1	0	**0**	**0**	**0**	**1**
0	1	1	1	**0**	**0**	**0**	**1**
0	1	1	2	**0**	**0**	**0**	**1**
0	1	2	0	**0**	**0**	**0**	**1**
0	1	2	1	**0**	**0**	**0**	**1**
0	1	2	2	**0**	**0**	**0**	**1**
0	2	0	0	-	-	-	-
0	2	0	1	**0**	**2**	**0**	**0**
0	2	0	2	**0**	**1**	**0**	**0**
0	2	1	0	**0**	**0**	**0**	**2**
0	2	1	1	**0**	**0**	**0**	**2**
0	2	1	2	**0**	**0**	**0**	**2**
0	2	2	0	**0**	**0**	**0**	**2**
0	2	2	1	**0**	**0**	**0**	**2**
0	2	2	2	**0**	**0**	**0**	**2**

$$= \quad A_1^0 A_0^2 \cdot \left[B_1^1 B_0^0 + B_1^1 B_0^1 + B_1^1 B_0^2 + B_1^2 B_0^0 + B_1^2 B_0^1 + B_1^2 B_0^2 \right] +$$

$$1. \left[A_1^0 A_0^1 \cdot \left(B_1^0 B_0^2 + B_1^1 B_0^0 + B_1^1 B_0^1 + B_1^1 B_0^2 + B_1^2 B_0^0 + B_1^2 B_0^1 + B_1^2 B_0^2 \right) \right]$$

B

A_1	A_0	B_1	B_0	Q_1	Q_0	R_1	R_0
1	0	0	0	-	-	-	-
1	0	0	1	1	0	0	0
1	0	0	2	0	1	0	1
1	0	1	0	0	1	0	0
1	0	1	1	0	0	1	0
1	0	1	2	0	0	1	0
1	0	2	0	0	0	1	0
1	0	2	1	0	0	1	0
1	0	2	2	0	0	1	0
1	1	0	0	-	-	-	-
1	1	0	1	1	1	0	0
1	1	0	2	0	2	0	0
1	1	1	0	0	1	0	1
1	1	1	1	0	1	0	0
1	1	1	2	0	0	1	1
1	1	2	0	0	0	1	1
1	1	2	1	0	0	1	1
1	1	2	2	0	0	1	1
1	2	0	0	-	-	-	-
1	2	0	1	1	2	0	0
1	2	0	2	0	2	0	1
1	2	1	0	0	1	0	2
1	2	1	1	0	1	0	1
1	2	1	2	0	1	0	0
1	2	2	0	0	0	1	2
1	2	2	1	0	0	1	2
1	2	2	2	0	0	1	2

$$Q_{1B} = 1.[\, A_1^1 A_0^0 B_1^0 B_0^1 + A_1^1 A_0^1 B_1^0 B_0^1 + A_1^1 A_0^2 B_1^0 B_0^1 \,] = 1.\,[\, A_1^1 B_1^0 B_0^1 . \,(\, A_0^0 + A_0^1 + A_0^2 \,)]$$

$$Q_{0B} = A_1^1 A_0^1 B_1^0 B_0^2 + A_1^1 A_0^2 B_1^0 B_0^1 + A_1^1 A_0^0 B_1^1 B_0^2$$

$$+1.\left[\begin{array}{c} A_1^1 A_0^0 B_1^0 B_0^2 + A_1^1 A_0^0 B_1^1 B_0^0 + A_1^1 A_0^0 B_1^0 B_0^1 + A_1^1 A_0^0 B_1^1 B_0^0 + A_1^1 A_0^1 B_1^1 B_0^1 + A_1^1 A_0^2 B_1^1 B_0^0 \\ + A_1^1 A_0^2 B_1^1 B_0^1 + A_1^1 A_0^0 B_1^1 B_0^2 \end{array} \right]$$

$$= A_1^1 B_1^0 \cdot [\, A_0^1 B_0^2 + A_0^2 B_0^1 + A_0^2 B_0^2 \,] + 1.[\, A_1^1 \{ A_0^0 (B_1^0 B_0^2 + B_1^1 B_0^0) + A_0^1 (B_1^0 B_0^1 + B_1^1 B_0^0 + B_1^1 B_0^1)$$

$$R_{1B} = 1.[\, A_1^1 A_0^0 B_1^1 B_0^0 + A_1^1 A_0^0 B_1^1 B_0^2 + A_1^1 A_0^0 B_1^2 B_0^0 + A_1^1 A_0^0 B_1^2 B_0^1 + A_1^1 A_0^0 B_1^2 B_0^2 +$$

$$A_0^1 B_1^1 B_0^2 + A_1^1 A_0^1 B_1^1 B_0^0 + A_1^1 A_0^1 B_1^2 B_0^1 + A_1^1 A_0^1 B_1^2 B_0^2 + A_1^1 A_0^2 B_1^2 B_0^0 + A_1^1 A_0^2 B_1^2 B_0^1 + A_1^1 A_0^2 B_1^2 B_0^2 \,]$$

$$= 1.[A_1^1 \{ A_0^0 (B_1^1 B_0^1 + B_1^1 B_0^2 + B_1^2 B_0^0 + B_1^2 B_0^1 + B_1^2 B_0^2) + A_0^1 (B_1^1 B_0^2 + B_1^2 B_0^0 + B_1^2 B_0^1$$

$$+ B_1^2 B_0^2) + A_0^2 B_1^2 (B_0^0 + B_0^1 + B_0^2) \}]$$

$$R_{0B} = A_1^1 A_0^2 B_1^1 B_0^0 + A_1^1 A_0^2 B_1^2 B_0^0 + A_1^1 A_0^2 B_1^2 B_0^1 + A_1^1 A_0^2 B_1^2 B_0^2$$

$$+1. \left[\begin{array}{c} A_1^1 A_0^0 B_1^0 B_0^2 + A_1^1 A_0^0 B_1^1 B_0^0 + A_1^1 A_0^0 B_1^1 B_0^2 + A_1^1 A_0^0 B_1^2 B_0^0 + A_1^1 A_0^0 B_1^2 B_0^1 + A_1^1 A_0^0 B_1^2 B_0^2 \\ +A_1^1 A_0^2 B_1^0 B_0^2 + A_1^1 A_0^2 B_1^1 B_0^1 \end{array} \right]$$

$$= A_1^1 [\, A_0^2 (\, B_1^1 B_0^0 + B_1^2 (B_0^0 + B_0^1 + B_0^2)) \,] +$$

$$1.[\, A_1^1 (\, A_0^0 B_1^0 B_0^2 + A_0^1 (B_1^1 (B_0^0 + B_0^2) + B_1^2 (B_0^0 + B_0^1 + B_0^2))) \,]$$

$$Q_{1C} = A_1^2 A_0^0 B_1^0 B_0^1 + A_1^2 A_0^1 B_1^0 B_0^1 + A_1^2 A_0^2 B_1^0 B_0^1 + 1.[A_1^2 A_0^0 B_1^0 B_0^2 + A_1^2 A_0^1 B_1^0 B_0^2 + A_1^2 A_0^2 B_1^0 B_0^2] \,]$$

$$= A_1^2 B_1^0 B_0^1 [\, A_0^0 + A_0^1 + A_0^2 \,] + 1.[\, A_1^2 B_1^0 B_0^2 (A_0^0 + A_0^1 + A_0^2) \,]$$

$$Q_{0C} = A_1^2 A_0^0 B_1^1 B_0^0 + A_1^2 A_0^1 B_1^1 B_0^0 + A_1^2 A_0^2 B_1^0 B_0^1 + A_1^2 A_0^2 B_1^1 B_0^0 + A_1^2 A_0^0 B_1^1 B_0^0$$

$$+1. \left[\begin{array}{c} A_1^2 A_0^0 B_1^1 B_0^1 + A_1^2 A_0^0 B_1^1 B_0^2 + A_1^2 A_0^0 B_1^2 B_0^0 + A_1^2 A_0^1 B_1^0 B_0^1 + A_1^2 A_0^1 B_1^1 B_0^1 + A_1^2 A_0^1 B_1^1 B_0^2 \\ +A_1^2 A_0^0 B_1^2 B_0^0 + A_1^2 A_0^0 B_1^2 B_0^1 + A_1^2 A_0^2 B_1^0 B_0^2 + A_1^2 A_0^2 B_1^1 B_0^1 + A_1^2 A_0^2 B_1^2 B_0^0 + A_1^2 A_0^2 B_1^2 B_0^1 \end{array} \right]$$

$$= A_1^2 [\, B_1^1 B_0^0 (\, A_0^0 + A_0^1 + A_0^2 \,) + A_0^2 B_0^0 (B_1^0 + B_1^1)] + 1.[\, A_1^2 (\, A_0^0 (B_1^1 (B_0^1 + B_0^2) + B_1^2 B_0^0) +$$

C

A$_1$	A$_0$	B$_1$	B$_0$	Q$_1$	Q$_0$	R$_1$	R$_0$
2	0	0	0	-	-	-	-
2	0	0	1	2	0	0	0
2	0	0	2	1	0	0	0
2	0	1	0	0	2	0	0
2	0	1	1	0	1	0	2
2	0	1	2	0	1	0	1
2	0	2	0	0	1	0	0
2	0	2	1	0	0	2	0
2	0	2	2	0	0	2	0
2	1	0	0	-	-	-	-
2	1	0	1	2	1	0	0
2	1	0	2	1	0	0	1
2	1	1	0	0	2	0	1
2	1	1	1	0	1	1	0
2	1	1	2	0	1	0	2
2	1	2	0	0	1	0	1
2	1	2	1	0	1	0	0
2	1	2	2	0	0	2	1
2	2	0	0	-	-	-	-
2	2	0	1	2	2	0	0
2	2	0	2	1	1	0	0
2	2	1	0	0	2	0	2
2	2	1	1	0	2	0	0
2	2	1	2	0	1	1	0
2	2	2	0	0	1	0	2
2	2	2	1	0	1	0	1
2	2	2	2	0	1	0	0

$$A \, {}^1_0 (B_0^1(\, B_1^0 + B_1^1 + B_1^2) + B_1^1 B_0^2 + B_1^2 B_0^0) + A_0^2(B_0^2(B_1^0 \, + \, B_1^1) + B_1^2(B_0^0 + B_0^1 + B_0^2)) \,) \,]$$

$$R_{1C} = A_1^2 A_0^0 B_1^2 B_0^1 + A_1^2 A_0^0 B_1^2 B_0^2 + A_1^2 A_0^1 B_1^2 B_0^2 + 1. \left[A_1^2 A_0^1 B_1^1 B_0^1 + A_1^2 A_0^2 B_1^1 B_0^2 \right]$$

$$= \, A_1^2 B_1^2 [\, A_0^0(B_0^1 + B_0^2) + A_0^1 B_0^2 \,] \, + \, 1.[A_1^2 B_1^1(\, A_0^1 B_0^1 + A_0^2 B_0^2 \,)]$$

$$R_{0C} = A_1^2 A_0^0 B_1^1 B_0^1 + A_1^2 A_0^1 B_1^1 B_0^2 + A_1^2 A_0^2 B_1^1 B_0^0 + A_1^2 A_0^2 B_1^2 B_0^0$$

$$+1. \left[A_1^2 A_0^0 B_1^1 B_0^2 + A_1^2 A_0^1 B_1^0 B_0^2 + A_1^2 A_0^1 B_1^1 B_0^0 + A_1^2 A_0^1 B_1^2 B_0^0 + A_1^2 A_0^1 B_1^2 B_0^2 + A_1^2 A_0^2 B_1^2 B_0^1 \right]$$

$$= A_1^2[\ B_1^1(A_0^0 B_0^1 + A_0^1 B_0^2) + B_0^0(\ A_0^2 B_1^1 + A_0^2 B_1^2)\ ]\ +\ 1.[A_1^2(\ B_0^2(A_0^0 B_1^1 + A_0^1 B_1^0) +$$

$$A_0^1 B_0^0(B_1^1\ +\ B_1^2\ ) + B_1^2(A_0^1 B_0^2 + A_0^2 B_0^1)\ )]$$

The logical expression for the quotient –

$Q_0 = Q_{0A} + Q_{0B} + Q_{0C}$

$Q_1 = Q_{1A} + Q_{1B} + Q_{1C}$

And the logical expression for the remainder –

$R_0 = R_{0A} + R_{0B} + R_{0C}$

$R_1 = R_{1A} + R_{1B} + R_{1C}$

11.10.2   Circuit Architecture of the Quantum Ternary Divider

Let's first construct the circuit for the 1-qutrit divider, where both the dividend and the divisor are the values of 1-qutrit. Figure 11.18 shows the circuit diagram of the quantum ternary 1-qutrit divider.

The circuit diagram of the quantum ternary 1-qutrit divider can be constructed using the expressions in Equation 11.1 and Equation 11.2. The values of the dividend and the divisor are gone through two quantum ternary 1-to-3 decoders and produce the corresponding outputs. Then using the equations construct the circuit diagram to get the quotient |Q> and remainder |R> outputs.

Now, using expressions Equation 11.3, Equation 11.4, and Equation 11.5, it is possible to construct the 2-qutrit quantum ternary divider. This architecture requires four decoders for the four input values (A_1, A_0, B_1, and B_0) and will produce 12 outputs (three outputs for each input). By now, it is known that how to work with decoder outputs. Figure 11.19 shows the general architecture of the 2-qutrit divider.

From the above general architecture, the circuit diagram can be constructed. The expression for Q_{1A}, Q_{1B}, Q_{1C}, Q_{0A}, Q_{0B}, Q_{0C}, R_{1A}, R_{1B}, R_{1C}, R_{0A}, R_{0B}, and R_{0C} was provided earlier. The decoders will provide the required input ternary values for the expressions.

11.10.3   The Working Procedures of the Quantum Ternary Divider

The logical expression is derived for the 1-qutrit and 2-qutrit quantum ternary divider from their operational truth table and the circuits are designed according to the logical expressions. Therefore, the circuits for the ternary division operations will work according to the logical expression directly. The working procedures are already explained in section 11.10.1.

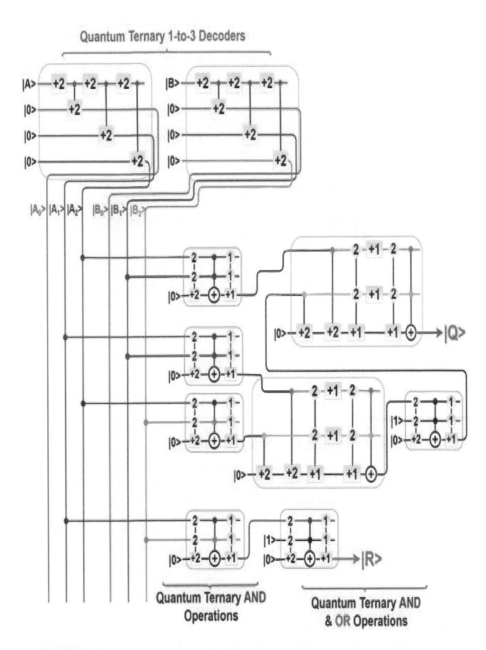

FIGURE 11.18
Circuit Diagram of the Quantum Ternary 1-Qutrit Divider

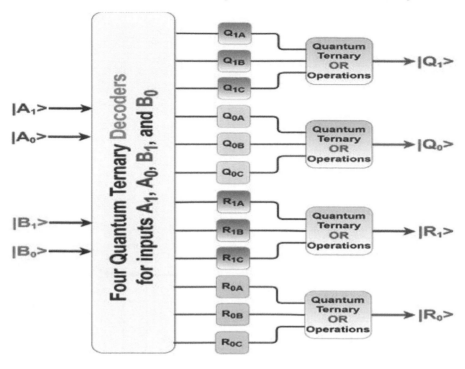

FIGURE 11.19
Circuit Diagram of the Quantum Ternary 2-Qutrit Divider

11.11 Multiple-Valued Quantum Comparator

A comparator that compares two input signals and gives output as which input is larger or smaller or equal. Multi-Valued Quantum 1-qubit comparator takes two single-qubit inputs, $|A>$ and $|B>$. The value of the inputs can be $|0>$, $|1>$, and $|2>$. This circuit gives three outputs $|X>$, $|Y>$ and $|Z>$ where $|X>$ is true when $|A>$ is smaller than $|B>$, $|Y>$ is true when $|A>$ is equal to $|B>$ and $|Z>$ is true when $|A>$ is greater than $|B>$. Here, Table 11.8 shows the truth table of the 1-qubit comparator.

From the truth table, the equations for $|X>$, $|Y>$ and $|Z>$ as –

$$|X> = A^0.B^1 + A^0.B^2 + A^1.B^2$$
$$|Y> = A^0.B^0 + A^1.B^1 + A^2.B^2$$
$$|Z> = A^1.B^0 + A^2.B^0 + A^2.B^1$$

Now, using the equations, it is possible to design the Multi-Valued Quantum 1-qubit comparator. Figure 11.20 display the block diagram of the quantum 1-qubit comparator. Here two inputs are $|A>$ and $|B>$, three outputs are $|X>$, $|Y>$ and $|Z>$.

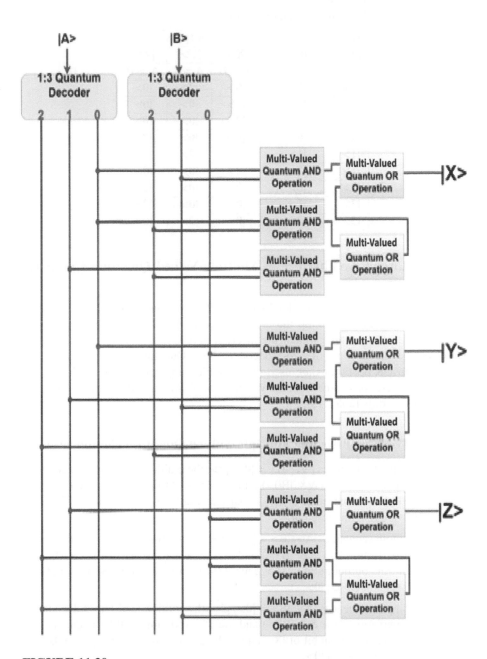

FIGURE 11.20
Block Diagram of Multi-Valued Quantum 1-Qubit Comparator

TABLE 11.8

1-Qubit Comparator Truth Table

Inputs		Outputs		
|A>	|B>	|X>	|Y>	|Z>
		A < B	A = B	A > B
|0>	|0>	|0>	|2>	|0>
|0>	|1>	|2>	|0>	|0>
|0>	|2>	|2>	|0>	|0>
|1>	|0>	|0>	|0>	|2>
|1>	|1>	|0>	|2>	|0>
|1>	|2>	|2>	|0>	|0>
|2>	|0>	|0>	|0>	|2>
|2>	|1>	|0>	|0>	|2>
|2>	|2>	|0>	|2>	|0>

11.11.1   Circuit Architecture of Multiple-Valued Quantum Comparator

The construction of the Multi-Valued Quantum 1-qubit comparator circuit is shown in Figure 11.21. In this circuit, the input, |A> and |B>, at first, pass through two 1-to-3 decoders separately, so that the inserted qubits, i.e., |0>, |1>, or |2> can be identified using the output of 1-to-3 decoders. Nine Quantum AND operations and six Quantum OR operations are used in a 1-qubit comparator where each input of these AND operations come from the various combinations of the output of a 1-to-3 decoder using truth Table 11.8.

11.11.2   Working Principle Multiple-Valued Quantum Comparator

1. For input sequences |A>, |B> = **|0>, |0>**, the multi-valued 1-to-3 decoder will perform (|A0> && |B0>) and the 4th AND line will be open. As a result, the output qubit of |Y> equal **|2>** and the remaining lines |X> to |Z> will remain closed **|0>**.

2. For input sequences |A>, |B> = **|0>, |1>**, the multi-valued 1-to-3 decoder will perform (|A0> && |B1>) and the 1st AND line will be open. As a result, the output qubit of |X> equal **|2>** and remaining lines |Y> to |Z> will remain closed **|0>**.

3. For input sequences |A>, |B> = **|0>, |2>**, the multi-valued 1-to-3 decoder will perform (|A0> && |B2>) and the 2nd AND line will be open. As a result, the output qubit of |X> equal **|2>** and remaining lines |Y> to |Z> will remain closed **|0>**.

4. For input sequences |A>, |B> = **|1>, |0>**, the multi-valued 1-to-3 decoder will perform (|A1> && |B0>) and the 7th AND line will be open. As a

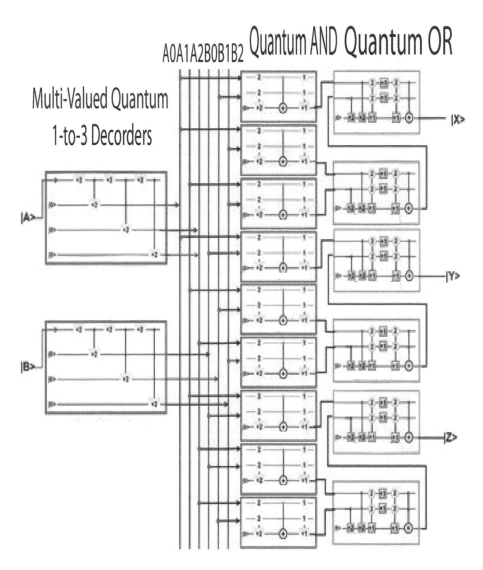

FIGURE 11.21
Circuit Architecture of Multi-Valued Quantum Comparator

result, the output qubit of |Z> equal |2> and remaining lines |X> to |Y> will remain closed |0>.

5. For input sequences |A>, |B> = |1>, |1> the multi-valued 1-to-3 decoder will perform (|A1> && |B1>) and the 5th AND line will be open. As a result, the output qubit of |Y> equal |2> and remaining lines |X> to |Z> will remain closed |0>.

6. For input sequences |A>, |B> = |1>, |2>, the multi-valued 1-to-3 decoder will perform (|A1> && |B2>) and the 3rd AND line will be open. As a result, the output qubit of |X> equal |2> and remaining lines |Y> to |Z> will remain closed |0>.

7. For input sequences A, B= |2>, |0>, the multi-valued 1-to-3 decoder will perform (|A2> && |B0>) and the 8th AND line will be open. As a result, the output qubit of |Z> equal |2> and remaining lines |X> to |Y> will remain closed |0>.

8. For input sequences |A>, |B> = |2>, |1>, the multi-valued 1-to-3 decoder will perform (|A2> && |B1>) and the 9th AND line will be open. As a result, the output qubit of |Z> equal |2> and remaining lines |X> to |Y> will remain closed |0>.

9. For input sequences |A>, |B> = |2>, |2>, the multi-valued 1-to-3 decoder will perform (|A2> && |B2>) and the 6th AND line will be open. As a result, the output qubit of |Y> equal |2> and remaining lines |X> to |Z> will remain closed |0>.

11.12   Summary

A quantum ternary Parallel Adder is a quantum circuit that finds the arithmetic sum of two ternary numbers that are more than one qutrit in length by operating on analogous pairs of qutrits in parallel. A Quantum ternary parallel adder consists of a chain of quantum ternary full-adders where the carry output of each quantum ternary full adder is output carry connected to the carry input of the next higher-order quantum ternary full-adder in the chain. A quantum ternary half adder's execution time is 373 microseconds whereas a quantum ternary full-adder operation's execution speed is 433 microseconds. All arithmetic operations in multi-valued quantum logic are explained in this chapter.

Bibliography

[1] Hallworth, R. P., & Heath, F. G. (1962). Semiconductor circuits for ternary logic. Proceedings of the IEE-Part C: Monographs, 109(15), 219-225.

[2] Dubrova, E. (1999, November). Multiple-valued logic in VLSI: challenges and opportunities. In Proceedings of NORCHIP (Vol. 99, No. 1999, pp. 340-350).

[3] Haghparast, M., Wille, R., & Monfared, A. T. (2017). Towards quantum reversible ternary coded decimal adder. Quantum Information Processing, 16(11), 1-25.

[4] Mandal, S. B., Chakrabarti, A., & Sur-Kolay, S. (2011, May). Synthesis techniques for ternary quantum logic. In 2011 41st IEEE International Symposium on Multiple-Valued Logic (pp. 218-223). IEEE.

12

Multiple-Valued Logic Operations in DNA Computing

12.1 Introduction

The multiple-valued system is more difficult to implement than the two-valued system. It is more costly, more space-consuming, requires more computational time, and generates more heat than the two-valued system. DNA computing can be used to implement the multiple-valued system. The reason DNA computers can solve complicated problems is that they generate multiple viable solutions at once. This is known as *parallel processing*. DNA computing is already used to implement two-valued operations. Now, this chapter will explain how DNA computing constructs the operations for multi-valued systems.

In this chapter, the most basic operations are implemented in DNA computing for the multi-valued system. Then the basic arithmetic operations will be described in the next chapter.

12.2 How Can Multiple-Valued Operations be Performed in DNA Computing?

In the two-valued DNA computing system, it is assumed that two DNA base sequences for the equivalent two binary values zero and one. But in the multiple-valued system (here working with the ternary system) the digits are 0, 1, and 2. Therefore now to take three DNA strands that will be equivalent to the three ternary digits.

The DNA sequences are ACCTAG = '0', CAAGCT = '1', and TGGATC = '2'. This is going to be different than the binary system. The operations are performed in DNA computing by conducting chemical reactions. And in binary logic, the bond is detected between DNA strands to perform operations using the DNase Enzyme (if the bond was not created the DNase enzyme would kill the base sequences).

In this Ternary DNA computing, the *fluorescent level* is used to detect the DNA sequence. Fluorescence is the temporary absorption of electromagnetic wavelengths from the visible light spectrum by fluorescent molecules, and the subsequent emission of light at a lower energy level. When it occurs in a living organism, it is sometimes called *Biofluorescence*. This causes the light that is emitted to be a different color than the light that is absorbed. Stimulating light excites an electron, raising energy to an unstable level. For each operation, the fluorescent level will be different and it will determine the output of the operation.

12.3 Performing Fundamental Operations in Ternary Logic in DNA Computing

In this section, the fundamental operations will be implemented in ternary logic in DNA computing. NOT, OR, NOR, AND, NAND, XOR, XNOR, etc. will be implemented in DNA computing with ternary logic system.

DNA ternary logic functions are those functions that have significance if a third value is acquainted with the DNA binary logic. Here, ACCTAG, CAAGCT, and TGGATC will represent 0, 1, and 2 in ternary logic which denotes the ternary levels for basic logic operations to represent false, undefined, and true, respectively [3]. The basic operations of DNA ternary logic can be defined as follows:

$$y_{OR} = \max(x, y)$$
$$y_{NOR} = \overline{\max(x, y)}$$
$$y_{AND} = \min(x, y)$$
$$y_{NAND} = \overline{\min(x, y)}$$
$$y_{XOR} = \text{sum}(x, y)$$
$$y_{XNOR} = \overline{\text{sum}(x, y)}$$
$$\text{Where, } x, y = \{0, \ 1, \ 2\}$$

But in this case, 0, 1, and 2 are used for simplicity. The actual values are the DNA strands which are ACCTAG, CAAGCT, and TGGATC as mentioned before.

The truth table of those DNA operations is shown in Table 12.1. Assume every operation is in the quantum system. For convenience, classical names are used.

The truth table of DNA Ternary AND, NAND, OR, NOR, XOR, and XNOR logic operations for ternary logic is depicted in the Table12.1. For AND logic operation its output value depends on the minimum value of its inputs. Similarly, in the case of the OR logic operation, its output value depends on the maximum value of its inputs. For the XOR operation, its output value is the sum of the value of its inputs.

Therefore, the outputs of DNA Ternary NAND, NOR, and XNOR logic operations become the inverted of quantum ternary AND, OR, and XNOR logic operations.

12.3.1 Ternary DNA-NOT Operation

It is known about the NOT operation principle that the NOT operation will simply invert the input value. Here in ternary DNA logic, the invert of ACCTAG is TGGATC and vice versa. And the invert value of CAAGCT is CAAGCT itself.

TABLE 12.1

Truth Table for Ternary AND, NAND, OR, NOR, XOR, XNOR Operations

Input 1	Input 2	AND	NAND	OR	NOR	XOR	XNOR
0	0	0	2	0	2	0	2
0	1	0	2	1	1	1	1
0	2	0	2	2	0	2	0
1	0	0	2	1	1	1	1
1	1	1	1	1	1	2	0
1	2	1	1	2	0	0	2
2	0	0	2	2	0	2	0
2	1	1	1	2	0	0	2
2	2	2	0	2	0	1	1

DNA-AND operation is defined as $Y_{DNOT}(X) = \overline{X}$, where X is the input from {ACCTAG, CAAGCT, TGGATC}.

12.3.1.1 Design Procedure of DNA Ternary NOT Operation

Figure 12.1 shows the architecture of the DNA Ternary NOT operation. As it is said earlier the architecture will not be complex in DNA ternary logic implementation. Here no base sequence is needed in the test tube. The fluorescent level will determine the output. The Annealing temperature should be more than 60°C for the DNA ternary NOT operations.

12.3.1.2 Working Principles of DNA Ternary NOT Operation

In ternary logic, there can be three types of NOT operation. It is defined as the positive inverter, negative inverter, and standard inverter. Table 12.2 shows the operations in the standard inverter. And the standard inverter in all operational circuits. In Figure 12.1, the fluorescent level is used here to detect the sequences and the logical value of the sequences. The operation in the ternary NOT operations can be shown as

$$y_0 = C_0(x) = \begin{cases} 2, x = 0 \\ 0, \ \ x \neq 0 \end{cases} \tag{12.1}$$

TABLE 12.2

Truth Table of DNA Ternary Standard NOT Operations

X	$\overline{X}$
ACCTAG	TGGATC
CAAGCT	CAAGCT
TGGATC	ACCTAG

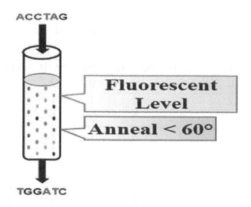

FIGURE 12.1
The Operational Diagram of DNA Ternary NOT Operation

$$y_1 = C_1(x) = \begin{cases} 2, & x \neq 2 \\ 0, & x = 2 \end{cases} \tag{12.2}$$

$$y_2 = C_2(x) = x' = 2 - x \tag{12.3}$$

Here equation 12.1 represents the operation of a positive ternary inverter (PTI), equation 12.2 represents the operation of the negative ternary inverter (NTI) and equation 12.3 represents the standard ternary inverter (STI) which will be used. In STI when the input sequence is ACCTAG, the STI generates the output TGGATC (which is equivalent to 2), if the input sequence is CAAGCT, the get generates the output CAAGCT (equivalent to 1), and if the input sequence is TGGATC, the get generates the output ACCTAG (which is equivalent to 0).

12.3.2   Ternary DNA-AND Operation

DNA ternary AND operation is defined as Y_{DTAND} = min(X, Y), where X and Y are the input from {0, 1, 2}. Table 12.3 shows the truth table of DNA ternary AND operations.

TABLE 12.3
Truth Table of DNA Ternary AND Operations

	ACCTAG	CAAGCT	TGGATC
ACCTAG	ACCTAG	ACCTAG	ACCTAG
CAAGCT	ACCTAG	CAAGCT	CAAGCT
TGGATC	ACCTAG	CAAGCT	TGGATC

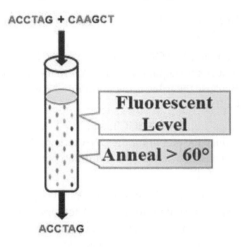

FIGURE 12.2
The Operational Diagram of DNA Ternary AND Operation

12.3.2.1 Design Procedure of DNA Ternary AND Operation

Figure 12.2 shows the architecture of the DNA ternary AND operation. Where two input sequences will add to the test tube and the fluorescence level will produce the output based on the input sequence. The annealing temperature is more than 60°C for the DNA ternary AND operations.

12.3.2.2 Working Principles of DNA Ternary AND Operation

Ternary AND functions are also similar to the binary "MINIMUM OF" function. The truth Table of DNA ternary AND operation is shown in Table 12.3. The following equation helps to get the AND operation's output for the ternary inputs.

$$y_{DTAND} = \min(x, y)$$

1. Let input values are, X = ACCTAG, and Y = ACCTAG. The fluorescence level will produce the output for the DNA ternary AND operation as AC-CTAG. Because this outputted value is the minimum between two input sequence values.

2. Let input values are, X = ACCTAG, and Y = CAAGCT. The fluorescence level will produce the output for the DNA ternary AND operation again as ACCTAG. Because the value is the minimum between two input sequence values is ACCTAG (equivalent to ternary digit 0).

3. Let input values are, X = ACCTAG, and Y = TGGATC. The fluorescence level will produce the output for the DNA ternary AND operation again as

ACCTAG. Because the value is the minimum between two input sequence values is ACCTAG.

4. Let input values are, X = CAAGCT, and Y = ACCTAG. The fluorescence level will produce the output for the DNA ternary AND operation again as ACCTAG. Because the value is the minimum between two input sequence values is ACCTAG.

5. Let input values are, X = CAAGCT, and Y = CAAGCT. The fluorescence level will produce the output for the DNA ternary AND operation again as CAAGCT. Because the value is the minimum between two input sequence values is CAAGCT.

6. For input values are, X = CAAGCT, and Y = TGGATC. The fluorescence level will produce the output for the DNA ternary AND operation again as CAAGCT. Because the value is the minimum between two input sequence values is CAAGCT (equivalent to ternary 1).

7. For input values are, X = TGGATC, and Y = ACCTAG. The fluorescence level will produce the output for the DNA ternary AND operation again as ACCTAG. Because the value is the minimum between two input sequence values is ACCTAG.

8. For input values are, X = TGGATC, and Y = CAAGCT. The fluorescence level will produce the output for the DNA ternary AND operation again as CAAGCT. Because the value is the minimum between two input sequence values is CAAGCT.

9. For input values are, X = TGGATC, and Y = TGGATC. The fluorescence level will produce the output for the DNA ternary AND operation again as TGGATC. Because the value is the minimum between two input sequence values is TGGATC (equivalent to ternary digit 2).

12.3.3 Ternary DNA NAND Operation

DNA ternary NAND operation is defined as $Y_{DTNAND} = \overline{\min(X, Y)}$, where X and Y are the input from {0, 1, 2}. The truth table of DNA ternary NAND operations is shown in Table 12.4.

TABLE 12.4
Truth Table of DNA Ternary NAND Operations

	ACCTAG	CAAGCT	TGGATC
ACCTAG	TGGATC	TGGATC	TGGATC
CAAGCT	TGGATC	CAAGCT	CAAGCT
TGGATC	TGGATC	CAAGCT	ACCTAG

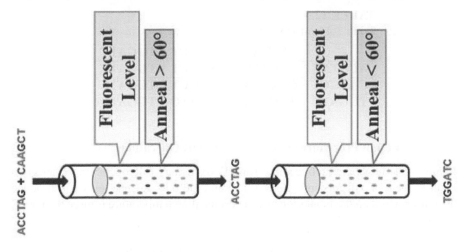

FIGURE 12.3

The Operational Diagram of DNA Ternary NAND Operation

12.3.3.1   Design Procedure of DNA Ternary NAND Operation

From Table 12.4, it is found that the DNA ternary NAND operation is the inverted output of the DNA ternary AND operation. The DNA ternary NOT operation is designed which inverts an input. Therefore, the DNA ternary NAND operation can be constructed by adding STI to the DNA ternary AND operation which is shown in Figure 12.3.

Figure 12.3 shows that the DNA ternary NAND is constructed using one DNA ternary AND operation followed by one DNA ternary NOT operation.

12.3.3.2   Working Principles of DNA Ternary NAND Operation

The working procedure of the DNA ternary NAND operation is discussed already. It is exactly the same as the DNA ternary AND operation, but the final value should be inverted to the result of the DNA ternary AND operation.

For example, if the input value is ACCTAG, CAAGCT then the DNA ternary AND will produce ACCTAG (the minimum value), and the final output will be the inverted value of ACCTAG, which is TGGATC.

12.3.4   Ternary DNA-OR Operation

DNA ternary OR operation is defined as $Y_{DTOR} = \max(X, Y)$, where X and Y are the input from {0, 1, 2}. Table 12.5 shows the truth table of DNA ternary OR operations.

TABLE 12.5

Truth Table of DNA Ternary OR Operations

	ACCTAG	CAAGCT	TGGATC
ACCTAG	ACCTAG	CAAGCT	TGGATC
CAAGCT	CAAGCT	CAAGCT	TGGATC
TGGATC	TGGATC	TGGATC	TGGATC

12.3.4.1   Design Procedure of DNA Ternary OR Operation

Figure 12.4 shows the architecture of the DNA ternary OR operation. Where two input sequences will mix into the test tube and the fluorescence level will produce the output based on the input sequence. The annealing temperature is more than 60°C for the DNA ternary OR operations.

12.3.4.2   Working Principles of DNA Ternary OR Operation

Ternary OR functions are also similar to the binary "MAXIMUM OF" function. The truth Table of DNA ternary OR operation is shown in table 12.5. The following equation helps to get the OR operation's output for the ternary inputs.

$$y_{DTOR} = \max(x, y)$$

1. Let input values are, X = ACCTAG, and Y = ACCTAG. The fluorescence level will produce the output for the DNA ternary OR operation as ACCTAG. Because this outputted value is the maximum (actually equal) between two input sequence values.

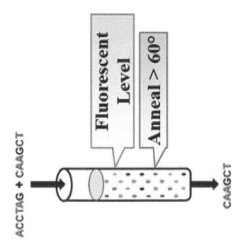

FIGURE 12.4

The Operational Diagram of DNA Ternary OR Operation

2. Let input values are, X = ACCTAG, and Y = CAAGCT. The fluorescence level will produce the output for the DNA ternary OR operation as CAAGCT. Because the maximum value between two inputs sequence values is CAAGCT (equivalent to ternary digit 1).

3. Let input values are, X = ACCTAG, and Y = TGGATC. The fluorescence level will produce the output for the DNA ternary OR operation as TGGATC. Because the maximum value between two inputs sequence values is TGGATC.

4. Let input values are, X = CAAGCT, and Y = ACCTAG. The fluorescence level will produce the output for the DNA ternary OR operation again as CAAGCT. Because the maximum value between two inputs sequence values is CAAGCT.

5. Let input values are, X = CAAGCT, and Y = CAAGCT. The fluorescence level will produce the output for the DNA ternary OR operation again as CAAGCT. Because the maximum value between two inputs sequence values is CAAGCT.

6. For input values are, X = CAAGCT, and Y = TGGATC. The fluorescence level will produce the output for the DNA ternary OR operation again as TGGATC. Because the maximum value between two inputs sequence values is TGGATC.

7. For input values are, X = TGGATC, and Y = ACCTAG. The fluorescence level will produce the output for the DNA ternary OR operation again as TGGATC. Because the maximum value between two inputs sequence values is TGGATC.

8. For input values are, X = TGGATC, and Y = CAAGCT. The fluorescence level will produce the output for the DNA ternary OR operation again as TGGATC. Because the maximum value between two inputs sequence values is TGGATC.

9. For input values are, X = TGGATC, and Y = TGGATC. The fluorescence level will produce the output for the DNA ternary OR operation again as TGGATC. Because the maximum value between two inputs sequence values is TGGATC.

12.3.5   Ternary DNA-NOR Operation

DNA ternary NOR operation is defined as $Y_{DTNOR} = \overline{max(X, Y)}$, where X and Y are the input from {0, 1, 2}. Table 12.6 shows the truth table of DNA ternary NOR operations.

12.3.5.1   Design Procedure of DNA Ternary NOR Operation

From Table 12.6 it is found that DNA ternary NOR operation is the inverted output of the DNA ternary OR operation. The DNA ternary NOT operation is designed

TABLE 12.6

Truth Table of DNA Ternary NOR Operations

	ACCTAG	CAAGCT	TGGATC
ACCTAG	TGGATC	CAAGCT	ACCTAG
CAAGCT	CAAGCT	CAAGCT	ACCTAG
TGGATC	ACCTAG	ACCTAG	ACCTAG

which inverts an input. Therefore, it is possible to construct the DNA ternary NOR operation adding STI to the DNA ternary OR operation which is shown in Figure 12.5.

Figure 12.5 shows that the DNA ternary NOR is constructed using one DNA ternary OR operation followed by one DNA ternary NOT operation.

12.3.5.2   Working Principles of DNA Ternary NOR Operation

By now the working procedure of the DNA ternary NOR operation is known. It is exactly the same as the DNA ternary OR operation, but the final output value should be inverted to the result of the DNA ternary OR operation.

For example, if the input value is ACCTAG, CAAGCT then the DNA ternary OR will produce CAAGCT (the maximum value), and the final output will be the inverted value of CAAGCT, which is also CAAGCT.

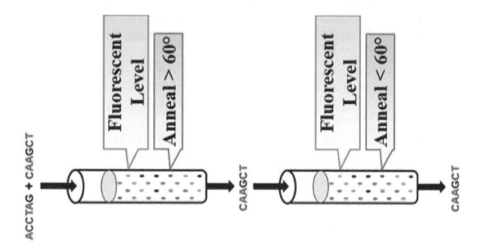

FIGURE 12.5

The Operational Diagram of DNA Ternary NOR Operation

TABLE 12.7
Truth Table of DNA Ternary XOR Operations

	ACCTAG	CAAGCT	TGGATC
ACCTAG	ACCTAG	CAAGCT	TGGATC
CAAGCT	CAAGCT	TGGATC	ACCTAG
TGGATC	TGGATC	ACCTAG	CAAGCT

12.3.6   Ternary DNA-XOR Operation

DNA ternary XOR operation is defined as Y_{DTXOR} = sum (X, Y), where X and Y are the input from {0, 1, 2}. Table 12.7 shows the truth table of DNA ternary XOR operations.

12.3.6.1   Design Procedure of DNA Ternary XOR Operation

Figure 12.6 shows the architecture of the DNA ternary XOR operation. Where two input sequences will mix to the test tube and the fluorescence level will produce the output based on the input sequence. The annealing temperature is more than 60°C for the DNA ternary XOR operations.

The ternary XOR's mechanism was described in the quantum ternary XOR section in chapter 10. The truth Table of DNA ternary XOR operation is shown in Table 12.7. The following equation helps to get the OR operation's output for the ternary inputs.

$$y_{DTXOR} = \text{sum}(x, y)$$

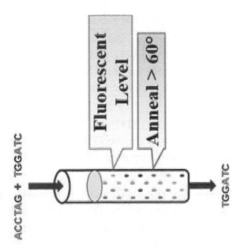

FIGURE 12.6
The Operational Diagram of DNA Ternary XOR Operation

The equation means that the output of the ternary XOR operation will be the sum value (the sum value during adding two bits).

1. Let input values are, X = ACCTAG, and Y = ACCTAG. The fluorescence level will produce the output for the DNA ternary XOR operation as AC-CTAG. Because this outputted value is the sum value when those two sequences will be added.

2. Let input values are, X = ACCTAG, and Y = CAAGCT. The fluorescence level will produce the output for the DNA ternary XOR operation as CAAGCT. Because this outputted value is the sum value when those two sequences will be added.

3. Let input values are, X = ACCTAG, and Y = TGGATC. The fluorescence level will produce the output for the DNA ternary XOR operation as TG-GATC. Because this outputted value is the sum value when those two sequences will be added.

4. Let input values are, X = CAAGCT, and Y = ACCTAG. The fluorescence level will produce the output for the DNA ternary XOR operation again as CAAGCT. Because this outputted value is the sum value when those two sequences will be added.

5. Let input values are, X = CAAGCT, and Y = CAAGCT. The fluorescence level will produce the output for the DNA ternary XOR operation again as TGGATC. Because this outputted value is the sum value when those two sequences will be added.

6. For input values are, X = CAAGCT, and Y = TGGATC. The fluorescence level will produce the output for the DNA ternary OR operation again as ACCTAG. Because this outputted value is the sum value when those two sequences will be added.

7. For input values are, X = TGGATC, and Y = ACCTAG. The fluorescence level will produce the output for the DNA ternary XOR operation again as TGGATC. Because this outputted value is the sum value when those two sequences will be added.

8. For input values are, X = TGGATC, and Y = CAAGCT. The fluorescence level will produce the output for the DNA ternary XOR operation again as ACCTAG. Because this outputted value is the sum value when those two sequences will be added.

9. For input values are, X = TGGATC, and Y = TGGATC. The fluorescence level will produce the output for the DNA ternary XOR operation again as CAAGCT. Because this outputted value is the sum value when those two sequences will be added.

TABLE 12.8

Truth Table of DNA Ternary NOR Operations

	ACCTAG	CAAGCT	TGGATC
ACCTAG	TGGATC	CAAGCT	ACCTAG
CAAGCT	CAAGCT	ACCTAG	TGGATC
TGGATC	ACCTAG	TGGATC	CAAGCT

12.3.7   Ternary DNA XNOR Operation

DNA ternary XNOR operation is defined as $Y_{DTNOR} = \overline{sum(X, Y)}$, where X and Y are the input from {0, 1, 2}. Table 12.8 shows the truth table of DNA ternary NOR operations.

12.3.7.1   Design Procedure of DNA Ternary XNOR Operation

From Table 12.8, it is found that DNA ternary XNOR operation is the inverted output of the DNA ternary XOR operation. The DNA ternary NOT operation is designed which inverts an input. Therefore it is possible to construct the DNA ternary XNOR operation by adding STI to the DNA ternary XOR operation which is shown in Figure 12.7.

Figure 12.7 shows that the DNA ternary XNOR is constructed using one DNA ternary XOR operation followed by one DNA ternary NOT operation.

12.3.7.2   Working Principles of DNA Ternary XNOR Operation

The DNA ternary XNOR operation is the same as the DNA ternary XOR operation, but the final output value should be inverted to the result of the DNA ternary XOR operation.

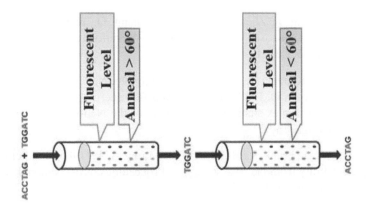

FIGURE 12.7

The Operational Diagram of DNA Ternary XNOR Operation

For example, if the input value is TGGATC, CAAGCT then the DNA ternary XOR will produce ACCTAG (the sum value), and the final output will be the inverted value of ACCTAG, which is also TGGATC. The truth table represents all possible truth values.

12.4 Summary

This chapter has presented the ternary logic in the DNA computing system that means establishing basic gates and fundamental operations in the ternary system. The circuit diagram of different DNA Computing operations in the multiple-valued system is discussed with necessary figures. As always, the design principles and working principles of the operations are shown. To be mentioned, DNA computing in multiple-valued logic is theoretically easier than in the two-valued logic system. Because no DNase Enzyme will be used to detect the DNA strands bond in the ternary system.

Bibliography

[1] Hallworth, R. P., & Heath, F. G. (1962). Semiconductor circuits for ternary logic. Proceedings of the IEE-Part C: Monographs, 109(15), 219-225.

[2] Dhande, A. P., & Ingole, V. T. (2005, March). Design and implementation of 2 bit ternary ALU slice. In Proc. Int. Conf. IEEE-Sci. Electron., Technol. Inf. Telecommun (Vol. 17).

[3] Zheng, X., Yang, J., Zhou, C., Zhang, C., Zhang, Q., & Wei, X. (2019). Allosteric DNAzyme-based DNA logic circuit: Operations and dynamic analysis. Nucleic Acids Research, 47(3), 1097-1109.

[4] Freier, S. M., Kierzek, R., Jaeger, J. A., Sugimoto, N., Caruthers, M. H., Neilson, T., & Turner, D. H. (1986). Improved free-energy parameters for predictions of RNA duplex stability. Proceedings of the National Academy of Sciences, 83(24), 9373-9377.

[5] Breslauer, K. J., Frank, R., Blöcker, H., & Marky, L. A. (1986). Predicting DNA duplex stability from the base sequence. Proceedings of the National Academy of Sciences, 83(11), 3746-3750.

13

Multiple-Valued Arithmetic Operations in DNA Computing

13.1 Introduction

Computing technologies are now built on the binary logic/number system, which is reliant on the existing transistors' basic on/off switching mechanism. With the exponential rise in data processing and storage requirements, there is a strong drive to transition to a higher radix logic/number system that can eliminate or mitigate many of the binary system's constraints. Moore's law is expected to be saturated, and the need to increase information density and processing speed in future micro and nanoelectronic circuits and systems provides a strong context and incentive for the beyond-binary logic system.

Multi-Valued logic (MVL) is considered a key enabler for next-generation and high-information-density digital electronics. While the conventional approaches to MVL systems rely on circuit-level design and implementation, the emergence of new electronic materials and device concepts now enables the MVL operation in a greatly reduced complexity and with intriguing new functionalities. The MVL system's viability in real-world applications is contingent on overcoming two major challenges: (i) developing an efficient mathematical approach for implementing the MVL logic using existing technology, and (ii) the availability of effective synthesis techniques.

In this chapter, previously designed robust arithmetic operations will be implemented in the ternary logic in DNA computing.

1. Multi-Valued DNA Half Adder
2. Multi-Valued DNA Full Adder
3. Multi-Valued DNA Half Subtractor
4. Multi-Valued DNA Full Adder
5. Multi-Valued DNA N-bit Parallel Adder
6. Multi-Valued DNA Carry Skip Adder
7. Multi-Valued DNA Carry Lookahead Adder
8. Multi-Valued DNA Multiplier
9. Multi-Valued DNA Divider
10. DNA Multi-Valued Comparator

DOI: 10.1201/9781003381938-13

TABLE 13.1

Truth Table of DNA Ternary Half-Adder Operation

A	B	Carry	Sum
ACCTAG	ACCTAG	ACCTAG	ACCTAG
ACCTAG	CAAGCT	ACCTAG	CAAGCT
ACCTAG	TGGATC	ACCTAG	TGGATC
CAAGCT	ACCTAG	ACCTAG	CAAGCT
CAAGCT	CAAGCT	ACCTAG	TGGATC
CAAGCT	TGGATC	CAAGCT	ACCTAG
TGGATC	ACCTAG	ACCTAG	TGGATC
TGGATC	CAAGCT	CAAGCT	ACCTAG
TGGATC	TGGATC	CAAGCT	CAAGCT

13.2   Multi-Valued DNA Half-Adder

A multi-valued or ternary half adder is a sort of adder, which is an electrical circuit that performs ternary number addition. The half adder may add two single ternary digits and output the result together with a carry value. It has two inputs, called A and B, and two outputs S (sum) and C (carry).

DNA Ternary Half Adder will add two molecular sequences from the ternary digit 0, 1, and 2 and produce two outputs as sum S and carry qutrit C.

The below truth table (Table 13.1) will show the outputs of a quantum ternary half adder and ACCTAG = '0', CAAGCT = '1', and TGGATC = '2'.

From the truth table, the logical expressions for the ternary sum and carry as –

   Sum = $A^0.B^2 + A^1.B^1 + A^2.B^0 + 1. (A^0.B^1 + A^1.B^0 + A^2.B^2)$

   Carry = $1. (A1.B2 + A^2.B^1 + A^2.B^2)$

So, using this equation, it is possible to construct the circuit diagram for the DNA ternary half-adder operation.

13.2.1   The Circuit Architecture of DNA Ternary Half-Adder Operation

To design a DNA ternary half adder, it is needed to use DNA decoders so that it is possible to decode the input bit into the corresponding bits. Two decoders will be used to decode two input sequences A and B that have to be added. They will produce 3-bit output. The circuit architecture of the DNA ternary half-adder is shown in Figure 13.1.

From the Figure 13.1, it is found that to construct the DNA ternary half-adder, it is needed to use two DNA ternary 1-to-3 decoders, eleven DNA ternary AND operations, and seven DNA ternary OR operations.

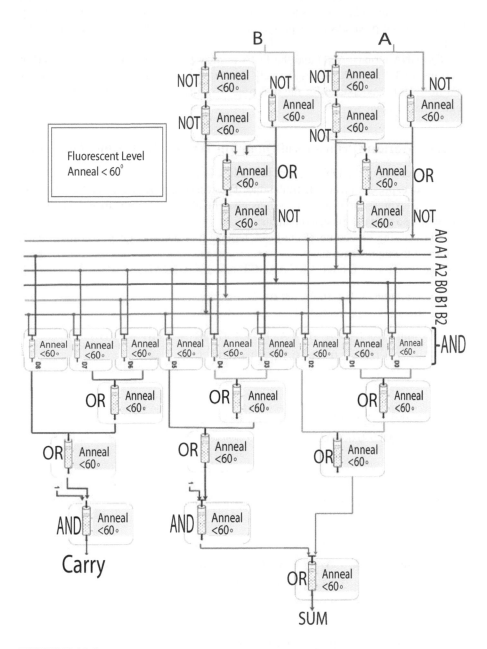

FIGURE 13.1
Circuit Architecture of DNA Ternary Half-Adder

The DNA combination of ternary AND and OR operations are nothing but the products-of-sum (which was shown in the logical expression of the DNA ternary half-adder) that produces the expected output that was mentioned in the truth table 13.1.

The DNA ternary AND and the DNA ternary OR operations were designed in sections 12.3.2 and 12.3.4, respectively. The architecture may seem very complex and difficult to understand for the first time. But the circuit diagram represents the logical expression.

13.2.2   Working Principles of Quantum Ternary Half-Adder Operation

The architecture of the DNA ternary half-adder is designed completely by following the logical expression from the truth table. The working procedures can be shown by the logical expression. If the expressions give the absolute outputs then the architecture will also give the expected outputs.

The working procedure of the DNA ternary half adder is given below for each pattern of the input sequences.

Assume that 0 = ACCTAG, 1 = CAAGCT, and 2 = TGGATC.

1.   For input A, B = 0, $A_0 = 2$, $A_1 = 0$, $A_2 = 0$

   $B_0 = 2$, $B_1 = 0$, $B_2 = 0$

   Sum $= A_2.B_0 + A_1.B_1 + A_0.B_2 + 1. (A_1.B_0 + A_0.B_1 + A_2.B_2)$

   $= 0.2 + 0.0 + 2.0 + 1. (0.2 + 2.0 + 0.0)$

   $= 0 + 1.0$

   $= 0$

   Carry $= 1. (A_2.B_1 + A_2.B_2 + A_1.B_2)$

   $= 1. (0.0 + 0.0 + 0.0)$

   $\qquad\qquad\qquad\qquad = 1.0$

   $\qquad\qquad\qquad\qquad = 0$

2.   For input A, B = 0, 1

   $A_0 = 2$, $A_1 = 0$, $A_2 = 0$

   $B_0 = 0$, $B_1 = 2$, $B_2 = 0$

   Sum $= A_2.B_0 + A_1.B_1 + A_0.B_2 + 1. (A_1.B_0 + A_0.B_1 + A_2.B_2)$

   $= 0.0 + 0.2 + 2.0 + 1.(0.2 + 2.2 + 0.0)$

   $= 0 + 1.2$

   $= 1$

   Carry $= 1. (A_2.B_1 + A_2.B_2 + A_1.B_2)$

   $= 1. (0.2 + 0.0 + 0.0)$

   $= 1.0$

   $= 0$

3. For input A, B = 0, 2

   $A_0 = 2, A_1 = 0, A_2 = 0$

   $B_0 = 0, B_1 = 0, B_2 = 2$

   Sum $= A_2.B_0 + A_1.B_1 + A_0.B_2 + 1. (A_1.B_0 + A_0.B_1 + A_2.B_2)$

   $= 0.0 + 0.0 + 2.2 + 1.(0.2 + 2.0 + 0.0)$

   $= 2 + 1.0$

   $= 2$

   Carry $= 1. (A_2.B_1 + A_2.B_2 + A_1.B_2)$

   $= 1. (0.0 + 0.2 + 0.2)$

   $= 1.0$

   $= 0$

4. For input A, B = 1, 0

   $A_0 = 0, A_1 = 2, A_2 = 0$

   $B_0 = 2, B_1 = 0, B_2 = 0$

   Sum $= A_2.B_0 + A_1.B_1 + A_0.B_2 + 1. (A_1.B_0 + A_0.B_1 + A_2.B_2)$

   $= 0.2 + 2.0 + 0.0 + 1.(2.2 + 0.0 + 0.0)$

   $= 0 + 1.2$

   $= 1$

   Carry $= 1. (A_2.B_1 + A_2.B_2 + A_1.B_2)$

   $= 1. (0.0 + 0.0 + 2.0)$

   $= 1.0$

   $= 0$

5. For input A, B = 1, 1

   $A_0 = 0, A_1 = 2, A_2 = 0$

   $B_0 = 0, B_1 = 2, B_2 = 0$

   Sum $= A_2.B_0 + A_1.B_1 + A_0.B_2 + 1. (A_1.B_0 + A_0.B_1 + A_2.B_2)$

   $= 0.0 + 2.2 + 0.0 + 1.(2.0 + 0.2 + 0.0)$

   $= 2 + 1.0$

   $= 2$

   Carry $= 1. (A_2.B_1 + A_2.B_2 + A_1.B_2)$

   $= 1. (0.2 + 0.0 + 2.0)$

   $= 1.0$

   $= 0$

6. For input A, B = 1, 2

   $A_0 = 0$, $A_1 = 2$, $A_2 = 0$

   $B_0 = 0$, $B_1 = 0$, $B_2 = 2$

   Sum $= A_2.B_0 + A_1.B_1 + A_0.B_2 + 1.(A_1.B_0 + A_0.B_1 + A_2.B_2)$

   $= 0.0 + 2.0 + 0.2 + 1.(2.0 + 0.0 + 0.2)$

   $= 0 + 1.0$

   $= 0$

   Carry $= 1.(A_2.B_1 + A_2.B_2 + A_1.B_2)$

   $= 1.(0.0 + 0.2 + 2.2)$

   $= 1.2$

   $= 1$

7. For input A, B = 2, 0

   $A_0 = 0$, $A_1 = 0$, $A_2 = 2$

   $B_0 = 2$, $B_1 = 0$, $B_2 = 0$

   Sum $= A_2.B_0 + A_1.B_1 + A_0.B_2 + 1.(A_1.B_0 + A_0.B_1 + A_2.B_2)$

   $= 2.2 + 0.0 + 0.0 + 1.(0.2 + 0.0 + 2.0)$

   $= 2 + 1.0$

   $= 2$

   Carry $= 1.(A_2.B_1 + A_2.B_2 + A_1.B_2)$

   $= 1.(2.0 + 2.0 + 0.0)$

   $= 1.0$

   $= 0$

8. For input A, B = 2, 1

   $A_0 = 0$, $A_1 = 0$, $A_2 = 2$

   $B_0 = 0$, $B_1 = 2$, $B_2 = 0$

   Sum $= A_2.B_0 + A_1.B_1 + A_0.B_2 + 1.(A_1.B_0 + A_0.B_1 + A_2.B_2)$

   $= 2.0 + 0.2 + 0.0 + 1.(0.0 + 0.2 + 2.0)$

   $= 0 + 1.0$

   $= 0$

   Carry $= 1.(A_2.B_1 + A_2.B_2 + A_1.B_2)$

   $= 1.(2.2 + 2.0 + 0.0)$

   $= 1.2$

   $= 1$

9. For input A, B = 2, 2

$A_0= 0, A_1=0, A_2=2$

$B_0= 0, B_1=0, B_2=2$

Sum $= A_2.B_0 + A_1.B_1 + A_0.B_2 + 1. (A_1.B_0 + A_0.B_1 + A_2.B_2)$

$= 2.0 + 0.0 +0.2 +1.(0.0 + 0.0 + 2.2)$

$= 0 +1.2$

$= 1$

Carry $= 1. (A_2.B_1 + A_2.B_2 + A_1.B_2)$

$= 1. (2.0 + 2.2 + 0.2)$

$= 1.2$

$= 1$

It is found that the designed architecture for the DNA ternary half-adder performs the half-adder computation for the DNA input sequences perfectly.

13.3   Multi-Valued DNA Full-Adder

In the DNA ternary full adder, there are a total of three inputs A, B, and C. Here the value of A and B can be ACCTAG CAAGCT, or TGGATC (i.e., 0, 1, or 2). But for C which acts as a carry input in the full-adder, can be either ACCTAG, or CAAGCT (i.e., 0 or 1; see the half-adder section; no carry was generated as 2). The reason is that the carry bit can never be 2(TGGATC). So, there are a total number of eighteen combinations of the input values. Table 13.2 shows the operations in a DNA ternary full-adder.

From the truth table 13.2, the logical expression can be derived to find the sum and carry outputs.

The logical expression for the sum and carry out can be derived as

Sum = A2.B0.C0 +A1.B0.C1 + A1. B1.C0+ A0.B1.C1 +A0.B2.C0 +A2.B2.C1+1 (A1. B0. C0+A0.B0.C1 +A0.B1. C0+A2.B1. C1+A2.B2. C0+A1.B2.C1)= (A2. B0+A1B1+A0B2).C0 + (A1. B0+A0.B1+A2.B2).C1+ 1((A1.B0+A0.B1+A2.B2). C0 + (A0B0+A2B1+A1B2).C1)

Carry = 1 (A0. B2. C0+A2.B1. C0+A2.B2.C0 +A1.B1. C1+B2.C1+A2.C1)= 1. ((A0. B2+A2.B1+A2B2).C0 + (A1. B1+ B2+A2).C1)

Here, A0, A1, and A2 refer that the value of input A might be 0, 1, and 2 (i.e. AC-CTAG, CAAGCT, and TGGATC), respectively. Similarly, B0, B1, and B2 refer that the value of B might be 0, 1, and 2, respectively. Again, the equations are nothing but the sum-of-products form.

TABLE 13.2

Truth Table of DNA Ternary Full-Adder Operation

A	B	C	Sum	C_{out}
ACCTAG	ACCTAG	ACCTAG	ACCTAG	ACCTAG
ACCTAG	ACCTAG	CAAGCT	CAAGCT	ACCTAG
ACCTAG	CAAGCT	ACCTAG	CAAGCT	ACCTAG
ACCTAG	CAAGCT	CAAGCT	TGGATC	ACCTAG
ACCTAG	TGGATC	ACCTAG	TGGATC	ACCTAG
ACCTAG	TGGATC	CAAGCT	ACCTAG	CAAGCT
CAAGCT	ACCTAG	ACCTAG	CAAGCT	ACCTAG
CAAGCT	ACCTAG	CAAGCT	TGGATC	ACCTAG
CAAGCT	CAAGCT	ACCTAG	TGGATC	ACCTAG
CAAGCT	CAAGCT	CAAGCT	ACCTAG	CAAGCT
CAAGCT	TGGATC	ACCTAG	ACCTAG	CAAGCT
CAAGCT	TGGATC	CAAGCT	CAAGCT	CAAGCT
TGGATC	ACCTAG	ACCTAG	TGGATC	ACCTAG
TGGATC	ACCTAG	CAAGCT	ACCTAG	CAAGCT
TGGATC	CAAGCT	ACCTAG	ACCTAG	CAAGCT
TGGATC	CAAGCT	CAAGCT	CAAGCT	CAAGCT
TGGATC	TGGATC	ACCTAG	CAAGCT	CAAGCT
TGGATC	TGGATC	CAAGCT	TGGATC	CAAGCT

13.3.1   The Architecture of DNA Ternary Full-Adder Operation

Figure 13.2 shows the sum output sequence generation part and Figure 13.3 shows the carry output generation part of the DNA ternary full-adder operation.

Like the DNA ternary half-adder, the circuit uses the equations. To get the ternary values from the input, decoders are used. For three input sequences, three decoders are used.

The DNA ternary decoder produces three output lines for each input value. Decoder for input A produces A0, A1, A2, for input B produce B0, B1, B2, and for input C produce C0, C1, C2. Only one output line is true (which gives TGGATC; see decoder section) and the other remains false (produce ACCTAG) for an input. From the truth table, it is seen that the sum of the full adder produces six true (i.e., TGGATC) values and six CAAGCT values. For these six true values, six ternary DNA-AND operations (DTAND) are designed and each of the DTAND operation outputs will DTOR (DNA ternary OR) to produce the true output when one of the outputs of the DTAND operation is true. There are also nine CAAGCT values in sum. For these six CAAGCT values, six DNA ternary AND operations (DTAND) are designed and each of the DTAND operation outputs will DTOR to produce the true (TGGATC) output when one of the outputs of the DTAND operation is true. But as the output CAAGCT is needed, DTAND performs the DTOR operation output with CAAGCT. DTAND operation produces the minimum value. So the output of the DTAND

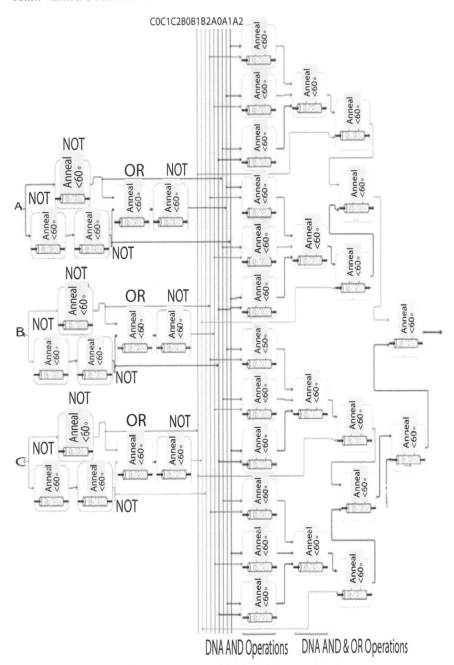

FIGURE 13.2
Circuit Diagram for DNA Ternary Full-Adder Sum Output Generation

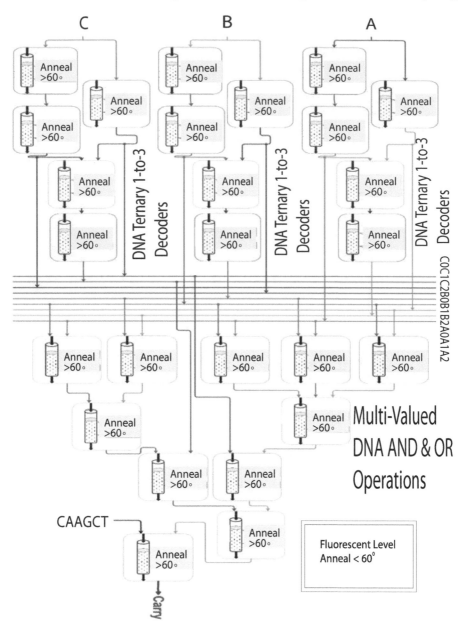

FIGURE 13.3
Circuit Diagram for DNA Ternary Full-Adder Carry Output Generation

operation will always CAAGCT. Finally, a DTOR operation is used to generate the maximum output among TGGATC, ACCTAG and ACCTAG, CAAGCT.

From the truth table 13.2, the carry of the full adder produces nine CAAGCT (i.e. 1) values. The design is constructed with nine DNA ternary AND operations (DTAND) and each of the DTAND operation's output will DTOR to produce the true (TGGATC) output when one of the outputs of the DTAND operation is true. But as the output CAAGCT is needed, DTAND is performed with the DTOR operation's output and CAAGCT. DTAND operation produces the minimum value. So the output of the DTAND operation will always be CAAGCT for one of the true inputs among nine DTAND operations. Again for one TGGATC value, one DTAND operation is used and the output of this DTAND will perform DTOR to produce the carry output.

13.3.2 Working Principles of Quantum Ternary Full-Adder Operation

A multiple-valued DNA Full-Adder contains three quantum bit sequences A, B, and C as input and two output DNA sequences, Sum and C_{out}. Here, the value of A and B can be any of the three DNA sequences from ACCTAG, CAAGCT, and TGGATC and for C, it can be ACCTAG and CAAGCT (in no situation carry input can be TGGATC). Here, three 1-to-3 multiple-valued DNA decoders are needed, seventeen multiple-valued DNA AND operations, and eleven multiple-valued DNA OR operations to get the result of Sum. Similarly, another seven multiple-valued DNA AND operations and five multiple-valued DNA OR operation are required to generate the output of C_{out}.

Let's have a look for one input pattern. When the input of A, B, and C are TGGATC, ACCTAG, and ACCTAG, then the decoder actives A2, B2, and C0 bypassing the DNA sequence TGGATC among them and ACCTAG through other lines. As a result, the first multiple-valued DNA AND operation is executed and produces the output TGGATC. This output pass through some multiple-valued DNA OR operation resulting in output TGGATC again. Then this output TGGATC performs a multiple-valued DNA AND operation with C0 which value is also TGGATC. The output then passes through two other multiple-valued DNA OR operations to produce the result of Sum which is TGGATC.

And for the output value of C_{out}, a similar procedure is followed. But before generating the result of C_{out}, one last multiple-valued DNA AND operation is executed with a fixed input sequence CAAGCT. And there is the value of C_{out} for this input combination is ACCTAG (i.e. 0).

13.4 Ternary DNA Half-Subtractor Operation

A ternary half subtractor is a sort of subtractor that conducts ternary number subtractions using an electrical circuit. The half subtractor may subtract two single ternary

TABLE 13.3

Truth Table of DNA Ternary Half-Subtractor
Operation

A	B	Borrow	Difference
ACCTAG	ACCTAG	ACCTAG	ACCTAG
ACCTAG	CAAGCT	CAAGCT	TGGATC
ACCTAG	TGGATC	CAAGCT	CAAGCT
CAAGCT	ACCTAG	ACCTAG	CAAGCT
CAAGCT	CAAGCT	ACCTAG	ACCTAG
CAAGCT	TGGATC	CAAGCT	TGGATC
TGGATC	ACCTAG	ACCTAG	TGGATC
TGGATC	CAAGCT	ACCTAG	CAAGCT
TGGATC	TGGATC	ACCTAG	ACCTAG

digits and output the result along with a borrow value. It has two inputs, called A and B, and two outputs D (difference) and B_{out} (borrow).

DNA Ternary Half Subtractor will subtract two DNA input sequences from the DNA ternary digit ACCTAG, CAAGCT, or TGGATC (i.e. 0, 1, and 2) and produce two outputs as DNA difference output D and DNA borrow output sequence B_{out}.

The truth table below (Table 13.3) will show the outputs of a DNA ternary half-subtractor.

From the truth table, the equations for the difference and borrow as –

Difference = $A^0.B^1 + A^1.B^2 + A^2.B^0 + 1. (A^0.B^2 + A^1.B^0 + A^2.B^1)$

Borrow = $1. (A^0.B^1 + A^0.B^2 + A^1.B^2)$

Here, A^0, A^1, and A^2 refer to the ACCTAG, CAAGCT, and TGGATC (i.e. 0, 1, and 2) which are the value of input A. Similarly, B^0, B^1, and B^2 refer to the ACCTAG, CAAGCT, and TGGATC (i.e. 0, 1, and 2) which are the value of input B.

13.4.1   Circuit Architecture of DNA Ternary Half-Subtractor Operation

To design a DNA ternary half subtractor, it is possible to use decoders to decode the input sequence into the corresponding input states. Two decoders will be used to decode two input sequences A and B that have to be subtracted. They will produce 3 output sequences. Where one output sequence will be active (i.e., having value TGGATC) and the other two will remain inactive (i.e. having value ACCTAG).

Figure 13.4 depicts the architecture of the DNA ternary half-subtractor operation. The Figure 13.4 shows that for the DNA ternary half-subtractor, it is needed to use two DNA ternary 1-to-3 decoders, eleven DNA ternary AND operations, and seven DNA ternary OR operations.

The DNA ternary AND and OR operations are nothing but the products-of-sum that produces the expected output that was mentioned in the above truth Table 13.3.

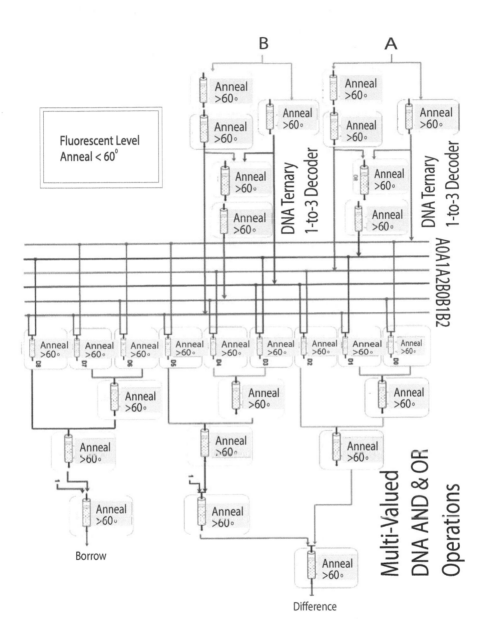

FIGURE 13.4
Circuit Diagram for DNA Ternary Half-Subtractor Operation

13.4.2   Working Principles of DNA Ternary Half-Subtractor Operation

The working principle of DNA ternary AND and OR operations are discussed already. The DNA ternary decoder operations are discussed also. Now, some inputs will be taken and simulate the circuit for the taken input sets. The result for every pattern with the logical expression that is derived for the DNA ternary half-subtractor is shown below.

The working procedure of the DNA ternary half subtractor is given below for each pattern of input bits.

1. For input A, B = 0,

    $A_0 = 2$, $A_1 = 0$, $A_2 = 0$

    $B_0 = 2$, $B_1 = 0$, $B_2 = 0$

    Difference = $A_0.B_1 + A_1.B_2 + A_2.B_0 + 1. (A_0.B_2 + A_1.B_0 + A_2.B_1)$

    = $2.0 + 0.0 + 0.2 + 1. (2.0 + 0.2 + 0.0)$

    = $0 + 1.0$

    = 0

    Borrow = $1. (A_0.B_1 + A_0.B_2 + A_1.B_2)$

    = $1. (2.0 + 2.0 + 0.0)$

    = 1.0

    = 0

2. For input A, B = 0, 1

    $A_0 = 2$, $A_1 = 0$, $A_2 = 0$

    $B_0 = 0$, $B_1 = 2$, $B_2 = 0$

    Difference = $A_0.B_1 + A_1.B_2 + A_2.B_0 + 1. (A_0.B_2 + A_1.B_0 + A_2.B_1)$

    = $2.2 + 0.0 + 0.0 + 1.(2.0 + 0.0 + 0.2)$

    = $2 + 1.0$

    = 2

    Borrow = $1. (A_0.B_1 + A_0.B_2 + A_1.B_2)$

    = $1. (2.2 + 2.0 + 0.0)$

    = 1.2

    = 1

3. For input A, B = 0, 2

    $A_0 = 2$, $A_1 = 0$, $A_2 = 0$

    $B_0 = 0$, $B_1 = 0$, $B_2 = 2$

    Difference = $A_0.B_1 + A_1.B_2 + A_2.B_0 + 1. (A_0.B_2 + A_1.B_0 + A_2.B_1)$

    = $2.0 + 0.2 + 0.0 + 1. (2.2 + 0.0 + 0.0)$

$= 0 + 1.2$

$= 1$

Borrow $= 1. (A_0.B_1 + A_0.B_2 + A_1.B_2)$

$= 1. (2.0 + 2.2 + 0.2)$

$= 1.2$

$= 1$

4. For input A, B = 1, 0

   $A_0 = 0, A_1 = 2, A_2 = 0$

   $B_0 = 2, B_1 = 0, B_2 = 0$

   Difference $= A_0.B_1 + A_1.B_2 + A_2.B_0 + 1. (A_0.B_2 + A_1.B_0 + A_2.B_1)$

   $= 0.0 + 2.0 + 0.2 + 1.(0.2 + 2.2 + 0.0)$

   $= 0 + 1.2$

   $= 1$

   Borrow $= 1. (A_0.B_1 + A_0.B_2 + A_1.B_2)$

   $= 1. (0.0 + 0.0 + 2.0)$

   $= 1.0$

   $= 0$

5. For input A, B = 1, 1

   $A_0 = 0, A_1 = 2, A_2 = 0$

   $B_0 = 0, B_1 = 2, B_2 = 0$

   Difference $- A_0.B_1 + A_1.B_2 + A_2.B_0 + 1. (A_0.B_2 + A_1.B_0 + A_2.B_1)$

   $= 0.2 + 2.0 + 0.0 + 1.(0.0 + 2.0 + 0.2)$

   $= 0 + 1.0$

   $= 0$

   Borrow $= 1. (A_0.B_1 + A_0.B_2 + A_1.B_2)$

   $= 1. (0.2 + 0.0 + 2.0)$

   $= 1.0$

   $= 0$

6. For input A, B = 1, 2

   $A_0 = 0, A_1 = 2, A_2 = 0$

   $B_0 = 0, B_1 = 0, B_2 = 2$

   Difference $= A_0.B_1 + A_1.B_2 + A_2.B_0 + 1. (A_0.B_2 + A_1.B_0 + A_2.B_1)$

$= 0.0 + 2.2 + 0.0 + 1.(0.2 + 2.0 + 0.0)$

$= 2 + 1.0$

$= 2$

Borrow $= 1. (A_0.B_1 + A_0.B_2 + A_1.B_2)$

$= 1. (0.0 + 0.2 + 2.2)$

$= 1.2$

$= 1$

7. For input A, B = 2, 0

$A_0 = 0, A_1 = 0, A_2 = 2$

$B_0 = 2, B_1 = 0, B_2 = 0$

Difference $= A_0.B_1 + A_1.B_2 + A_2.B_0 + 1. (A_0.B_2 + A_1.B_0 + A_2.B_1)$

$= 0.0 + 0.0 + 2.2 + 1.(0.0 + 0.2 + 2.0)$

$= 2 + 1.0$

$= 2$

Borrow $= 1. (A_0.B_1 + A_0.B_2 + A_1.B_2)$

$= 1. (0.0 + 0.0 + 0.0)$

$= 1.0$

$= 0$

8. For input A, B = 2, 1

$A_0 = 0, A_1 = 0, A_2 = 2$

$B_0 = 0, B_1 = 2, B_2 = 0$

Difference $= A_0.B_1 + A_1.B_2 + A_2.B_0 + 1. (A_0.B_2 + A_1.B_0 + A_2.B_1)$

$= 0.2 + 0.0 + 2.0 + 1.(0.0 + 0.0 + 2.2)$

$= 0 + 1.2$

$= 1$

Borrow $= 1. (A_0.B_1 + A_0.B_2 + A_1.B_2)$

$= 1. (0.0 + 0.0 + 0.0)$

$= 1.0$

$= 0$

9. For input A, B = 2, 2

$A_0 = 0, A_1 = 0, A_2 = 2$

$B_0 = 0, B_1 = 0, B_2 = 2$

Difference $= A_0.B_1 + A_1.B_2 + A_2.B_0 + 1. (A_0.B_2 + A_1.B_0 + A_2.B_1)$

$= 0.0 + 0.2 + 2.0 + 1.(0.2 + 0.0 + 2.0)$

$= 0 + 1.0$

$= 0$

Borrow $= 1. (A_0.B_1 + A_0.B_2 + A_1.B_2)$

$= 1. (0.0 + 0.2 + 0.2)$

$= 1.0$

$= 0$

The circuit reflects those two expressions for the difference and borrow outputs. Therefore, the results will be the same for the inputs to the circuit shown in Figure 13.4.

13.5   Multiple-Valued DNA Full Subtractor

Multi-Valued full subtractor performs subtraction operation taking borrow DNA sequence in consideration, thus it solves the problem that half subtractor cannot overcome. In multi-valued full subtractor, there are a total of three inputs A, B, and C. Here the value of A and B can be ACCTAG (0), CAAGCT (1), and TGGATC (2). But for C which acts as a carry input in full subtractor, can be either ACCTAG (0) or CAAGCT (1). The reason is that the carry bit can never be TGGATC (2). Hence, there are a total number of 18 combinations of the input values. The truth table of multi-valued full subtractors is given in Table 13.4.

From the truth Table, it is possible to construct the following K-maps and find the equations for the difference (Table 13.5) and borrow (Table 13.6).

C= ACCTAG (0) C= CAAGCT (1)

C= ACCTAG (0) C= CAAGCT

$D = A0.B1.C0 + A1.B2.C0 + A2.B0.C0 + A0.B0.C1 + A1.B1.C1 + A2.B2.C1 \quad 1.$
$(A0.B2.C0 + A1.B0.C0 + A2.B1.C0 + A0.B1.C1 + A1.B2.C1 + A2.B0.C1)$
$= (A0.B1 + A1.B2 + A2.B0).C0 + (A0.B0 + A1.B1 + A2.B2).C1 \quad 1. ((A0.B2 + A1.B0 + A2.B1).C0 + (A0.B1 + A1.B2 + A2.B0).C1)$

$B_{out} = 1.(A0.B1.C0 + A0.B2.C0 + A1.B2.C0 + A0.C1 + B2.C1 + A1.B1.C1)$
$= 1.((A0.B1 + A0.B2 + A1.B2).C0 + (A0 + B2 + A1.B1).C1)$

Now, using the equations of D (difference) and B_{out} (Borrow), the circuits of

TABLE 13.4

Truth Table of Multi-Valued DNA Full Subtractor

Inputs			Outputs	
A	**B**	**C**	**D**	**Bout**
ACCTAG	ACCTAG	ACCTAG	ACCTAG	ACCTAG
ACCTAG	CAAGCT	ACCTAG	TGGATC	CAAGCT
ACCTAG	TGGATC	ACCTAG	CAAGCT	CAAGCT
CAAGCT	ACCTAG	ACCTAG	CAAGCT	ACCTAG
CAAGCT	CAAGCT	ACCTAG	ACCTAG	ACCTAG
CAAGCT	TGGATC	ACCTAG	TGGATC	CAAGCT
TGGATC	ACCTAG	ACCTAG	TGGATC	ACCTAG
TGGATC	CAAGCT	ACCTAG	CAAGCT	ACCTAG
TGGATC	TGGATC	ACCTAG	ACCTAG	ACCTAG
ACCTAG	ACCTAG	CAAGCT	TGGATC	CAAGCT
ACCTAG	CAAGCT	CAAGCT	CAAGCT	CAAGCT
ACCTAG	TGGATC	CAAGCT	ACCTAG	CAAGCT
CAAGCT	ACCTAG	CAAGCT	ACCTAG	ACCTAG
CAAGCT	CAAGCT	CAAGCT	TGGATC	CAAGCT
CAAGCT	TGGATC	CAAGCT	CAAGCT	CAAGCT
TGGATC	ACCTAG	CAAGCT	CAAGCT	ACCTAG
TGGATC	CAAGCT	CAAGCT	ACCTAG	ACCTAG
TGGATC	TGGATC	CAAGCT	TGGATC	CAAGCT

TABLE 13.5

K Map for Difference

A/B	ACCTAG	CAAGCT	TGGATC	ACCTAG	CAAGCT	TGGATC
ACCTAG		TGGATC	CAAGCT	TGGATC	CAAGCT	
CAAGCT	CAAGCT		TGGATC		TGGATC	CAAGCT
TGGATC	TGGATC	CAAGCT		CAAGCT		TGGATC

TABLE 13.6

K Map for Borrow

A/B	ACCTAG	CAAGCT	TGGATC	ACCTAG	CAAGCT	TGGATC
ACCTAG		CAAGCT	CAAGCT	CAAGCT	CAAGCT	CAAGCT
CAAGCT			CAAGCT			CAAGCT
TGGATC						CAAGCT

multi-valued DNA Full Subtractor have to be designed. At the starting, three 1-to-3 DNA decoders are needed to find the corresponding input of the full subtractor. Inputs can be DNA sequence ACCTAG (0), CAAGCT (1), and TGGATC (2).

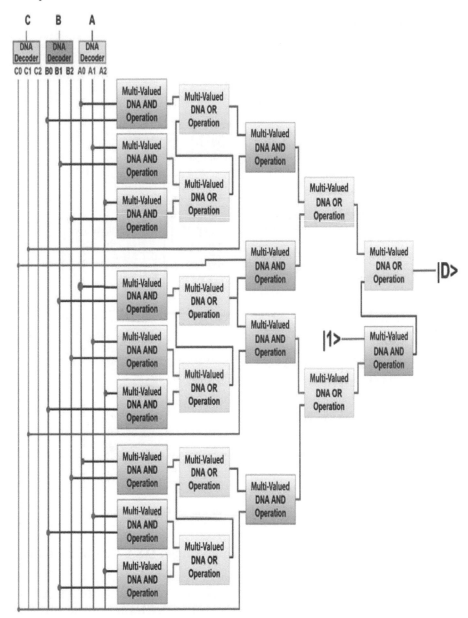

FIGURE 13.5
Block Diagram of Multi-Valued DNA Full Subtractor (Difference)

Figure 13.5 (Difference) and Figure 13.6 (Borrow) show the block diagram of multi-valued DNA Full Subtractor.

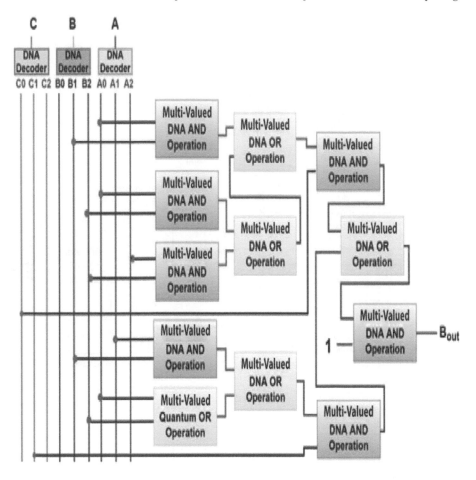

FIGURE 13.6
Block Diagram of Multi-Valued DNA Full Subtractor (Borrow)

13.5.1 Circuit Architecture of Multiple-Valued DNA Full Subtractor

The ternary decoder produces 3 outputs line for each input value. A 3-molecular Full subtractor (A, B, C) is used. Decoder for input A produces A0, A1, A2, for input B produce B0, B1, B2 and for input C produce C0, C1. Only one output line is true, TGGATC (2) and other remains false, ACCTAG (0) for an input. From the truth table it is seen that the difference of the full subtractor produce six true, TGGATC (2) values and six CAAGCT (1) values. For these six true values, six ternary AND (DNA AND) gates are designed and each of the DNA AND gate output will DNA OR to produce the true output when one of the output of the DNA AND gate is true. There are also six CAAGCT(1) values in difference. For this six CAAGCT (1) values, six ternary AND gate (DNA AND) are designed and each of the DNA AND gate output will DNA OR to produce the true, TGGATC (2) output when one of the output of the DNA AND gate is true (TGGATC). But as the output CAAGCT (1) is needed, so DNA AND operation is performed the DNA OR gate's output with CAAGCT (1). DNA AND gate produce the minimum value. So, the output of the DNA AND gate will always CAAGCT (1). Finally, a DNA OR gate is used to generate the maximum output among TGGATC (2), ACCTAG (0) and ACCTAG (0), CAAGCT (1) (Figure 13.7).

From the truth table, it is seen that the borrow of the full subtractor produces eight CAAGCT (1) values. After minimization using k map, the carry function produces six CAAGCT (1) values. Six ternary AND gate (DNA AND) gates are designed and each of the DNA AND gate outputs will DNA OR to produce the true, TGGATC(2) output when one of the outputs of the DNA AND gate is true. But the output CAAGCT (1) is needed, so DNA AND operation is performed between the DNA OR gate output and CAAGCT (1). DNA AND gate produce the minimum value. So, the output of the DNA AND gate will always CAAGCT (1) for one of the true inputs among the six DNA AND gate (Figure 13.8). So, Figure 13.7 shows the circuit architecture of multi-valued DNA Full Subtractor for Difference and Figure 13.8 shows the circuit architecture of multi-valued DNA Full Subtractor for Borrow.

13.5.2 Working Principle of Multiple-Valued DNA Full Subtractor

A ternary DNA Full-Subtractor contains 3 DNA input sequences as input and 2 output DNA sequences one for sum and another for carry. To understand the working principle of DNA Full Subtractor, let's consider the following case:

When the input sequence A, B = ACCTAG (0), and C = CAAGCT (1), DNA decoder generate A0, A1, A2 = (2, 0, 0), B0, B1, B2 = (2, 0, 0) and C0, C1, C2 = (0, 2, 0). The 10th AND gate is connected to the A0 and B0 and C1 line. As a result, the 10th AND gate value, that evaluates A0.B0.C1, will be true (TGGATC). The 13th AND gate is connected to the OR gate, so the output of the OR gate will be true (TGGATC). This output will go through the last OR gate and generate TGGATC (2) as difference (D).

In the full-subtractor borrow circuit, it is seen that none of the AND operations and OR operations are connected to A0, B0 and C1 lines. As a result, none of the AND operations and OR operation value will be true. Thus, they produce ACCTAG (0). This truth value will TAND with CAAGCT (1), and produce minimum ACCTAG (0) as output. This output will go through the OR gate and generate ACCTAG (0) as borrow (Bout).

13.6 Multiple-Valued DNA Parallel Adder

A single DNA ternary complete adder adds two single DNA sequence values and one carry input sequence. A DNA ternary Parallel Adder, on the other hand, is a DNA

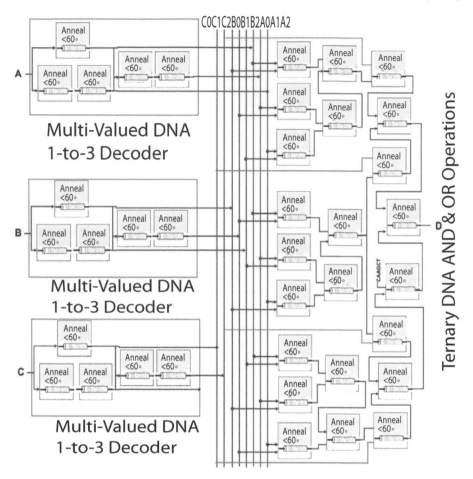

FIGURE 13.7
Circuit Architecture of Multi-Valued DNA Full Subtractor (Difference)

architecture that finds the arithmetic sum of two ternary numbers that are more than one bit in length by operating on analogous pairs of bits in parallel. It consists of a chain of DNA ternary full-adders, with each DNA ternary full adder's output carry sequence connected to the carry input of the next higher-order DNA ternary full-adder in the chain. An n-molecular parallel adder requires n DNA ternary full-adders to operate. So two DNA ternary full-adders are required for a two-bit number, four DNA ternary full adders are required for a four-bit value, and so on.

As always before performing the operation, it is needed to encode the ternary values to the equivalent DNA sequences and after the operation, the outputted result needs to be decoded to get the actual value.

13.6.1 General Organizations of DNA Multiple-Valued Parallel Adder

Figure 13.9 shows the general organizations of a DNA ternary 4-molecular sequence parallel adder. It is seen that to perform the ternary addition of two numbers of 4-molecular, four DNA ternary full-adders are required connected in a manner where the carry output of the 1st ternary full adder is the carry input of the 2nd ternary full adder, and the carry output of the 2nd ternary full adder is the carry input of the 3rd ternary full adder, and the carry output of the 3rd ternary adder is the carry input of the 4th ternary full adder. Finally, the carry output of the 4th ternary full-adder will be the most significant bit sequence of the ternary addition result in the DNA computing system.

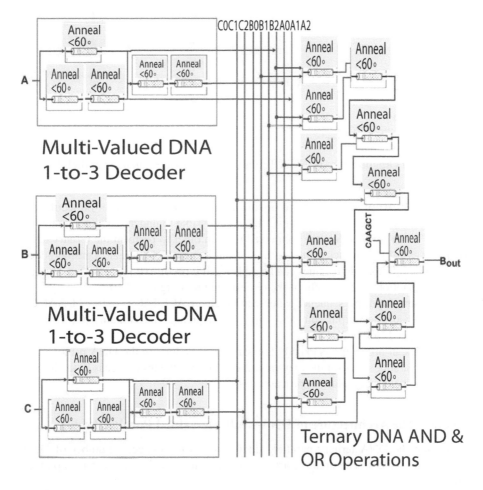

FIGURE 13.8
Circuit Architecture of Multi-Valued DNA Full Subtractor (Borrow)

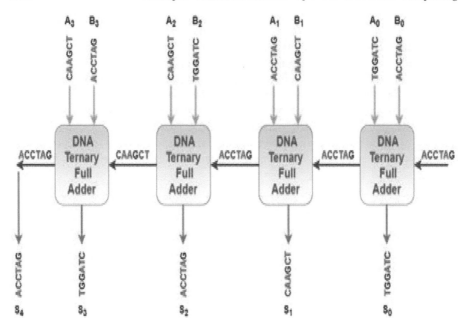

FIGURE 13.9
The General Organizations of DNA Ternary 4-Molecular Sequence Parallel Adder

Figure 13.10 shows the general organizations of a DNA ternary n-molecular sequence parallel adder. From Figure 13.10, it is possible to understand how n number of DNA ternary full-adders are connected to each other to perform the ternary addition process of two values of n-bits. In an n-molecular DNA ternary parallel adder, any number of bits length values can be added. All it needs to do is to design a circuit for that operation with the help of the circuit shown in Figure 13.10.

The carry input bit sequence value can be either ACCTAG (i.e. 0) or CAAGCT (i.e. 1), where the output sum and carry bit sequence can contain ACCTAG, or CAAGCT, or TGGATC.

The design procedure is exactly the same as it is done in the DNA n-molecular sequence parallel adder for the two-valued system. But the difference is in the logical operation- the ternary addition process.

13.6.2   The Architecture of DNA Multiple-valued Parallel Adder

To design the DNA ternary parallel adder it needs to connect a number of DNA ternary full-adders. The architecture of the DNA ternary full-adder was explained already.

The circuit diagram for sum output was shown in Figure 13.2, and the carry output was shown in Figure 13.3. All it needs to do is to connect the DNA full-

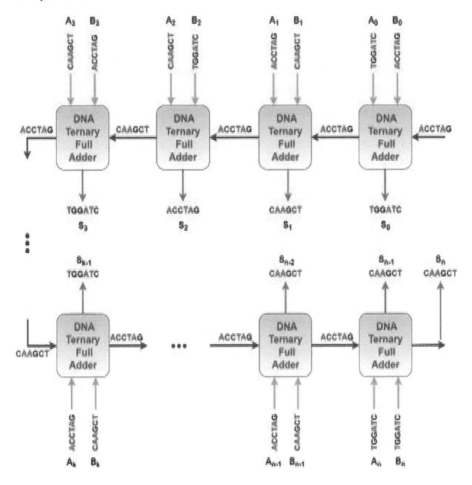

FIGURE 13.10
General Organizations of DNA Ternary n-Molecular Sequence Parallel Adder

adders in a way the general organization of the DNA ternary parallel adder is shown. The carry output will be the carry input sequence to the next DNA ternary full-adder.

13.6.3 The Working Principles of DNA Multiple-Valued Parallel Adder

The working mechanism of the DNA ternary full-adder is described already. There two sequences (the addend and the augend) are added which were only single-bit input.

The 4-bit sequence DNA ternary parallels adder operational procedure is explained below by taking an example.

Suppose two ternary values, A = 2021 and B = 1201 are to be added. Therefore, A_0 = CAAGCT, A_1 = TGGATC, A_2 = ACCTAG, and A_3 = TGGATC. In the same way, B_0 = CAAGCT, B_1 = ACCTAG, B_2 = TGGATC, and B_3 = CAAGCT.

Now, let's see how the DNA 4-bit ternary parallel adder perform the addition operation and generates the expected output.

1. The first DNA ternary parallel will get A_0 = CAAGCT, B_0 = CAAGCT, and the carry input sequence C_0 = ACCTAG. And this adder will produce the sum sequence value S_0 = TGGATC and the carry output as ACCTAG. This carry output will work as the carry input to the next DNA full-adder.

2. The 2nd DNA ternary parallel will get A_1 = TGGATC, B_1 = ACCTAG, and the carry input sequence C_1 = ACCTAG (from step 1). And this adder will produce the sum sequence value S_1 = TGGATC and the carry output as ACCTAG. This carry output will work as the carry input to the next DNA full-adder.

3. The 3rd DNA ternary parallel will get A_2 = ACCTAG, B_2 = TGGATC, and the carry input sequence C_2 = ACCTAG (from step 2). And this adder will produce the sum sequence value S_2 = TGGATC and the carry output as ACCTAG. This carry output will work as the carry input to the next DNA full-adder.

4. The 4th DNA ternary parallel will get A_3 = TGGATC, B_2 = CAAGCT, and the carry input sequence C_3 = ACCTAG (from step 3). And this adder will produce the sum sequence value S_3 = ACTAG and the carry output as CAAGCT. This carry output will be the value of S_4.

Therefore, after decoding the value, S = 10222 is obtained, which is the correct summation value of those two inputs.

13.7 Multiple-Valued DNA Carry-Lookahead Adder

In the DNA ternary carry-lookahead adder, all the carry outputs will be calculated without performing the *FullAdderCarryValue ()* operations. That means the carry output will be calculated using combinational circuits. Remember that, no carry will be generated as TGGATC, therefore it will not be a problem to determine the carry output values in the ternary system.

13.7.1 General Organizations of DNA Ternary Carry-Lookahead Adder

Figure 13.11 shows the general architecture of the 4-molecular sequence carry-lookahead adder. The combinational circuits consist of DNA ternary AND and OR operations which are shown in the previous chapters. This representation is the ternary representation of the DNA 4-molecular carry-lookahead adder.

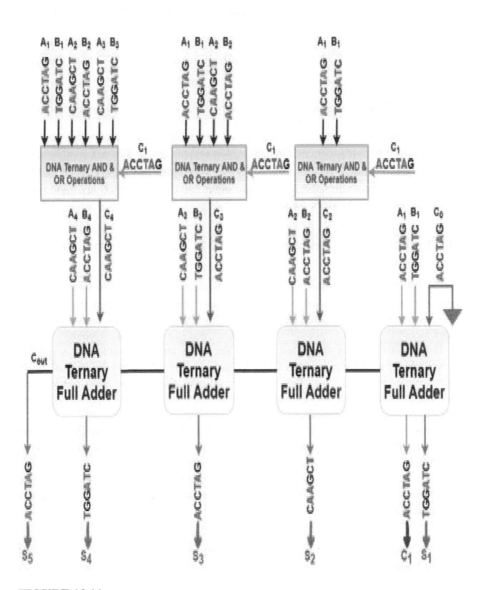

FIGURE 13.11
General Organizations of DNA Ternary 4-Molecular Sequence Carry-Lookahead Adder

Figure 13.12 shows the general architecture of the n-molecular sequence carry-lookahead adder to perform the addition of n-bit inputs, and all operations will be executed at the same time after determining the value of C_1. This will reduce the total computational time.

Figure 13.13 and Figure 13.14 show the combinational circuit to determine the value of C_2 and C_3.

In Figure 13.13, it is seen that the DNA ternary AND and OR operations find out the value of C2 using the value of P_1, G_1, and C_1.

In Figure 13.14, it is seen that the combinational circuit of the DNA ternary AND and OR operations find out the value of C3 using the value of P_1, P_2, G_1, G_2, and C_1. Construct the combinational circuit for the other carry values in ternary logic using the logical expression shown for them.

13.7.2   The Circuit Architecture of DNA Ternary Carry-Lookahead Adder

The value of C_i is determined to perform the addition operation concurrently. With the previously gained knowledge and the general organization shown in Figure 13.12, it is possible to construct the circuit diagram for the DNA ternary carry-lookahead adder.

Figure 13.15 shows the architecture of the DNA ternary 3-molecular sequence carry-lookahead adder where the DNA ternary 3-molecular carry-lookahead adder is constructed using the most familiar universal operations in the DNA ternary system. That means the architecture contains the DNA ternary XOR operations, AND operations, and OR operations.

Figure 13.15 shows how to calculate the P_i and G_i to determine the carry bit sequences. P_i can be determined by A_i **DTXOR** B_i and G_i can be determined by A_i **DTAND** B_i. The architecture does the same thing in the ternary system.

It is already seen that the value of C_2 and C_3 can be determined easily. C_2 can be determined from $P_1C_1 + G_1$, and C_3 can be determined from $P_2C_2 + G_2$. The values are determined and connected them accordingly.

13.7.3   The Working Principles of Quantum Ternary Carry-Lookahead Adder

The working principles of the DNA ternary carry-lookahead adder are straightforward. The equations are devised and implemented in the circuit. As a result, the circuit should provide the desired outcome if the equations are right. However, let's take an example, it is possible to simulate the inputs to the circuit and observe how the circuit behaves.

Suppose, A = 120 and B = 101.

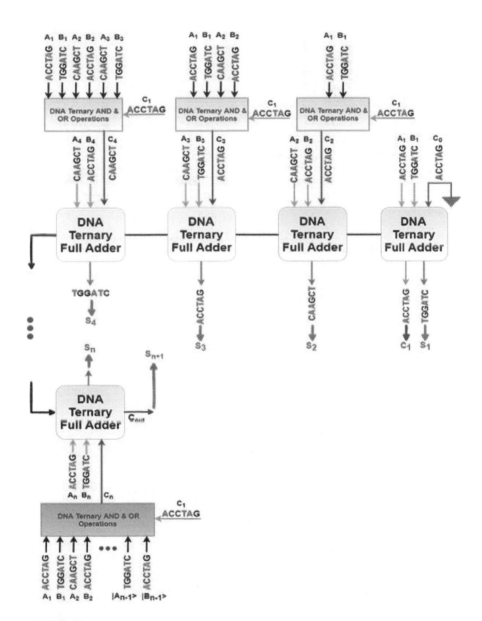

FIGURE 13.12
The General Organizations of DNA Ternary *n*-Molecular Carry-Lookahead Adder

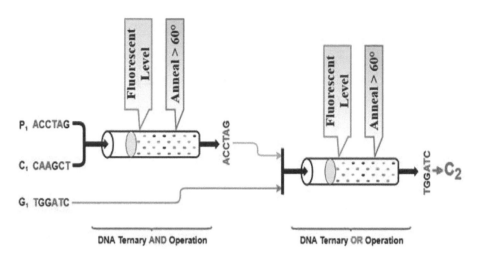

FIGURE 13.13
The Combinational Circuit to Determine the Value of C_2

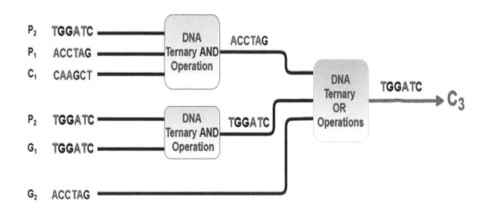

FIGURE 13.14
The Combinational Circuit to Determine the Value of C_3

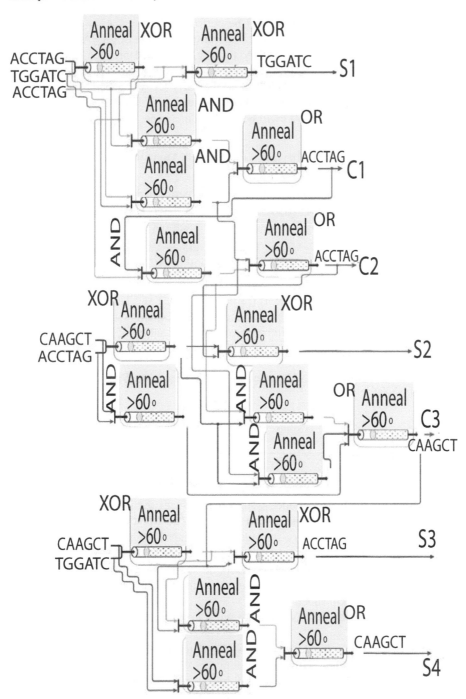

FIGURE 13.15
The Circuit Diagram of DNA Ternary 3-Molecular Sequence Carry-Lookahead Adder

1. The 1st addition operation will be performed between A_1 = ACCTAG, and B1 = CAAGCT. And for 1st operation the value of C_0 =ACCTAGA. The 1st ternary DTXOR (DNA Ternary XOR) will produce CAAGCT. And the 2nd ternary DTXOR will produce also CAAGCT. Therefore the value of S_1 will be CAAGCT. And the carry C_1 will generate using the ternary DTAND and ternary DTOR operations, which will be ACCTAG. Thus the C_1 = ACCTAG will be used to determine all the values of other C_i.

2. The value of P_1 will be 1, and the value of G_1 will be 0. Therefore, the value of C_2 = ACCTAG ($P_1C_1 + G_1$).

3. Now the 2nd addition operation will be performed between A_2 = TGGATC, and B_2 = ACCTAG. And C_2 = ACCTAG. Therefore, the value of S_2 =TGGATC. Now the value of C_3 is determined.

4. Here, the value of P_2 will be TGGATC, and the value of G_2 will be ACCTAG again. Therefore, the value of C_3 will be ACCTAG.

5. Now the 3rd addition operation will be performed between A_3 = CAAGCT, and B_3 = CAAGCT. And C_3 = ACCTAG. Therefore, the value of S_3 = TGGATC. And the value of C_4 = ACCTAG (see the truth table for ternary full-adder) will be the most significant bit of the result.

6. Therefore, the value of S_4 will be ACCTAG.

So, after decoding the DNA sequences for the sum value, S = 0221, which is the expected value.

13.8   Multi-Valued DNA Carry-Skip Adder

DNA carry-skip adder is an adder system that reduces the propagation delay of a DNA n-molecular parallel adder with low effort. The worst-case delay can be decreased by merging several carry-skip adders to create a DNA block-carry-skip adder.

The working mechanism of the DNA carry-skip adder is learned. The circuit is constructed for it. Now in this section, it will be learned in the DNA ternary system. But the operational change will be needed to perform the addition of ternary qutrits in the ternary carry-skip adder.

Figure 13.16 shows the basic structure of a 4-molecular sequence DNA ternary carry-skip adder.

Here, BP stands for block propagate which determines whether a carry will generate as ACCTAG (in the two-valued system the ACCTAG = 1) or not. If there will be no carry output as 1, a multiplexer will bypass the 1st carry input C_0 as the input for the next DNA full-adder operation. And, if there will be generated a carry as 1 within these four DNA full-adder operations, this carry skips adder will work just like the DNA parallel adder.

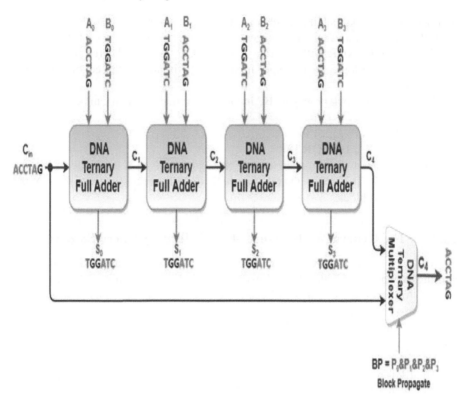

FIGURE 13.16
Basic Structure of a 4-Molecular Sequence DNA Ternary Carry-Skip Adder

The question is how to determine the block propagate in the DNA ternary system? Will this also same as the DNA two-valued system? If the block propagate is 1 (which means CAAGCT in the ternary system) the carry will not change! In the DNA full-adder operation the carry qubit was generated as ACCTAG (i.e. 1) only when the input addend and augend both were 1 (i.e. ACCTAG). Otherwise, the carry was always TGGATC (i.e. 0). Now in the ternary system carry generates as 1 (i.e. CAAGCT) only two bits are added either 1 and 2, or 2 or 1 (in the ternary system 1 = CAAGCT, and 2 = TGGATC). And for another input pattern carry can generate as 1 is if both inputs are 2. Therefore, avoiding these three input patterns, the block propagates when no carry will be generated as 1.

So, to perform DNA carry-skip adder in the ternary system, it is needed to consider 2 (TGGATC) also as the block propagates along with the 1 (CAAGCT). So, the DNA ternary carry-skip adder will bypass the value of C_0 as the final carry output of the block if the value of BP is either 1 (CAAGCT) or 2 (TGGATC). When the BP is 0 (ACCTAG) then this carry-skip adder will work as the DNA parallel adder. And one thing that can be noted is that to optimize the result, it is needed add a condition with the system, that is, if the input value both are 2 (TGGATC) the block will not perform the carry-skip addition, rather it will perform the regular parallel addition as well. Because this input set will also generate a carry output sequence 1 (CAAGCT) but the ternary DTXOR will generate 1 (CAAGCT) as well. Therefore, this input set will break the logic for the other input set if it is not set the condition by the algorithm.

Therefore, the block propagate BP in the DNA ternary system can be determined as–

$BP_i = P_0 . P_1 . P_2 . \ldots . P_i$

Where, $P_i = (A_i$ **DTXOR** $B_i)$

All operations will be for the ternary bit sequence. And if the value of BP is 1 (CAAGCT) or 2 (TGGATC), the DNA multiplexer will bypass the carry input of the block to the next block.

13.8.1   The General Architecture of DNA Ternary Carry-Skip Adder

Figure 13.17 shows the architecture of the DNA 4-bit block carry-skip adder. According to Figure 13.17, the block propagate will generate first, if its BP = CAAGCT or TGGATC then the circuit will bypass the first carry input sequence to the next DNA ternary full-adder using a DNA ternary 3-to-1 multiplexer.

The circuit diagram of the DNA ternary full-adder was shown already. Figure 13.17 shows how to connect the components accordingly to construct a 4-bit DNA ternary carry-skip adder.

The BP is generated as TGGATC (i.e. 2), and that is why the DNA sequence AC-CTAG which is the carry input to the block is bypassed through the DNA multiplexer as the carry-out of the block.

13.8.2   The Working Principles of DNA Ternary Carry-Skip Adder

To make the thing easier, as always, it is possible to understand the working procedure of the DNA ternary carry-skip adder through example.

Suppose, the addition will be performed of two inputs A and B, where A = 0120100102 and B = 2001021020.

A: 0120100102

B: 2001021020

S: 2121121122

In the ternary system for such cases, carry does not generate as 1. Therefore determining carry for such cases will increase the computational time.

For the above example, block propagate (BP) can be calculated as –

$BP = \prod_{i=0}^{n-1} P_i$

And, $P_i = A_i$ **DTXOR** B_i, in ternary logic.

If BP = 1 or 2, then the carry of the last adder operation will be the carry output of the 1st adder operation. This means no carry is been generated as 1. Let's calculate BP for the above example.

$BP = 2 . 2 . 1 . 1 . 2 . 1 . 1 . 2 . 1 . 2 = \mathbf{1}$

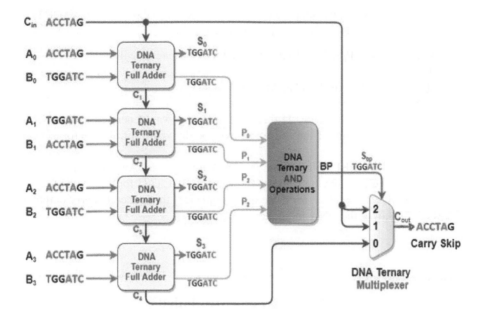

FIGURE 13.17

General Organizations of a 4-Molecular DNA Ternary Carry-Skip Adder

And suppose the inputs are A = 20002020220, and B = 02220202002, then the block propagate will be –

BP = 2.2.2.2.2.2.2.2.2.2.2 = **2**

Therefore, in both cases, no carry will be generated as 1. Therefore, the last carry output will be 0 which was the 1st carry input of this operation.

Note that, to explain in an easier way, the ternary digits are used instead of the equivalent DNA sequence. It needs to consider, 0 = ACCTAG, 1 = CAAGCT, and 2 = TGGATC.

Now let's take an example for the designed 4-bit DNA ternary carry-skip adder. As the circuit is for 4-bit, suppose A = 1012, and B = 0200. So, A_0 = TGGATC, A_1 = CAAGCT, A_2 = ACCTAG, and A_3 = CAAGCT. And B_0 = ACCTAG, B_1 = ACCTAG, B_2 = TGGATC, and B_3 = ACCTAG.

1. 1st DNA full-adder will receive inputs A_0 = TGGATC, B_0 = ACCTAG, and C_0 = ACCTAG. Remember the carry will determine only if the value of BP is not 1 or 2. Therefore the sum will be TGGATC, and the value of P_0 = TGGATC.

2. For 2nd DNA full-adder operation, input sequences are A_1 = CAAGCT, B_1 = ACCTAG. Therefore, P_1 = CAAGCT.

3. For 3rd DNA full-adder operation, input sequences are A_2 = ACCTAG, B_2 = TGGATC. Therefore, P_2 = TGGATC.

4. For 4th DNA full-adder operation, input sequences are A_3 = CAAGCT, B_3 = ACCTAG. Therefore, P_3 = CAAGCT.

All the values of P_i will be found. So the block propagate will be,
BP = P_0 . P_1 . P_2 . P_3
    = TGGATC . CAAGCT . TGGATC . CAAGCT
    = CAAGCT
Here, BP = CAAGCT (i.e. 1) therefore, no carry was generated as CAAGCT. And that is why the sum of the operations will be the values of all P_i's for each A_i and B_i.

And the value of BP will activate the DNA multiplexer. And consequently, the carry output for C_4 will be the value of C_0 which is ACCTAG for this example.

This circuit can be tested for more input sequences that generate carry output as CAAGCT in any of the DNA full adder operations.

13.9   Multi-Valued DNA Ternary Multiplier

A multiplier is a mechanism that multiplies two numbers together. In DNA logic, the multiplier is defined. The DNA multiplier has been demonstrated. The architecture was built using DNA logic. The two-valued system was the focus of everything that was accomplished. However, now in a multi-valued system, the DNA multiplier in the ternary system must be designed.

13.9.1   How Does The DNA Ternary Multiplier Work?

DNA Binary Multiplication rules were just like the decimal multiplication rules. The DNA ternary multiplier will take two input sequences and produce a product value and a carry value. The next multiplication will consider the carry to determine the product as well. Figure 13.18 shows the general working block of a DNA ternary multiplier.

To establish the DNA ternary multiplier it is needed to have to start from the core of the ternary multiplication. That means it is possible to understand how 1-bit multiplication can be formed. It is already shown in quantum ternary multiplication. Table 13.7 shows the truth table for 1-molecular DNA ternary multiplication. From the truth table (Table 13.7), the logical expression for the product and the carry can be derived.

The derived logical expression from the truth Table can be shown as –
Product = $A^1.B^2 + A^2.B^1 + 1. (A^1.B^1 + A^2.B^2)$
Carry = $1. (A^2.B^2)$
Here, A^1 means the value of A is 1 (CAAGCT), similarly A^2 means that the value of A is 2 (TGGATC). The same thing goes for the others as well.

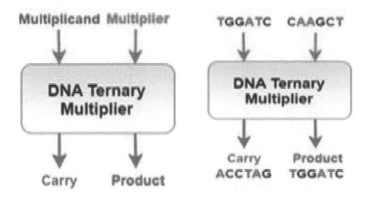

FIGURE 13.18
The General Working Block of a DNA Ternary Multiplier

From the equation, it is possible to generate the 1-molecular DNA ternary multiplier.

13.9.2 The Architecture of the DNA Ternary 2 × 2 Multiplier

From the logical expression, a DNA ternary 1-bit multiplier can be constructed which is shown in Figure 13.19. This 1-bit multiplier can perform the multiplication operation as shown in the truth table of the DNA multiplication rules. Figure 13.19 shows that two decoders decode the given input values for the multiplicand and multiplier. They produce the corresponding outputs, and the product and the carry output by using the outputs of the decoders as the input of the DNA ternary AND and OR operations. The circuit reflects the equations for the product and carry.

TABLE 13.7
Truth Table for 1-Molecular DNA Ternary Multiplication

MultiplicandA	MultiplierB	ProductP	CarryC	ResultR
ACCTAG	ACCTAG	ACCTAG	ACCTAG	00
ACCTAG	CAAGCT	ACCTAG	ACCTAG	00
ACCTAG	TGGATC	ACCTAG	ACCTAG	00
CAAGCT	ACCTAG	ACCTAG	ACCTAG	00
CAAGCT	CAAGCT	CAAGCT	ACCTAG	01
CAAGCT	TGGATC	TGGATC	ACCTAG	02
TGGATC	ACCTAG	ACCTAG	ACCTAG	00
TGGATC	CAAGCT	TGGATC	ACCTAG	02
TGGATC	TGGATC	CAAGCT	CAAGCT	11

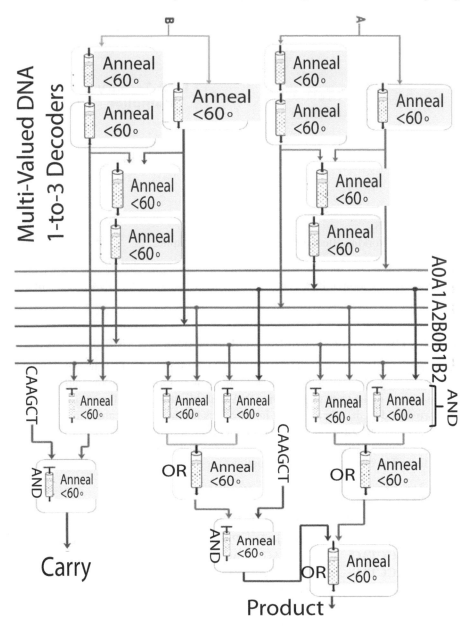

FIGURE 13.19
The Circuit Diagram of the DNA Ternary 1-Molecular Multiplier

So, this was just to perform 1-bit multiplication which will produce only one bit as the product, and just for one case, it will produce a carry as CAAGCT (i.e. 1).

Figure 13.20 shows the architecture of the DNA 2x2 qutrit ternary multiplier where the multiplicand and the multiplier are in 2-bit in size.

In Figure 13.20, two inputs, the multiplicand $A = A_1A_0$, and multiplier $B = B_1B_0$ are performing multiplication operations and generating the product of A and B as $P = P_4P_3P_2P_1P_0$. It is found that to perform the multiplication operation of the inputs of 2-bit size, four 1-bit DNA ternary multipliers are needed. These multipliers will perform the 1-bit multiplication operation. But to get the final product outputs, it is needed to connect five DNA ternary half adders (which will add two sequence values) and two DNA ternary full adders (which will also add a carry input sequence along with two input sequences) accordingly to Figure 13.20. The circuit diagram for the DNA ternary half-adder and DNA ternary full-adder was given already. So, connect every component and construct the circuit to perform the DNA ternary 2x2 multiplication operation.

13.9.3 The Working Procedure of the Quantum Ternary 2×2 Multiplier

The architecture of the DNA 2x2 ternary multiplier itself shows the multiplication process of values A = 11, and B = 20, which produces the product P = 00220. The operational behavior of the designed circuit for the DNA ternary 2x2 multiplier is shown in Figure 13.20.

Suppose, it is needed to multiply two ternary input values, where A = 02, and B = 10. After multiplication operations, these inputs will produce $P = P_4P_3P_2P_1P_0$. So, the value of $A_0 = $ TGGATC, and $A_1 = $ ACCTAG. And the value of $B_0 = $ ACCTAG, and $B_1 = $ CAAGCT. The operations will be described from right to left in Figure 13.20.

1. The 1st DNA ternary 1-bit multiplier will get the input values of A_0 and B_0. Here, $|A_0> = $ TGGATC, and $B_0 = $ ACCTAG. Therefore, this multiplier will generate two outputs $m_0 = $ ACCTAG and $c_0 = $ ACCTAG (see Table 13.7). Here, m_0 is the multiplication product, and c_0 is the carry value. The value of m_0 will be the least significant bit of the final product. Therefore, $P_0 = 0$, and the value of c_0 will be work as an input to the 1st DNA ternary half-adder shown in Figure 13.20 (from the right side of the Figure).

2. The 2nd DNA ternary 1-bit multiplier will get the input values of A_1 and B_0. Here, $A_1 = $ ACCTAG, and $B_0 = $ ACCTAG. Therefore, this multiplier will generate two outputs $m_1 = $ ACCTAG and $c_1 = $ ACCTAG (see Table 13.7). The value of m_1 will be work as the 2nd input value to the 1st DNA ternary half-adder. And the value of c_1 will be work as an input to the 3rd DNA ternary half-adder.

3. The 3rd DNA ternary 1-bit multiplier will get the input values of A_0 and B_1. Here, $A_0 = $ TGGATC, and $B_1 = $ CAAGCT. And, this multiplier will generate two outputs $m_2 = $ TGGATC and $c_2 = $ ACCTAG (see Table 13.7). The value of m_2 will be work as an input value to the 5th DNA ternary

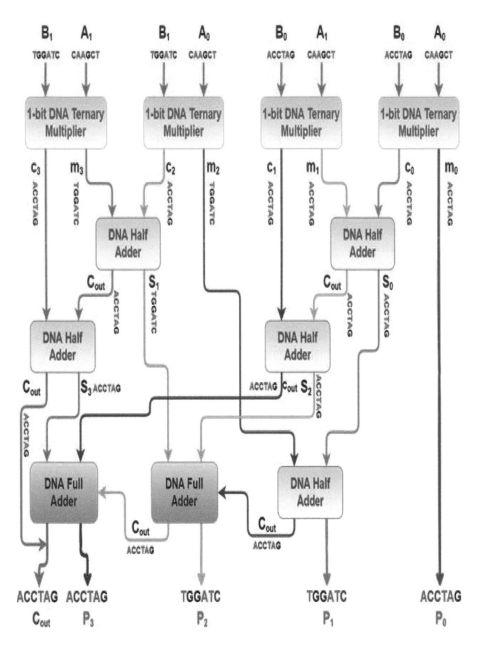

FIGURE 13.20
The Architecture of the DNA Ternary 2×2 Multiplier

half-adder. And the value of c_2 will work as an input to the 2nd DNA ternary half-adder.

4. The 4th DNA ternary 1-bit multiplier will get the input values of A_1 and B_1. Here, A_1 = ACCTAG, and B_1 = CAAGCT. And, this multiplier will generate two outputs m_3 = ACCTAG and c_3 = ACCTAG (see Table 13.7). The value of m_3 will be work as another input value to the 2nd DNA ternary half-adder. And the value of c_3 will work as an input to the 4th DNA ternary half-adder.

5. Now, the 1st DNA ternary half-adder will get inputs c_0 = ACCTAG (from step 1), and m_1 = ACCTAG (from step 2). So, this half-adder will produce two outputs. Here, from the 1st half adder S_0 = ACCTAG, and the carry output will also be ACCTAG. The value of S_0 will be work as another input of the 5th DNA ternary half-adder. And the carry value from this half-adder will work as another input value of the 3rd DNA ternary half-adder (according to Figure 13.20).

6. The 2nd DNA ternary half-adder will get inputs c_2 = ACCTAG (from step 3), and m_3 = ACCTAG (from step 4). So, this half-adder will produce two outputs, S_1 = ACCTAG, and the carry output will also be ACCTAG (see the truth Table of DNA ternary half-adder). The value of S_1 will be work as an input of the 1st DNA ternary full-adder. And the carry value from this half-adder will work as another input value of the 4th DNA ternary half-adder (according to Figure 13.20).

7. The 3rd DNA ternary half-adder will get inputs c_1 = ACCTAG (from step 2), and also ACCTAG from the carry output of the 1st half-adder (from step 5). So, this half-adder will produce two outputs, S_2 = ACCTAG, and the carry output will also be ACCTAG (see the truth Table of DNA ternary half-adder). The value of S_2 will be work as an input of the 1st DNA ternary full-adder. And the carry value from this half-adder will work as an input value of the 2nd DNA ternary full-adder (according to Figure 13.20).

8. The 4th DNA ternary half-adder will get inputs c_3 = ACCTAG (from step 4), and also ACCTAG from the carry output of the 2nd half-adder (from step 6). So, this half-adder will produce two outputs, S_3 = ACCTAG, and the carry output will also be ACCTAG (see the truth Table of DNA ternary half-adder). The value of S_3 will be work as an input of the 2nd DNA ternary full-adder. And the carry value from this half-adder produces the most significant bit of the final product value.

9. The 5th DNA ternary half-adder will get inputs S_0 = ACCTAG (from step 5), and m_2 = TGGATC from step 3. So, this half-adder will produce the value of P_1 = TGGATC. And a carry out value ACCTAG which will work as an carry input to the 1st DNA full-adder.

10. Now, the 1st DNA ternary full-adder operations can be performed. It gets three inputs as, S_1 = ACCTAG, S_2 = ACCTAG, and ACCTAG (from step

9). Therefore the sum of the addition will be ACCTAG which is the value of P_2. And the carry will also be ACCTAG which will be the carry input to the 2nd DNA ternary full-adder.

11. The last operation is the 2nd DNA ternary full-adder. It gets the three inputs as - S_3 = ACCTAG, ACCTAG from step 7, and also ACCTAG from step 10. Therefore it will also produce sum ACCTAG and carry output as ACCTAG. The sum output will be the value of P_3 and the carry output will be the value of P_4. And therefore, both the value ae obtained as ACCTAG.

So, finally, all the product values rae obtained.

Here, the input values are A = 02, and B = 10. And the product values are as- P_0 = ACCTAG, P_1 = TGGATC, P_2 = ACCTAG, P_3 = ACCTAG, and P_4 = ACCTAG. Therefore, the multiplication result for the two input values after decoding the DNA sequences to their equivalent ternary value P = 00020 is found. Which is absolutely correct. That is entirely accurate.

13.10   Multi-Valued DNA Divider

Figure 13.21 shows the general block of the DNA ternary divider. Here, the inputs and outputs are the DNA base sequences that are equivalent to the ternary values. As always, for the DNA ternary system – ACCTAG = "0", CAAGCT = "1", and TGGATC = "2".

The following truth table (Table 13.8) is for the 1-molecular DNA ternary divider. This is exactly the same as shown in the quantum ternary 1-qutrit divider, but the values here are DNA sequences instead of quantum ternary bits.

TABLE 13.8

Truth Table for 1-Molecular DNA Ternary Divider

DividendA	DivisorB	QuotientQ	RemainderR
ACCTAG	ACCTAG	-	-
ACCTAG	CAAGCT	ACCTAG	ACCTAG
ACCTAG	TGGATC	ACCTAG	ACCTAG
CAAGCT	ACCTAG	-	-
CAAGCT	CAAGCT	CAAGCT	ACCTAG
CAAGCT	TGGATC	ACCTAG	CAAGCT
TGGATC	ACCTAG	-	-
TGGATC	CAAGCT	TGGATC	ACCTAG
TGGATC	TGGATC	CAAGCT	ACCTAG

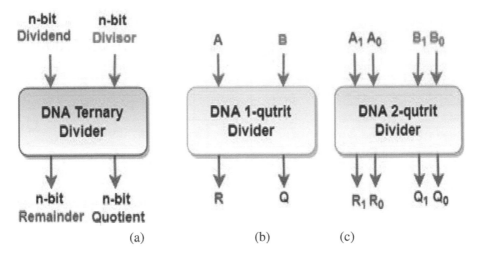

FIGURE 13.21
The General Working Block of the DNA Ternary Divider; (a) n-Bit DNA Ternary Divider, (b) 1-Bit DNA Ternary Divider, and (c) 2-Bit DNA Ternary Divider

The derived logical expression from the truth table can be shown as –
Quotient $= A^2. B^1 + 1.(A^1. B^1 + A^2. B^2)$
Remainder $= 1.(A^1. B^2)$ Here, A^1 means the value of A is 1 (CAAGCT), similarly A^2 means that the value of A is 2 (TGGATC). The same thing goes for the others as well.
From the above equation, it is possible to construct the DNA architecture for the DNA ternary 1-bit divider which is shown in Figure 13.22.

13.10.1 The Construction of the DNA Ternary Divider

Figure 13.22 shows the architecture of the DNA ternary 1-bit divider. Where two DNA 1-to-3 ternary divider is used to decode the value of the inputs A and B. And the rest of the structure is constructed using the logical expression for the 1-bit DNA ternary divider.

It is needed to construct the general architecture for the DNA ternary system according to those logical expressions. Figure 13.23 shows the general architecture of the DNA ternary 2-bit divider is depicted.

Figure 13.23 shows that four DNA ternary 1-to-3 decoders are required. And they will generate twelve outputs containing the values of the input bit sequences. And then a large number of DNA ternary AND and DNA ternary OR operations will produce the outputs of the division operations.

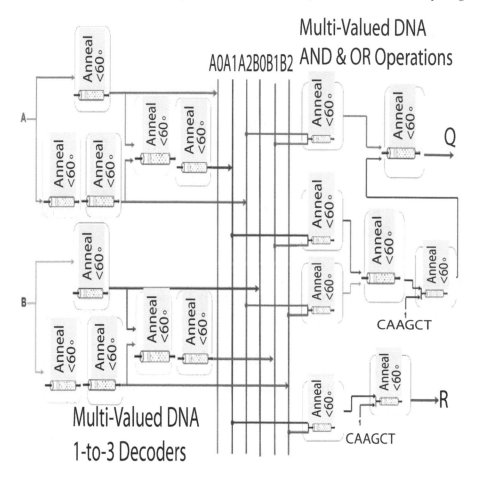

FIGURE 13.22
Circuit Architecture of the DNA Ternary 1-Bit Divider

13.11 Multiple-Valued DNA Comparator

A comparator that compares two input signals and gives output as which input is larger or smaller or equal. Multi-Valued DNA 1-bit comparator takes two single DNA sequence inputs, A and B. The value of the inputs can be **ACCTAG(0)**, **CAAGCT(1)** and **TGGATC (2)**. This circuit gives three outputs X, Y, Z where X is true when A is smaller than B, Y is true when A is equal to B and Z is true when A is greater than B. Here, Table 13.9 shows the truth table of DNA 1-bit comparator.

From the truth table, the equations for X, Y, and Z as:

$$X = A^0.B^1 + A^0.B^2 + A^1.B^2$$

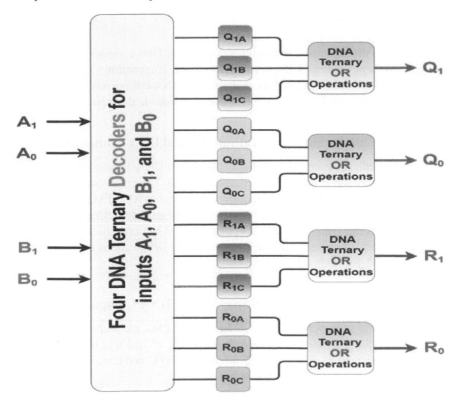

FIGURE 13.23
The General Architecture of the DNA Ternary 2-Bit Divider

TABLE 13.9
1-Bit Comparator Truth Table

Inputs		Outputs		
A	B	X	Y	Z
		A < B	A = B	A > B
ACCTAG	ACCTAG	ACCTAG	TGGATC	ACCTAG
ACCTAG	CAAGCT	TGGATC	ACCTAG	ACCTAG
ACCTAG	TGGATC	TGGATC	ACCTAG	ACCTAG
CAAGCT	ACCTAG	ACCTAG	ACCTAG	TGGATC
CAAGCT	CAAGCT	ACCTAG	TGGATC	ACCTAG
CAAGCT	TGGATC	TGGATC	ACCTAG	ACCTAG
TGGATC	ACCTAG	ACCTAG	ACCTAG	TGGATC
TGGATC	CAAGCT	ACCTAG	ACCTAG	TGGATC
TGGATC	TGGATC	ACCTAG	TGGATC	ACCTAG

$$Y = A^0.B^0 + A^1.B^1 + A^2.B^2$$
$$Z = A^1.B^0 + A^2.B^0 + A^2.B^1$$

Now, using the equations, the Multi-Valued DNA 1-bit comparator is designed. Figure 13.24 shows the block diagram of DNA 1-bit comparator where nine multi-valued DNA AND operations and six multi-valued DNA OR operations are used. At the beginning two 1-to-3 DNA decoders are used to decode the inputs.

13.11.1   Circuit Architecture Multiple-Valued DNA Comparator

Figure 13.25 depicts the creation of the Multi-Valued DNA 1-qubit comparator circuit. The inputs A and B are passed through two 1-to-3 Decoders separately in this circuit so that the inserted DNA molecular sequence, i.e. **ACCTAG**, **CAAGCT** or **TGGATC**, which represent "1," "2," and "3," can be detected using the outputs of the decoders. In the 1-qubit comparator, nine DNA AND operations and six DNA OR operation are employed, with each input coming from various combinations of the output of the 1-to-3 decoders.

13.11.2   Working Principle Multiple-Valued DNA Comparator

1. For input sequences A, B =**ACCTAG**, **ACCTAG**, the multi-valued 1-to-3 decoder will perform (A0 && B0) and the 4th line will be open. As a result, the output sequence of Y equal **TGGATC** and remaining lines, X and Z will remain closed, **i.e. ACCTAG**.

2. For input sequences A, B =**ACCTAG**, **CAAGCT**, the multi-valued 1-to-3 decoder will perform (A0 && B1) and the 1st line will be open. As a result, the output sequence of X equal **TGGATC** and remaining lines, Y and Z will remain closed, **i.e. ACCTAG**.

3. For input sequences A, B =**ACCTAG**, **TGGATC** , the multi-valued 1-to-3 decoder will perform (A0 && B2) and the 2nd line will be open. As a result, the output sequence of X equal **TGGATC** and remaining lines, Y and Z will remain closed, **i.e. ACCTAG**.

4. For input sequences A, B =**CAAGCT**, **ACCTAG**, the multi-valued 1-to-3 decoder will perform (A1 && B0) and the 7th line will be open. As a result, the output sequence of Z equal **TGGATC** and remaining lines, X and Y will remain closed, **i.e. ACCTAG**.

5. For input sequences A, B =**CAAGCT**, **CAAGCT**, the multi-valued 1-to-3 decoder will perform (A1 && B1) and the 5th line will be open. As a result, the output sequence of Y equal **TGGATC** and remaining lines, X and Z will remain closed, **i.e. ACCTAG**.

6. For input sequences A, B =**CAAGCT**, **TGGATC** , the multi-valued 1-to-3 decoder will perform (A1 && B2) and the 3rd line will be open. As a result, the output sequence of X equal **TGGATC** and remaining lines, Y and Z will remain closed, **i.e. ACCTAG**.

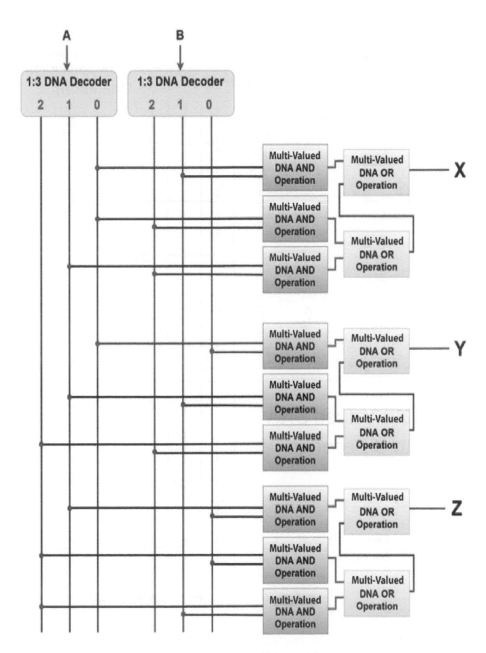

FIGURE 13.24
Block Diagram of Multi-Valued DNA 1-Bit Comparator

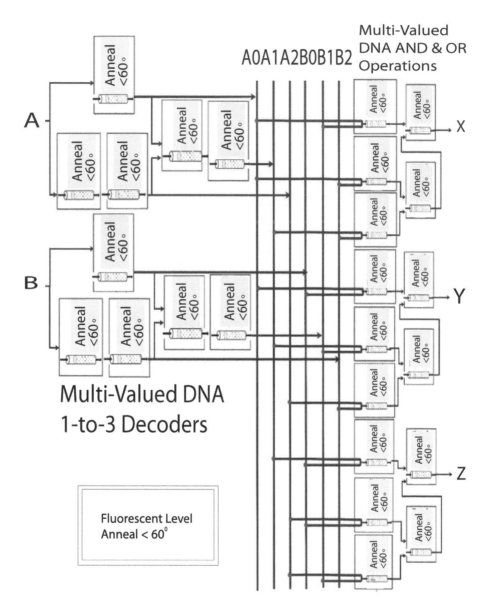

FIGURE 13.25
Circuit Architecture of Multi-Valued DNA Comparator

7. For input sequences A, B =**TGGATC** , **ACCTAG**, the multi-valued 1-to-3 decoder will perform (A2 && B0) and the 8th line will be open. As a result, the output sequence of Z equal **TGGATC** and remaining lines, X and Y will remain closed, **i.e. ACCTAG**.

8. For input sequences A, B =**TGGATC** , **CAAGCT**, the multi-valued 1-to-3 decoder will perform (A2 && B1) and the 9th line will be open. As a result, the output sequence of Z equal **TGGATC** and remaining lines, X and Y will remain closed, **i.e. ACCTAG**.

9. For input sequences, A, B =**TGGATC** , **TGGATC**, the multi-valued 1-to-3 decoder will perform (A2 && B2) and the 6th line will be open. As a result, the output sequence of Y equal **TGGATC** and the remaining lines, X and Z will remain closed, **i.e. ACCTAG**.

13.12   Summary

The ternary digits 0, 1, and 2 can be represented in the DNA system as ACCTAG, CAAGCT, and TGGATC, respectively. The challenges to implement the arithmetic operations in ternary logic systems are to develop an efficient mathematical approach for implementing the ternary logic using existing technology and the availability of effective synthesis techniques. A DNA ternary Parallel Adder is a DNA architecture that finds the arithmetic sum of two ternary numbers that are more than one bit in length by operating on analogous pairs of bits in parallel. A DNA ternary carry-lookahead adder finds out all the carries with combinational circuits by only determining the first carry, in a consequence, all the DNA full-adders can be executed concurrently. The total required time to perform the addition of two n-bit ternary values in the DNA ternary parallel adder is $(20n + 8)$ hours. The total execution time to perform a 2×2 bit multiplication operation in the DNA ternary 2×2 multiplier is 100 hours.

Bibliography

[1] Hallworth, R. P., & Heath, F. G. (1962). Semiconductor circuits for ternary logic. Proceedings of the IEE-Part C: Monographs, 109(15), 219-225.

[2] Mandal, S. B., Chakrabarti, A., & Sur-Kolay, S. (2011, May). Synthesis techniques for ternary quantum logic. In 2011 41st IEEE International Symposium on Multiple-Valued Logic (pp. 218-223). IEEE.

[3] Nie, H., Han, X., He, B., Sun, L., Chen, B., Zhang, W., ... & Kong, H. (2019, November). Deep sequence-to-sequence entity matching for heterogeneous en-

tity resolution. In Proceedings of the 28th ACM International Conference on Information and Knowledge Management (pp. 629-638).

[4] Dhande, A. P., & Ingole, V. T. (2005, March). Design and implementation of 2 bit ternary ALU slice. In Proc. Int. Conf. IEEE-Sci. Electron., Technol. Inf. Telecommun (Vol. 17).

[5] Giri, S., & Saraswathi, M. N. (2012). Implementation of combinational circuits using ternary multiplexer. International Journal of Computational Engineering Research, 2(2), 457-463.

14

Multiple-Valued Arithmetic Operations in Quantum-DNA Computing

14.1 Introduction

The combination of quantum computing and DNA computing is a new creation that is introduced in this book with multi-valued logic. This is actually the combination of quantum physics and molecular biology. So, it can be said that the combination is actually quantum molecular biology. Then established two types of cross-platform environment – (i) Quantum-DNA computing system, and (ii) DNA-Quantum computing system where both can be said as a form of quantum molecular biology but both are not the same. The arithmetic operations on both the Quantum-DNA and DNA-Quantum systems are done after solving all the prerequisites to develop the cross-platform environment between Quantum and DNA computing. Everything is done for the two-valued logic system. Now, as in the multiple-valued system, it is possible to build the cross-platform environment to construct the arithmetic circuits.

The underlying concept and functioning principles will remain the same, however, each component's logical operation will diverge. So, let's go further to learn everything about the multiple-valued arithmetic operations in Quantum-DNA computing.

14.2 Establishing Quantum-DNA Cross-Platform for the Ternary Logic System

The system was divided into four parts. The first part contained the Quantum system, then a quantum cache memory where the qubits were stored, then the data conversion unit to convert the quantum data into DNA base sequences, and the last part was the DNA system.

In the multiple-valued logic system, the general organization of the Quantum-DNA cross-platform for the multiple-valued logic system is depicted in Figure 14.1. Every ternary operation in the Quantum-DNA system will contain –

1. Multiple-valued Quantum System,

DOI: 10.1201/9781003381938-14

2. Multiple-valued Quantum Cache Memory,

3. Data Conversion Unit, and

4. Multiple-valued DNA System.

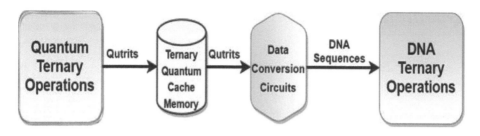

FIGURE 14.1
The General Organization Multiple-Valued Quantum-DNA Computing to Perform Ternary Logical Operations

14.3   Multiple-Valued Quantum-DNA Half Adder

A Multi-Valued half adder is a type of adder or electronic circuit that performs ternary number addition. Multi-Valued Half Adder is discussed two times in previous chapters in this part. Now, a half adder is designed with quantum qubits as input, but the outputs of this half adder will produce a DNA sequence. The truth table of the DNA-Quantum Half adder is shown in Table 14.1.

TABLE 14.1
Truth Table of Multi-Valued Quantum-DNA Half Adder

Inputs		Outputs	
|A>	|B>	Carry	Sum
|0>	|0>	ACCTAG	ACCTAG
|0>	|1>	ACCTAG	CAAGCT
|0>	|2>	ACCTAG	TGGATC
|1>	|0>	ACCTAG	CAAGCT
|1>	|1>	ACCTAG	TGGATC
|1>	|2>	CAAGCT	ACCTAG
|2>	|0>	ACCTAG	TGGATC
|2>	|1>	CAAGCT	ACCTAG
|2>	|2>	CAAGCT	CAAGCT

From the truth table, it is found in equations for the sum and carry as –

Sum = $A^0.B^2 + A^1.B^1 + A^2.B^0 + 1. (A^0.B^1 + A^1.B^0 + A^2.B^2)$

Carry = $1. (A^1.B^2 + A^2.B^1 + A^2.B^2)$

To design a Quantum-DNA ternary half adder, quantum decoders will be used so that it is possible to decode the input qubits into the corresponding bits. Except for the two Quantum decoders at the input level, other operations will be conducted using DNA operations.

14.3.1 Multiple-Valued Quantum-DNA Half Adder

For performing Quantum-DNA multi-valued half adder operation, first it is needed to operate two Quantum 1-to-3 ternary decoders. The purpose of using this decoder is to take three inputs. If the 1-to-3 decoders are not used, it will not be possible to access all combinations of these three inputs. Two 1-to-3 decoders will perform nine combinations. These nine Quantum qubit outputs of the Quantum decoder are temporarily stored in Quantum cache memory. Then they will be converted into DNA sequences with the help of NMR relaxation. After the conversion, they act as the inputs of the remaining DNA operations and these combinations then will go through DNA ternary AND operation and DNA ternary OR operations to perform sum and carry. The circuit of the Multi-Valued DNA half adder is displayed in Figure 14.2.

14.3.2 Working Principle of Multiple-Valued Quantum-DNA Half Adder

A 2-input ternary Quantum-DNA half-adder consists of 2 DNA input sequences as input and 2 output qubits, one for Sum and another for Carry.

1. When the input $|A>$, $|B>$ = $|0>$ (0), Quantum decoders generate A0, A1, A2 = (2, 0, 0) and B0, B1, B2 = (2, 0, 0). None of the AND operations (AND1-AND9) are connected to the A0 and B0 line. As a result, none of the AND gate values will be true. So the output of Sum and Carry both are ACCTAG (0).

2. When the input $|A> = |0>$ and $|B> =|1>$, Quantum decoders generate A0, A1, A2 = (2, 0, 0) and B0, B1, B2 = (0, 2, 0). AND4 is connected to the A0 and B1 line. As a result, the AND4 value (TGGATC) is also AND with CAAGCT (1)to produces minimum sum CAAGCT. Here, Carry is ACCTAG (0).

3. When the input $|A> = |0>$ and $|B> =|2>$, Quantum decoders generate A0, A1, A2 = (2, 0, 0) and B0, B1, B2 = (0, 0, 2). AND1 is connected to the A0 and B2 line. AND1 operation values will be true (TGGATC) and produces minimum as sum TGGATC and minimum as carry ACCTAG (0).

4. When the input $|A> = |1>$ and $|B> =|0>$, Quantum decoders generate

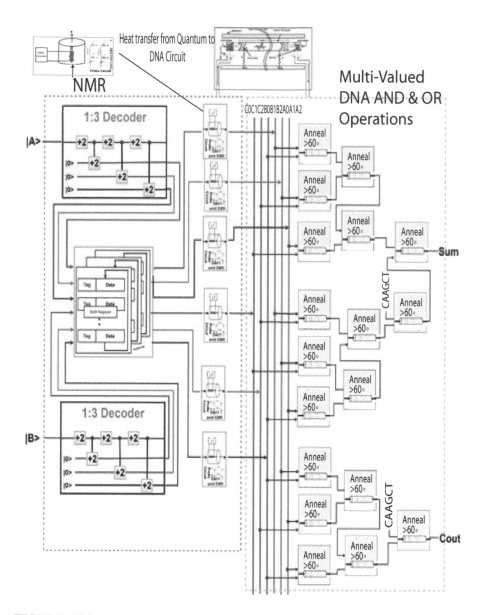

FIGURE 14.2
Circuit Architecture of Multi-Valued Quantum-DNA Half Adder

A0, A1, A2 = (0, 2, 0) and B0, B1, B2 = (2, 0, 0). AND5 operation is connected to A1 and B0 lines. As a result, AND5 output will be true (TGGATC). So AND5 value (TGGATC) is also AND with CAAGCT (1)and produces minimum as sum CAAGCT and carry is ACCTAG (0).

5. When the input |A>, |B> = |1>, Quantum decoders generate A0, A1, A2 = (0, 2, 0) and B0, B1, B2 = (0, 2, 0). DNA decoders generate A0, A1, A2 = (2, 0, 0) and B0, B1, B2 = (0, 0, 2). AND2 is connected to the A1 and B1 line. AND2 operation value will be true (TGGATC) and produce minimum as sum TGGATC and minimum as carry ACCTAG (0).

6. When the input |A> = |1>, and |B> = |2>, Quantum decoders generate A0, A1, A2 = (0, 2, 0) and B0, B1, B2 = (0, 0, 2). AND7 gate is connected to A1 and B2 lines. As a result, AND7 gate value will be true (TGGATC). So, the output of differences is ACCTAG (0) and AND7 value (TGGATC) is also AND with CAAGCT (1) and produces minimum as borrow CAAGCT.

7. When the input |A> = |2>, and |B> = |0>, Quantum decoders generate A0, A1, A2 = (0, 0, 2) and B0, B1, B2 = (2, 0, 0). AND3 is connected to the A2 and B0 line. AND3 operation values will be true (TGGATC) and produce minimum as sum TGGATC and minimum as carry ACCTAG (0).

8. When the input |A> = |2>, and |B> = |1>, Quantum decoders generate A0, A1, A2 = (0, 0, 2) and B0, B1, B2 = (0, 2, 0). AND7 gate is connected to A2 and B1 lines. As a result, AND8 gate value will be true (TGGATC). So, the output of differences is ACCTAG (0) and AND8 value (TGGATC) is also AND with CAAGCT (1)and produces minimum as borrow CAAGCT.

9. When the input |A>, |B> = |2>, Quantum decoders generate A0, A1, A2 = (0, 0, 2) and B0, B1, B2 = (0, 0, 2). AND6 and AND9 operation is connected to A2 and B2 lines. As a result, AND6 output will be true (TGGATC). So AND6 value (TGGATC) is also AND with CAAGCT (1)and produces a minimum sum CAAGCT. AND9 gate value will be true (TGGATC). So AND9 value (TGGATC) is also AND with CAAGCT (1) and produces a minimum as borrow CAAGCT.

14.4   Multiple-Valued Quantum-DNA Full Adder

A Multiple-Valued Quantum-DNA full adder removes the drawback of a Multiple-Valued Quantum-DNA half adder as it includes carry bit in the addition operation. Hence, there are a total of three Quantum inputs |A>, |B>, and |C> and this circuit will generate the output in DNA sequence. The truth table of Quantum-DNA Full Adder is shown in Table 14.2.

TABLE 14.2
Truth Table of Multi-Valued Quantum-DNA
Full Adder

Inputs			Outputs	
$\|A\rangle$	$\|B\rangle$	$\|C\rangle$	Carry	Sum
$\|0\rangle$	$\|0\rangle$	$\|0\rangle$	ACCTAG	ACCTAG
$\|0\rangle$	$\|1\rangle$	$\|0\rangle$	ACCTAG	CAAGCT
$\|0\rangle$	$\|2\rangle$	$\|0\rangle$	ACCTAG	TGGATC
$\|1\rangle$	$\|0\rangle$	$\|0\rangle$	ACCTAG	CAAGCT
$\|1\rangle$	$\|1\rangle$	$\|0\rangle$	ACCTAG	TGGATC
$\|1\rangle$	$\|2\rangle$	$\|0\rangle$	CAAGCT	ACCTAG
$\|2\rangle$	$\|0\rangle$	$\|0\rangle$	ACCTAG	TGGATC
$\|2\rangle$	$\|1\rangle$	$\|0\rangle$	CAAGCT	ACCTAG
$\|2\rangle$	$\|2\rangle$	$\|0\rangle$	CAAGCT	CAAGCT
$\|0\rangle$	$\|0\rangle$	$\|1\rangle$	ACCTAG	CAAGCT
$\|0\rangle$	$\|1\rangle$	$\|1\rangle$	ACCTAG	TGGATC
$\|0\rangle$	$\|2\rangle$	$\|1\rangle$	CAAGCT	ACCTAG
$\|1\rangle$	$\|0\rangle$	$\|1\rangle$	ACCTAG	TGGATC
$\|1\rangle$	$\|1\rangle$	$\|1\rangle$	CAAGCT	ACCTAG
$\|1\rangle$	$\|2\rangle$	$\|1\rangle$	CAAGCT	CAAGCT
$\|2\rangle$	$\|0\rangle$	$\|1\rangle$	CAAGCT	ACCTAG
$\|2\rangle$	$\|1\rangle$	$\|1\rangle$	CAAGCT	CAAGCT
$\|2\rangle$	$\|2\rangle$	$\|1\rangle$	CAAGCT	TGGATC

It is known that

Sum = A2.B0.C0 +A1.B0.C1 + A1.B1.C0+ A0.B1.C1 +A0.B2.C0 + A2.B2.C1 +
1. (A1.B0.C0 + A0.B0.C1 + A0.B1.C0 + A2.B1.C1 + A2.B2.C0 + A1.B2.C1)
Carry = 1 (A0.B2.C0+A2.B1.C0+A2.B2.C0 +A1.B1.C1+B2.C1+A2.C1)
  = 1. ((A0.B2+ A2.B1+A2B2).C0 + (A1.B1+ B2+A2).C1)

14.4.1 Circuit Architecture of Multiple-Valued Quantum-DNA Full Adder

The ternary decoder produces 3 outputs line for each input value. A 3-bit Full adder (A, B, C) is used. Quantum Decoder for input $|A\rangle$, produce A0, A1, A2, for input $|B\rangle$ produce B0, B1, B2 and for input $|C\rangle$ produce C0, C1 (Figure 8.4.6(a)). Only one output line is true, $|2\rangle$ and the other remains false, $|0\rangle$ for an input. The outputs of these decoders are temporarily stored in Quantum cache memory. Then they convert into DNA sequences with the help of NMR relaxation. After the conversion, they act as the inputs of the remaining DNA operations

From the truth table it is seen that, the sum of the full adder produces six true TGGATC (2) values and six TGGATC values. For these six true values, six ternary AND gate (TAND) are designed and each of the DNA AND gate outputs will ternary

OR to produce the true output when one of the outputs of the ternary AND gate is true. There are also six 1 values in sum. For these six 1 values, six ternary AND gate (DNA AND) are designed and each of the DNA AND gate output will DNA OR to produce the true, TGGATC (2) output when one of the outputs of the DNA AND gate is true. But the output CAAGCT (1) is needed, so the DNA AND operation will be performed between the DNA OR gate output and CAAGCT (1). DNA AND gate produce the minimum value. So, the output of the DNA AND gate will always CAAGCT (1). Finally, a DNA OR gate is used to generate the maximum output among TGGATC (2), ACCTAG (0) and ACCTAG (0), CAAGCT (1) (Figure 14.3).

From the truth table, carry of the full adder produces nine CAAGCT (1) values. After minimization, the carry function produces nine CAAGCT (1) values and one TGGATC (2) value. Nine ternary AND gates (DNA AND) are designed and each of the DNA AND gate outputs will DNA OR to produce the true (TGGATC) output when one of the outputs of the DNA AND gate is true. But as it is needed the output CAAGCT (1), DNA AND operation is performed between the DNA OR gate output with CAAGCT (1). TAND gate produces the minimum value. So, the output of the DNA AND gate will always CAAGCT (1) for one of the true inputs among nine DNA AND gate. Again, for one TGGATC (2) value, one DNA AND gate is used and the output of this DNA AND will perform DNA OR produce the carry (Figure 14.4).

So, Figure 14.3 shows the circuit architecture of multi-valued Quantum-DNA Full Adder for Sum and Figure 14.4 shows the circuit architecture of multi-valued Quantum-DNA Full Adder for Carry.

14.4.2 Working Principle of Multiple-Valued Quantum-DNA Full Adder

A ternary Quantum-DNA Full-Adder contains 3 Quantum input qubits as input and 2 output DNA sequences, one for sum and another for carry. To understand the working principle of Quantum-DNA Full Adder, let's consider the following case.

When the input $|A>$, $|B> = |0>$, and $C = |1>$, Quantum decoder generates $|A0>$, $|A1>$, $|A2> = (2, 0, 0)$, $|B0>$, $|B1>$, $|B2> = (2, 0, 0)$ and $|C0>$, $|C1>$, $|C2> = (0, 2, 0)$. The 13th AND gate are connected to the $|A0>$ and $|B0>$ and $|C1>$ line. As a result, the 7th AND gate value, that evaluates A0.B0.C1, will be true (TGGATC). It is seen that the 13th AND gate is connected to the OR gate, so the output of the OR gate will be true (TGGATC). This true value (TGGATC) is also AND with CAAGCT (1)and produces minimum CAAGCT (1) as output. This output will go through the OR gate and generate CAAGCT (1) as sum.

In the full-adder carry circuit, it is seen that none of the AND operations and OR operations are connected to $|A0>$, $|B0>$ and $|C1>$ lines. As a result, none of the AND operations and OR operations value will be true. Thus, they produce ACCTAG (0). This truth value will TAND with CAAGCT (1), and produce minimum ACCTAG (0) as output. This output will go through the OR gate and generate ACCTAG (0) as carry.

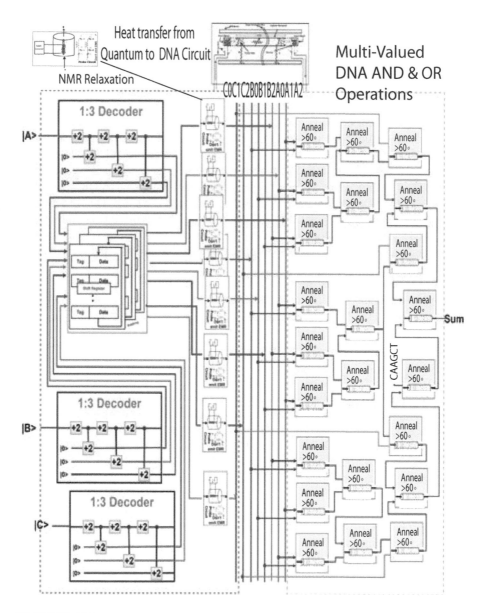

FIGURE 14.3
Circuit Architecture of Multi-Valued Quantum-DNA Full Adder (Sum)

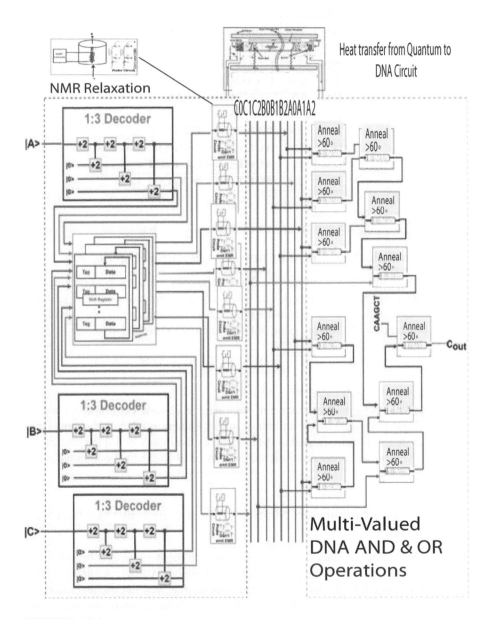

FIGURE 14.4
Architecture of Multi-Valued Quantum-DNA Full Adder (Carry)

TABLE 14.3
Truth Table of Quantum-DNA Ternary
Half Subtractor

Inputs		Outputs	
$|A>$	$|B>$	**Borrow**	**D**
$|0>$	$|0>$	ACCTAG	ACCTAG
$|0>$	$|1>$	CAAGCT	TGGATC
$|0>$	$|2>$	CAAGCT	CAAGCT
$|1>$	$|0>$	ACCTAG	CAAGCT
$|1>$	$|1>$	ACCTAG	ACCTAG
$|1>$	$|2>$	CAAGCT	TGGATC
$|2>$	$|0>$	ACCTAG	TGGATC
$|2>$	$|1>$	ACCTAG	CAAGCT
$|2>$	$|2>$	ACCTAG	ACCTAG

14.5   Multiple-Valued Quantum-DNA Half Subtractor

A ternary half subtractor is a type of subtractor, an electronic circuit that performs the subtractions of ternary numbers. The Quantum-Quantum half subtractor can subtract two single ternary digits which are provided as Quantum qubit and provide the output, difference and a borrow value in DNA sequence representation.
The difference and borrow are –

   Difference $= A^0.B^1 + A^1.B^2 + A^2.B^0 + 1.(A^0.B^2 + A^1.B^0 + A^2.B^1)$

   Borrow $= 1.(A^0.B^1 + A^0.B^2 + A^1.B^2)$

The general design of a ternary half subtractor is already explained in chapters 11 and 13. However, the major difference in this Quantum-DNA half subtractor is that the quantum decoder is used at the starting as the inputs as a qubit and except that the other operations are designed using DNA gates as the final output will be in DNA sequences.

14.5.1   Circuit Architecture Multiple-Valued Quantum-DNA Half Subtractor

The ternary decoder produces 3 outputs line for each input value. A 2-bit subtractor (A, B) is used. Quantum decoder for input $|A>$ produces A0, A1, A2 and for input $|B>$ produces B0, B1, B2. Only one output line is true, $|2>$ and the other remains false, $|0>$ for an input (Figure 14.5). The outputs of these decoders are temporarily stored in Quantum cache memory. Then they convert into DNA sequences with the help of NMR relaxation. After the conversion, they act as the inputs of the remaining DNA operations

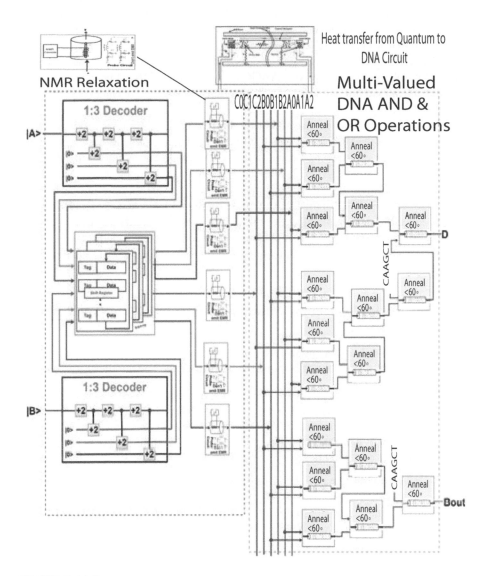

FIGURE 14.5
Circuit Architecture of Multi-Valued Quantum-DNA Half Subtractor

From the truth table it is seen that the difference of the subtractor produces three true, TGGATC (2), values and three CAAGCT (1) values. For these three true values, three ternary AND gates (DNA AND) are designed and each of the DNA AND gate outputs will DNA OR to produce the true output when one of the outputs of the DNA AND gate is true. There are also three CAAGCT (1) values in difference. For these three CAAGCT (1) values, three ternary AND gates (DNA AND) are designed and each of the DNA AND gate output will TOR to produce the true, TGGATC (2) output when one of the outputs of the TAND gate is true. But as to produce the output CAAGCT (1), so TAND operation is performed between the TOR gate output and CAAGCT (1). DNA AND gate produce the minimum value. So, the output of the DNA AND gate will always CAAGCT (1). Finally, a DNA OR gate is used to generate the maximum output among TGGATC (2), ACCTAG (0) and ACCTAG (0), CAAGCT (1).

From the truth table, it is seen that the borrow of the subtractor produces three CAAGCT (1) values. For these three CAAGCT (1) values, three ternary AND gates (DNA AND) are designed and each of the DNA AND gate outputs will DNA OR to produce the true, TGGATC (2) output when one of the outputs of the DNA AND gate is true. But as to produce the output CAAGCT (1), DNA AND operation is needed to perform between the DNA OR gate output with CAAGCT (1). DNA AND gate produce the minimum value. So, the output of the DNA AND gate will always CAAGCT (1) for one of the true inputs among three DNA AND gate.

Figure 14.5 shows the circuit architecture of multi-valued Quantum-DNA Half Subtractor.

14.5.2 Working Principle of Multiple-Valued Quantum-DNA Half Subtractor

A 2-input multi-valued Quantum-DNA half-subtractor contains 2 Quantum bits as input and 2 DNA sequences as outputs, one for difference and another for borrow.

1. When the input qubit $|A\rangle$, $|B\rangle = |0\rangle$, quantum decoders generate A0, A1, A2 = (2, 0, 0) and B0, B1, B2 = (2, 0, 0). None of the AND gates (D0-D8) are connected to the A0 and B0 line. As a result, none of the AND gate value will be true. So, the output of differences and borrow both are ACCTAG (0).

2. When the input qubit $|A\rangle = |0\rangle$ and $|B\rangle = |1\rangle$, quantum decoders generate A0, A1, A2 = (2, 0, 0) and B0, B1, B2 = (0, 2, 0). D0 and D6 AND gates are connected to the A0 and B1 line. As a result, D0 and D6 AND gates value will be true (TGGATC). So, the output of differences is TGGATC and D6 value (TGGATC) is also AND with CAAGCT (1) and produces a minimum as borrow CAAGCT.

3. When the input qubit $|A\rangle = |0\rangle$ and $|B\rangle = |2\rangle$, quantum decoders generate A0, A1, A2 = (2, 0, 0) and B0, B1, B2 = (0, 0, 2). D3 and D7 AND gates are connected to the A0 and B2 line. As a result, D3 and D7 AND gates values will be true (TGGATC). So D3 value (TGGATC) is also AND with CAAGCT (1) and produces minimum as difference CAAGCT and D7 value (TGGATC) is also AND with CAAGCT (1) and produces minimum as borrow CAAGCT.

4. When the input qubit $|A\rangle = |1\rangle$ and $|B\rangle = |0\rangle$, quantum decoders generate A0, A1, A2 = (0, 2, 0) and B0, B1, B2 = (2, 0, 0). D4 gate is

connected to A1 and B0 line. As a result, D4 AND gate value will be true (TGGATC). So D4 value (TGGATC) is also AND with CAAGCT (1) and produces minimum as difference CAAGCT and borrow is ACCTAG (0).

5. When the input qubit |A>, |B> = |1>, quantum decoders generate A0, A1, A2 = (0, 2, 0) and B0, B1, B2 = (0, 2, 0). None of the AND gates (D0-D8) are connected to A1 and B1 lines. As a result, none of the AND gate value will be true. So, the output of differences and borrow both are ACCTAG (0).

6. When the input qubit |A> = |1>, and |B> = |2>, quantum decoders generate A0, A1, A2 = (0, 2, 0) and B0, B1, B2 = (0, 0, 2). D1 and D8 AND gates are connected to A1 and B2 lines. As a result, D1 and D8 AND gates values will be true (TGGATC). So the output of differences is TGGATC and D8 value (TGGATC) is also AND with CAAGCT (1) and produces minimum as borrow CAAGCT.

7. When the input qubit |A> = |2>, and |B> = |0>, quantum decoders generate A0, A1, A2 = (0, 0, 2) and B0, B1, B2 = (2, 0, 0). D2 gate is connected to A1 and B0 line. As a result, D2 AND gate value will be true (TGGATC). So, the output of differences is TGGATC and borrow is ACCTAG (0).

8. When the input qubit |A> =|2>, and |B> = |1>, quantum decoders generate A0, A1, A2 = (0, 0, 2) and B0, B1, B2 = (0, 2, 0). The D5 gate is connected to the A2 and B1 line. As a result, D5 AND gate value will be true (TGGATC). So D5 value (TGGATC) is also AND with CAAGCT (1) and produces minimum as difference CAAGCT and borrow is ACCTAG (0).

9. When the input qubit |A>, |B> = |2>, quantum decoders generate A0, A1, A2 (0, 0, 2) and B0, B1, B2 = (0, 0, 2). None of the AND gates (D0-D8) are connected to the A2 and B2 lines. As a result, none of the AND gate values will be true. So, the output of differences and borrow both are ACCTAG (0).

14.6 Multiple-Valued Quantum-DNA Full Subtractor

Multi-Valued full subtractor performs subtraction operation taking borrow qubit in consideration, thus it solves the problem that half subtractor cannot overcome. In a Multi-Valued full subtractor, there are a total of three Quantum inputs |A>, |B>, and |C>. The truth Table of Multi-Valued Quantum-DNA full subtractor is given in Table 14.4.

From the truth table, the equations for the difference and borrow as −

$D = A0.B1.C0 + A1.B2.C0 + A2.B0.C0 + A0.B0.C1 + A1.B1.C1 + A2.B2.C1. 1.$
$(A0.B2.C0 + A1.B0.C0 + A2.B1.C0 + A0.B1.C1 + A1.B2.C1 + A2.B0.C1)$

TABLE 14.4

Truth Table of Multi-Valued Quantum-DNA
Full Subtractor

Inputs			Outputs	
|A>	|B>	|C>	D	Bout
|0>	|0>	|0>	ACCTAG	ACCTAG
|0>	|1>	|0>	TGGATC	CAAGCT
|0>	|2>	|0>	CAAGCT	CAAGCT
|1>	|0>	|0>	CAAGCT	ACCTAG
|1>	|1>	|0>	ACCTAG	ACCTAG
|1>	|2>	|0>	TGGATC	CAAGCT
|2>	|0>	|0>	TGGATC	ACCTAG
|2>	|1>	|0>	CAAGCT	ACCTAG
|2>	|2>	|0>	ACCTAG	ACCTAG
|0>	|0>	|1>	TGGATC	CAAGCT
|0>	|1>	|1>	CAAGCT	CAAGCT
|0>	|2>	|1>	ACCTAG	CAAGCT
|1>	|0>	|1>	ACCTAG	ACCTAG
|1>	|1>	|1>	TGGATC	CAAGCT
|1>	|2>	|1>	CAAGCT	CAAGCT
|2>	|0>	|1>	CAAGCT	ACCTAG
|2>	|1>	|1>	ACCTAG	ACCTAG
|2>	|2>	|1>	TGGATC	CAAGCT

$= (A0.B1 + A1.B2 + A2.B0).C0 + (A0.B0 + A1.B1 + A2.B2).C1. 1. ((A0.B2 + A1.B0 + A2.B1).C0 + (A0.B1 + A1.B2 + A2.B0).C1)$

B_{out} $= 1. (A0. B1. C0+A0.B2.C0 + A1.B2.C0 + A0.C1 +B2.C1+A1.B1.C1)$
$= 1. ((A0. B1+A0.B2+A1.B2).C0 + (A0+B2+A1.B1).C1)$

The main difference in this Quantum-DNA full subtractor is that a Quantum Decoder is used at the beginning because the inputs will be in Quantum qubit, and the other operations will be designed using DNA gates because the final output will be in a DNA sequence.

14.6.1 Circuit Architecture of Multiple-Valued Quantum-DNA Full Subtractor

The ternary decoder produces 3 outputs line for each input value. The 3-bit Full subtractor (A, B, C) is used. Quantum Decoder for input |A> produce A0, A1, A2, for input |B> produce B0, B1, B2 and for input |C> produce C0, C1. Only one output line is true, |2> and the other remains false, |0> for an input. The outputs of these decoders are temporarily stored in Quantum cache memory. Then they convert into DNA sequences with the help of NMR relaxation. After the conversion, they act as the inputs of the remaining DNA operations.

From the truth table, it is seen that the difference of the full subtractor produces six true, TGGATC (2) values and six CAAGCT (1) values. For these six true values, six ternary AND gate (DNA AND) are designed and each of the DNA AND gate output will DNA OR to produce the true output when one of the outputs of the DNA AND gate is true. There are also six CAAGCT (1) values in difference. For these six CAAGCT (1) values, six ternary AND gates (DNA AND) are designed and each of the DNA AND gate output will DNA OR to produce the true, TGGATC (2) output when one of the outputs of the DNA AND gate is true (TGGATC). But as the output CAAGCT (1) is needed, DNA AND operation are performed between the DNA OR gate output with CAAGCT (1). DNA AND gate produce the minimum value. So, the output of the DNA AND gate will always CAAGCT (1). Finally, a DNA OR gate is used to generate the maximum output among TGGATC (2), ACCTAG (0) and ACCTAG (0), CAAGCT (1) (Figure 14.6).

From the truth table, it is seen that the borrow of the full subtractor produces eight CAAGCT (1) values. After minimization using the k map, the carry function produces six CAAGCT (1) values. Six ternary AND gate (DNA AND) are designed and each of the DNA AND gate outputs will DNA OR to produce the true, TGGATC (2) output when one of the outputs of the DNA AND gate is true. But as the output CAAGCT (1) is needed, it is needed to perform DNA AND operation between the DNA OR gate output with CAAGCT (1). DNA AND gate produce the minimum value. So, the output of the DNA AND gate will always CAAGCT (1) for one of the true inputs among six DNA AND gate (Figure 14.7).

Figure 14.6 shows the circuit architecture of multi-valued Quantum-DNA Full Subtractor (Difference) and Figure 14.7 shows the circuit architecture of multi-valued Quantum-DNA Full Subtractor (Borrow).

14.6.2 Working Principle of Multiple-Valued Quantum-DNA Full Subtractor

A ternary Quantum-DNA Full-Subtractor contains 3 quantum input qubits as input and 2 output DNA sequences one for sum and another for carry. To understand the working principle of Quantum-DNA Full Subtractor, let's consider the following case:

When the input $|A>$, $|B> = |0>$, and C = $|1>$, Quantum decoder generates $|A0>$, $|A1>$, $|A2> = (2, 0, 0)$, $|B0>$, $|B1>$, $|B2> = (2, 0, 0)$ and $|C0>$, $|C1>$, $|C2> = (0, 2, 0)$. The 10th AND gate are connected to the $|A0>$ and $|B0>$ and $|C1>$ line. As a result, the 10th AND gate value, that evaluates A0.B0.C1, will be true (TGGATC). It is seen that the 13th AND gate is connected to the OR gate, so the output of the OR gate will be true (TGGATC). This output will go through the last OR gate and generate TGGATC (2) as difference (D).

In the full-subtractor borrow circuit, it is seen that none of the AND operations and OR operations are connected to $|A0>$, $|B0>$ and $|C1>$ lines. As a result, none of the AND operations and OR operation values will be true. Thus, they produce ACCTAG (0). This truth value will TAND with CAAGCT (1), and produce

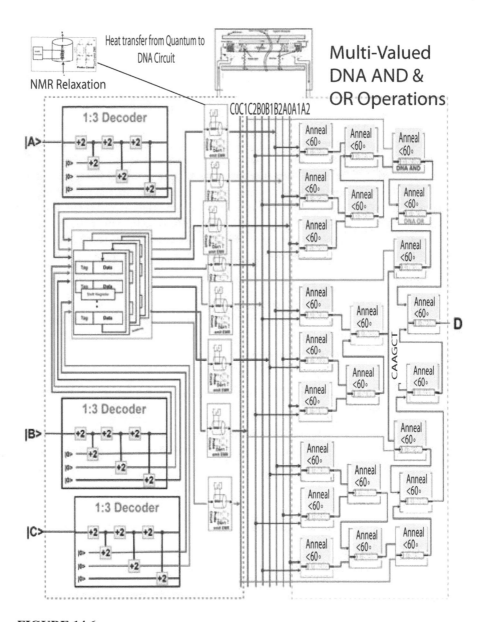

FIGURE 14.6

Circuit Architecture of Multi-Valued Quantum-DNA Full Subtractor (Difference)

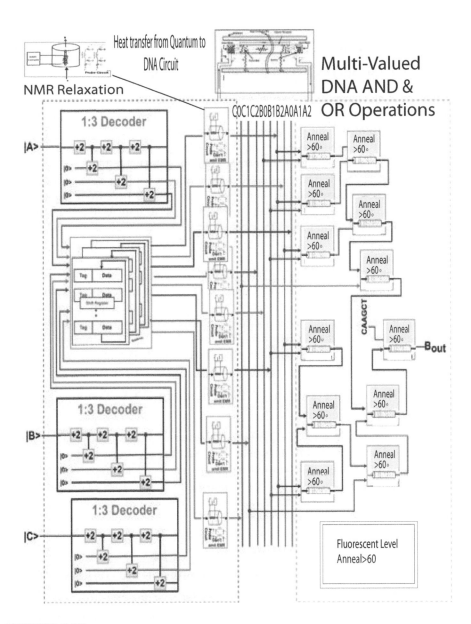

FIGURE 14.7
Circuit Architecture of Multi-Valued Quantum-DNA Full Subtractor (Borrow)

minimum ACCTAG (0) as output. This output will go through the OR gate and generate ACCTAG (0) as borrow (Bout).

14.7   Quantum-DNA Ternary Parallel Adder

The general organizations of Quantum-DNA 4-qutrit ternary parallel adder is given which is shown in Figure 14.8. The Figure shows that three ternary full-adder operations are performed in the quantum system and the last ternary full-adder operation is performed in the DNA system. A quantum ternary cache memory is used to store the qutrit output and the qutrit input for the last full-adder operation.

The qubits are converted into the DNA base sequences using the data conversion circuits. Besides, a heat transfer circuit is used to transfer heat from the quantum system to a cooling system. And again, it is possible to transfer the required heat from the quantum system to the DNA system as well.

Figure 14.9 shows the general organizations of Quantum-DNA ternary N-qutrit parallel adder.

The Figure 14.9 shows, N-1 ternary full-adder operations are performed in the quantum system and the last ternary full-adder operation is performed in the DNA system. A quantum ternary cache memory is used to store the qutrit output and the qutrit input for the last full-adder operation. The qutrits are converted into the DNA base sequences using the data conversion circuits. Besides, a heat transfer circuit is used to transfer heat from the quantum system to a cooling system.

14.7.1   The Circuit Design of the Quantum-DNA Ternary Parallel Adder

The general organization of the Quantum-DNA ternary 4-qutrit and N-qutrit parallel adder is shown in Figures 14.8 and 14.9. And the architecture contains quantum ternary full-adders, quantum ternary cache memory, data conversion circuits, DNA ternary full adders, and the heat conductance circuit. Every component was constructed earlier. The quantum ternary cache memory has been designed in the previous section of this chapter. NMR relaxation or trap ions can be used as a data conversion circuit to convert the qutrits into the equivalent DNA sequences.

14.7.2   The Working Principles of Quantum-DNA Ternary Parallel Adder

There is nothing new to understanding the working procedure. All its needed to do is to combine two systems. In the 1st three quantum ternary full-adders in the quantum system, the input and output mapping will be according to the quantum ternary parallel adder operations. Then the NMR relaxation or trap ions will receive the qutrits from a quantum ternary cache memory. It will convert the qutrits to the equivalent

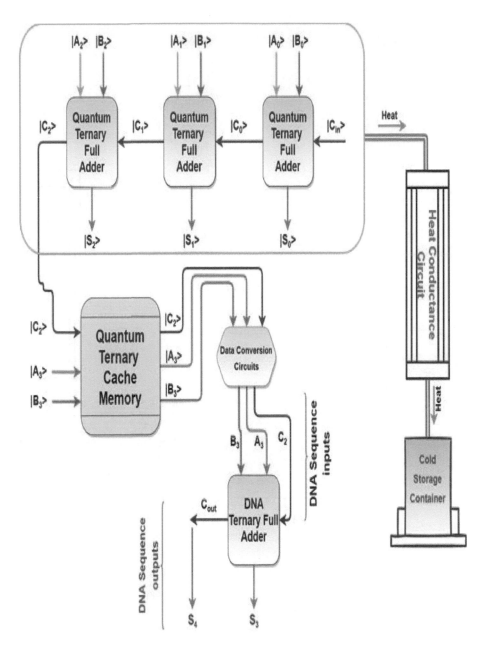

FIGURE 14.8
The General Organizations of Quantum-DNA Ternary 4-Qutrit Parallel Adder

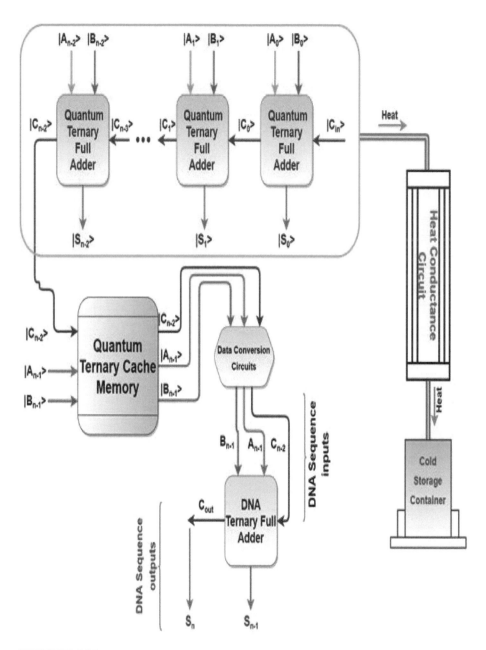

FIGURE 14.9
The General Organizations of Quantum-DNA Ternary N-Qutrit Parallel Adder

DNA sequences. Then the converted DNA sequences will be the inputs to the DNA ternary full-adder. And eventually, the DNA ternary full-adder will generate the output of the last ternary full-adder operation.

Let, the addend and augend qutrits are $|A>$ and $|B>$, respectively. Then, for a 4-qutrit input the input $|A>$ will be $|A_3A_2A_1A_0>$ and the input $|B>$ will be $|B_3B_2B_1B_0>$. Let's take some value for those inputs and observe the behavior of the circuit.

Suppose $|A> = |1020>$ and $|B> = |1100>$. Therefore, the values of $|A_0> = |0>$, $|A_1> = |2>$, $|A_2> = |0>$, and $|A_3> = |1>$. And the values of $|B_0> = |0>$, $|B_1> = |0>$, $|B_2> = |1>$, and $|B_3> = |1>$.

1. For the inputs $|A_0>$, and $|B_0>$ both is $|0>$. And the carry input qubit $|C_{in}>$ is also $|0>$ - The first quantum ternary full-adder will get all the inputs as $|0>$. Therefore, it will produce output $|0>$ as the sum and also $|0>$ as the carry output qubit. This carry will be working as the carry input of the next quantum ternary full-adder.

2. The next input qutrits to the second quantum full-adder are $|A_1> = |2>$, $|B_1> = |0>$, and the carry input $|C_0>$ is also $|0>$. Therefore, the second quantum ternary full-adder will generate $|2>$ as the sum output and again $|0>$ as the carry output of $|C_1>$. The value of $|C_1>$ will be delivered to the next quantum ternary full-adder as a carry input.

3. The input to the third quantum ternary full-adder are $|A_2> = |0>$, $|B_1> = |1>$, and the carry input $|C_0>$ is again $|0>$. Therefore, the third quantum ternary full-adder will generate $|1>$ as the sum output and also again $|0>$ as the carry output of $|C_2>$.

4. Now it has done with the quantum part. The carry output of the very last quantum ternary full-adder operation will be stored in a quantum ternary cache memory. And the next addend and augend qutrits are also stored in the quantum ternary cache memory.

5. From cache memory, the values $|A_3> = |1>$, $|B_3> = |1>$, and $|C_2> = |0>$ will be delivered to the NMR relaxation or trap ions. From here the qutrit values will be converted into the equivalent DNA sequences. The converted values will be $A_3 = CAAGCT$, $B_3 = CAAGCT$, and $C_2 = ACCTAG$. These values will be the inputs to the last ternary full-adder which is in DNA system.

6. The final ternary full-adder is in the DNA system. It will receive the value as $A_3 = CAAGCT$, $B_3 = CAAGCT$, and $C_2 = ACCTAG$. Therefore it will generate the output as TGGATC (i.e. 2 in ternary number) for the sum, and ACCTAG for the last carry output which will be the most significant qutrit of the addition result eventually.

Therefore, the expected result is obtained from the designed architecture for the Quantum-DNA ternary parallel adder.

14.8   Quantum-DNA Ternary Carry-Lookahead Adder

To develop Quantum-DNA carry-lookahead adder by dividing the whole system into two segments – one for the quantum system and the other for the DNA system, as before, this can be done in two ways. Dividing the system for N-qutrit into (N-1) qutrit in the quantum system and the last one qutrit in the DNA system. Or (N-X) qutrit in the quantum system and X qutrit in the DNA system. Again let us see how to construct it in the 1st way which is much simpler. Figure 14.10 shows the general organization of a 4-qutrit Quantum-DNA ternary carry-lookahead adder.

Here, among 4-qutrit input addition operations, the first 3 operations are occurring in the quantum system, and the last bit addition operation is occurring in the DNA system. The quantum ternary cache memory store the qutrit values of the most significant qutrits of the inputs and the carry value obtained by the combinational operations. The heat transfer circuit transfers the heat from the quantum system, and the DNA system can get the necessary heat from the quantum system.

14.8.1   The Circuit Architecture of Quantum-DNA Ternary Carry-Lookahead Adder

Figure 14.11 shows the circuit diagram of a Quantum-DNA 3-qutrit ternary carry-lookahead adder.

The circuit diagram is almost familiar except for the added quantum ternary cache memory as an intermediary quantum qutrit storage device, NMR relaxation at zero Kelvin temperature for data conversion from quantum qutrit to DNA ternary value equivalent sequence, and the heat conductance circuit to transfer heat from the quantum system.

The working mechanism of all those systems was explained in previous sections and chapters. Figure 14.11 depicted to understand the working procedures of the 3-qutrit Quantum-DNA ternary carry-lookahead adder.

14.8.2   The Working Principles of Quantum-DNA Ternary Carry-Lookahead Adder

The working procedures of the first part of the Quantum-DNA system which is the quantum ternary system are the same as those learned in the working procedures of the quantum ternary carry-lookahead adder. As the circuit diagram shows, first 2-qutrit addition and the carry input qutrit for the last addition operation is occurring in the quantum system. Then carry qutrit along with the other two input qutrits stored in the quantum ternary cache memory. From the cache memory, they are passing through the NMR relaxation process to become the input of the DNA ternary full-adder when the DNA system is ready. Then the DNA system will perform the DNA ternary full-adder operation and will generate the final sum outputs.

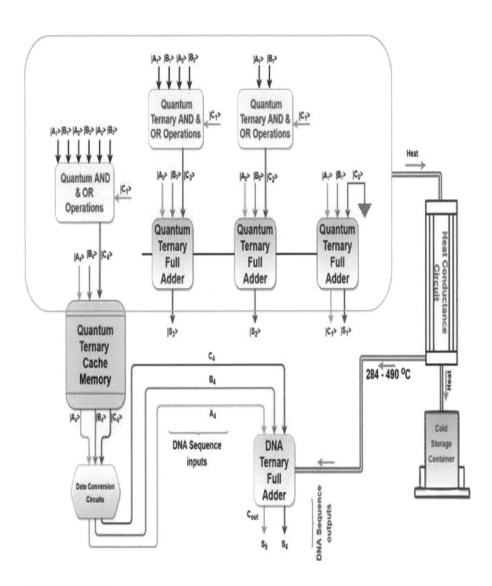

FIGURE 14.10
The General Organizations of Quantum-DNA Ternary 4-Qutrit Carry-Lookahead Adder

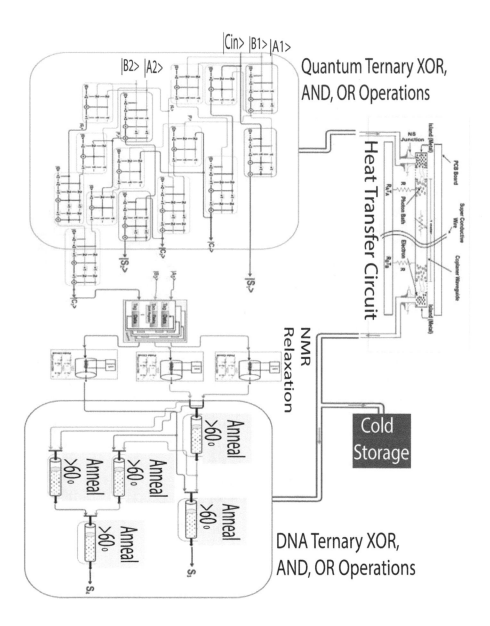

FIGURE 14.11
The Circuit Diagram of Quantum-DNA Ternary 3-Qutrit Carry-Lookahead Adder
Operation

14.9 Quantum-DNA Ternary Carry-Skip Adder

In the ternary carry-skip adder, a block of ternary full-adders is created, and then determine the block propagates before operating the actual ternary full-adder operations. If the block propagates BP is equal |1> or |2> (for quantum computing in ternary logic), then the output carry will be the same as the carry input given to the block.

Here, block propagate,

$BP_i = |P_0> . |P_1> . |P_2> . .... . |P_i>$

Where, $|P_i> = (A_i$ **QTXOR** $B_i)$

This is how to reduce the time delay to propagate the carry from one block to another. For more explanation, chapter 11 and chapter 13 are recommended again to observe the architecture of both quantum ternary carry-skip adder and DNA ternary carry-skip adder properly. The ternary carry-skip adder is exactly the same as the ternary parallel adder when the BP is not equal |1> or |2>. Therefore, it is needed to design n number of ternary full-adders of an n-qutrit block. And if BP is equal |1> or |2>, it is possible to bypass the value of the carry input qutrit of the block to the carry output using a 3-to-1 ternary multiplexer.

14.9.1 The General Organizations of Quantum-DNA Ternary Carry-Skip Adder

The general overview of the circuit diagram for the Quantum-DNA ternary carry-skip adder is depicted in Figure 14.12.

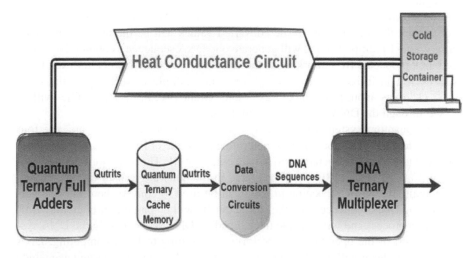

FIGURE 14.12
The Front-View of the Quantum-DNA Ternary Carry-Skip Adder

From the above Figure 14.12, it is easy to understand that the Quantum-DNA ternary carry-skip adder is going to be built by dividing it into two parts – one part is in the quantum system which consists of connected quantum ternary parallel adders, and the second part is in the DNA system which consists of a 3x1 DNA ternary multiplexer that will bypass the carry output of the block.

And a heat transfer circuit is connected to the quantum system to transfer the massive produced heat into the cooler. Moreover, it is needed to transfer the necessary heat to the DNA system to operate the DNA system.

Figure 14.13 shows another front view of a 4-qutrit Quantum-DNA ternary carry-skip adder. Where the quantum system includes the four quantum ternary full-adders. The first carry input to the block and carry output of the block are stored in a quantum ternary cache memory so that they can be passed to the next operations.

The data conversion unit converts the data and delivers the DNA sequences to the DNA system which is a DNA 3-to-1 ternary multiplexer as the inputs. And the DNA multiplexer does the rest of the operations. It bypasses the carry output sequence based on the selection input. And as it is already known, it passes the carry input from the block to the next block only if the value of the BP (block propagate) is either $|1>$ or $|2>$. Otherwise, the ternary 3-to-1 multiplexer will bypass the last ternary full adder's carry output as the block's carry output.

14.9.2   The Circuit Architecture of Quantum-DNA Ternary Carry-Skip Adder

Figure 14.14 shows an inside view of the Quantum-DNA ternary 4-qutrit carry-skip adder operation. The value of P_i is determined first, and the block propagate is determined by the quantum ternary AND operations accordingly. The value of block propagates BP, and the two carry qutrits $|C_0>$ and $|C_4>$ (as the Figure 14.14 shows) will be stored to the quantum ternary cache memory. And the data conversion unit will convert them accordingly. The converted values will be the input of the DNA ternary 3-to-1 multiplexer and will generate one DNA sequence as output. Note that, the block propagate BP is working as the select input of the DNA multiplexer. If BP is CAAGCT or TGGATC (i.e. 1 or 2), then the multiplexer will bypass the value of C_0 (1st carry input to the block) as output. Otherwise, the value of $|C_4>$ will be passed through the DNA ternary 3-to-1 multiplexer.

The Quantum system will contain four quantum ternary full-adders, the four quantum ternary XOR operations to determine the value of P_i. It will also have three quantum ternary AND operations to get the output of BP. The three qutrit values (values of $|C_0>$, $|C_4>$, and $|BP>$) will be stored in the quantum ternary cache memory. Here again, the NMR relaxation process can be used to convert them into the corresponding DNA sequences. So, the stored qutrits will be passing through the NMR relaxation process and will become the equivalent DNA sequences. In the DNA system, the rest of the operations are performed according to DNA computing. DNA ternary multiplexer will pass the output according to the select input sequence which

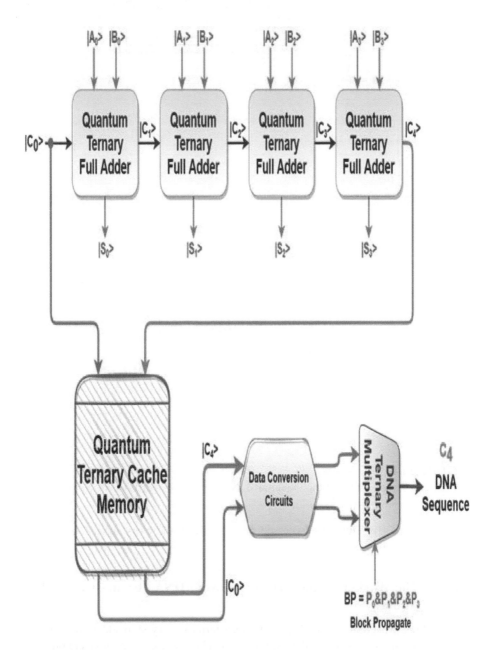

FIGURE 14.13
The Front-View of the Quantum-DNA Ternary 4-Qutrit Carry-Skip Adder

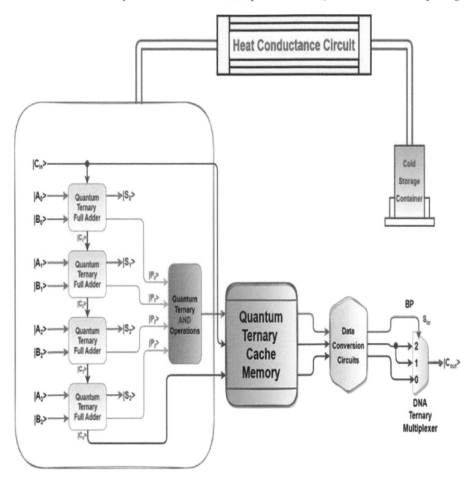

FIGURE 14.14
The Circuit Architecture of the Quantum-DNA Ternary 4-Qutrit Carry-Skip Adder

is the value of BP. Thus, the bypassed output will be the carry input for the next cascaded block if existed.

14.9.3 The Working Principles of Quantum-DNA Ternary Carry-Skip Adder

The working procedures of the Quantum-DNA ternary carry-skip adder include four main units: quantum system, data storing to cache memory, data conversion, and the DNA system. The operations in the quantum ternary system were discussed in chapter 11. Besides, the working procedures of the DNA circuit have been discussed already in chapter 13. Now it's easy to understand the working principle of quantum-DNA circuits with some of the input sets |A> and |B>.

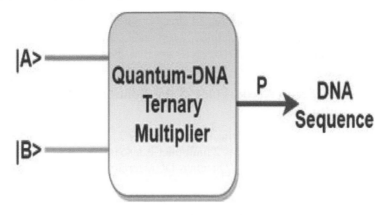

FIGURE 14.15
The General Working Block of a Quantum-DNA Ternary Multiplier

14.10 Quantum-DNA Ternary Multiplier

In this section, the architecture of the Quantum-DNA ternary multiplier operation will be developed. A 2×2 qutrit Quantum-DNA ternary multiplier will be designed. The general working block of the Quantum-DNA ternary multiplier is shown in Figure 14.15.

Figure 14.16 shows the general working block of a Quantum-DNA ternary 2×2 multiplication operation. Where, two inputs $|A\rangle$ and $|B\rangle$ are two-qutrit values in size, and they produce five DNA sequences product values.

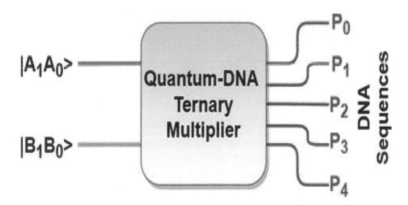

FIGURE 14.16
The General Working Block of a Quantum-DNA Ternary 2x2 Multiplier

The 1-qutrit multiplier is constructed already to implement the ternary multiplication rules. Using the 1-qutrit multiplier, it is possible to design the larger qutrit size ternary multiplication.

14.10.1 The Circuit Architecture of the Quantum-DNA Ternary 2 × 2 Multiplier

In chapter 11, the architecture of the quantum ternary 2 × 2 multiplier is constructed. In chapter 13, the DNA ternary 2 × 2 multiplier is also constructed. So, it is possible to design the architecture of the Quantum-DNA ternary multiplier for two qutrits multiplier operations.

Figure 14.17 shows an architecture of the Quantum-DNA ternary 2 × 2 multiplier. The architecture divides the multiplier into two systems – the Quantum system and the DNA system. Quantum system, which includes four 1-qutrit quantum ternary multiplier operations, and four quantum ternary half-adders. The outputs from this quantum system are stored in the quantum cache memory. And here, trap ions are used as the data conversion units. After converting the qutrits into the equivalent DNA sequences, the DNA sequences are entered into the DNA system. Where the DNA system includes one ternary DNA half-adder and two ternary DNA full-adders. Finally, the outputs of the product are obtained in the DNA base sequence form.

The heat transfer circuit is not included; the total system will be destroyed immediately due to the presence of the massive heat from the quantum system. So let's try another circuit that will include the heat conductance circuit as well.

The complete architecture of a Quantum-DNA ternary 2x2 multiplier is shown in 14.18. The heat conductance circuit is added and through the heat conductance circuit the generated massive heat from the quantum system will be transferred to cold storage. And notice, this time the NMR relaxation process is used to convert the qutrits into the corresponding DNA sequences. To be mentioned, here the NMR relaxation process will be performed at zero kelvin temperature. The NMR relaxation process and its working principle are described in the earlier chapter. It is also known that the working procedure of the heat conductance circuit. And all the other operations were also described earlier.

14.10.2 The Working Procedure of the Quantum-DNA Ternary 2 × 2 Multiplier

The operational behavior of the designed architecture for the Quantum-DNA ternary 2x2 multiplier is shown in Figure 14.17.

Suppose, it is needed to multiply two ternary input values in the Quantum-DNA multiplier, where $|A\rangle = |02\rangle$, and $|B\rangle = |10\rangle$. After multiplication operations these inputs will produce $P = P_4 P_3 P_2 P_1 P_0$, where these product values are in the form of DNA sequences. So, the value of $|A_0\rangle = |2\rangle$, and $|A_1\rangle = |0\rangle$. And the value of $|B_0\rangle = |0\rangle$, and $|B_1\rangle = |1\rangle$. The operations will be described from right to left of Figure 14.17.

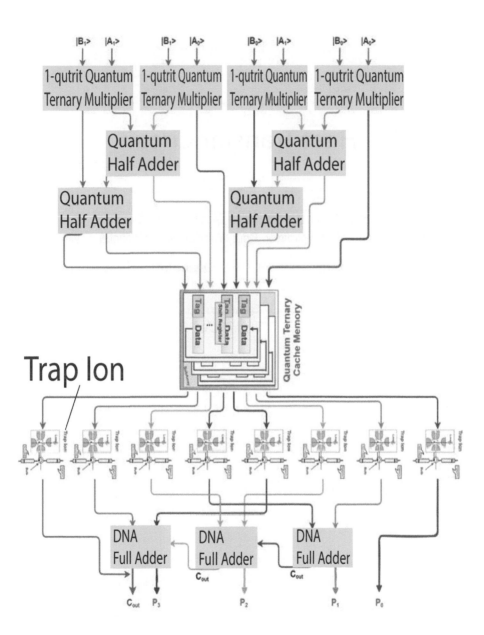

FIGURE 14.17
The Architecture of a Quantum-DNA Ternary 2 × 2 Multiplier

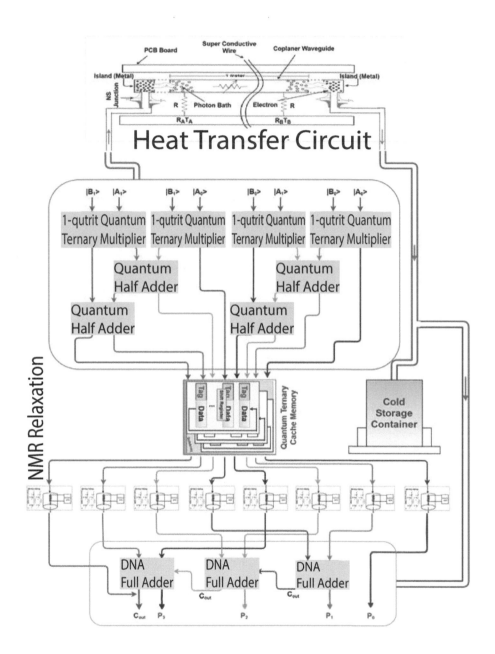

FIGURE 14.18
The Complete Circuit Architecture of a Quantum-DNA Ternary 2 × 2 Multiplier

1. In the quantum system, the 1st quantum ternary 1-qutrit multiplier will get the input values of $|A_0>$ and $|B_0>$. Here, $|A_0> = |2>$, and $|B_0> = |0>$. Therefore, this multiplier will generate two outputs $|m_0> = |0>$ and $|c_0> = |0>$. Here, $|m_0>$ is the multiplication product, and $|c_0>$ is the carry value. The value of $|m_0>$ will be the least significant qutrit of the final product. Therefore, the value of $|m_0>$ can be passed to the quantum ternary cache memory. And the value of $|c_0>$ will be work as an input to the 1st quantum ternary half-adder shown in Figure 14.17 (from the right side of the Figure).

2. The 2nd quantum ternary 1-qutrit multiplier will get the input values of $|A_1>$ and $|B_0>$. Here, $|A_1> = |0>$, and $|B_0> = |0>$. Therefore, this multiplier will generate two outputs $|m_1> = |0>$ and $|c_1> = |0>$. The value of $|m_1>$ will be work as the 2nd input value to the 1st quantum ternary half-adder. And the value of $|c_1>$ will be work as an input to the 3rd quantum ternary half-adder.

3. The 3rd quantum ternary 1-qutrit multiplier will get the input values of $|A_0>$ and $|B_1>$. Here, $|A_0> = |2>$, and $|B_1> = |1>$. And, this multiplier will generate two outputs $|m_2> = |2>$ and $|c_2> = |0>$. The value of $|m_2>$ will be stored to the quantum ternary cache memory which will work as an input to the 1st DNA ternary half adder eventually. And the value of $|c_2>$ will work as an input to the 2nd quantum ternary half-adder.

4. The 4th quantum ternary 1-qutrit multiplier will get the input values of $|A_1>$ and $|B_1>$. Here, $|A_1> = |0>$, and $|B_1> = |1>$. And, this multiplier will generate two outputs $|m_3> = |0>$ and $|c_3> = |0>$. The value of $|m_3>$ will be work as another input value to the 2nd quantum ternary half-adder. And the value of $|c_3>$ will work as an input to the 4th quantum ternary half-adder.

5. Now, the 1st quantum ternary half-adder will get inputs $|c_0> = |0>$ (from step 1), and $|m_1> = |0>$ (from step 2). So, this half-adder will produce two outputs. Here, from the 1st half adder, $|S_0> = |0>$, and the carry output will also be $|0>$. The value of $|S_0>$ will be stored in the quantum ternary cache memory and it will work as another input of the 1st DNA ternary half-adder further. And the carry value from this half-adder will work as another input value of the 3rd quantum ternary half-adder (according to Figure 14.17).

6. The 2nd quantum ternary half-adder will get inputs $|c_2> = |0>$ (from step 3), and $|m_3> = |0>$ (from step 4). So, this half-adder will produce two outputs, $|S_1> = |0>$, and the carry output will also be $|0>$ (see the truth table of ternary half-adder). The value of $|S_1>$ will be stored in the cache memory and it will work as an input of the 1st DNA ternary full-adder eventually. And the carry value from this half-adder will work as another input value of the 4th quantum ternary half-adder (see Figure 14.17).

7. The 3rd quantum ternary half-adder will get inputs $|c_1\rangle = |0\rangle$ (from step 2), and also $|0\rangle$ from the carry output of the 1st half-adder (from step 5). So, this half-adder will produce two outputs, $|S_2\rangle = |0\rangle$, and the carry output will also be $|0\rangle$ (see the truth table of ternary half-adder). The value of $|S_2\rangle$ will be stored in the cache memory and it will work as an input of the 1st DNA ternary full-adder. And the carry value from this half-adder will also be stored in the cache memory and will work as an input value of the 2nd DNA ternary full-adder (according to Figure 14.17).

8. The 4th quantum ternary half-adder will get inputs $|c_3\rangle = |0\rangle$ (from step 4), and also $|0\rangle$ from the carry output of the 2nd half-adder (from step 6). So, this half-adder will produce two outputs, $|S_3\rangle = |0\rangle$, and the carry output will also be $|0\rangle$ (see the truth table of ternary half-adder). The value of $|S_3\rangle$ will be stored in the cache memory and will work as an input of the 2nd DNA ternary full-adder. And the carry value from this half-adder produces the most significant DNA sequence value of the final product value, and thus it will be stored in the cache memory as well.

9. The quantum system operations are done. All the outputs are stored from the quantum system to the quantum ternary cache memory. Now, from the cache memory, all the values will be passed through the NMR relaxation process to convert them into their equivalent DNA base sequences.

10. In the DNA system, The 1st DNA ternary half-adder will get inputs S_0 = ACCTAG (from the converted value from step 5), and also m_2 = TG-GATC from step 3. So, this half-adder will produce the value of P_1 = TGGATC. And a carry out value ACCTAG which will work as a carry input to the 1st DNA full-adder.

11. Now, the 1st DNA ternary full-adder operations can be performed. It gets three inputs as, S_1 = ACCTAG, S_2 = ACCTAG, and also ACCTAG (from step 9). Therefore the sum of the addition will be ACCTAG which is the value of P_2. And the carry will also be ACCTAG which will be the carry input to the 2nd DNA ternary full-adder.

12. The last operation is the 2nd DNA ternary full-adder. It gets the three inputs as - S_3 = ACCTAG, ACCTAG from the converted value of step 7, and also ACCTAG from step 10. Therefore, it will also produce sum ACCTAG and carry output as ACCTAG. The sum output will be the value of P_3 and the carry output will be the value of P_4. And therefore, both the value as ACCTAG is obtained.

13. And the value of P_0 is obtained from the converted DNA sequence of $|m_0\rangle = |0\rangle$ from step 1. Therefore, the value of P_0 as ACCTAG is gained.

So, finally, all the product values are P_0 = ACCTAG, P_1 = TGGATC, P_2 = ACCTAG, P_3 = ACCTAG, and P_4 = ACCTAG. After decoding the DNA sequences, the product of the ternary multiplication operation of P = 00020 is obtained, which is absolutely correct for the given input.

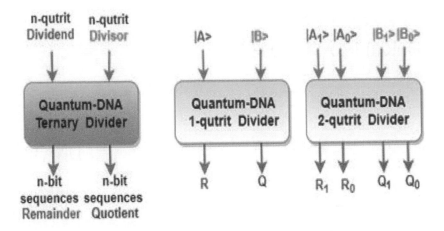

FIGURE 14.19
The General Block Diagrams of Quantum-DNA Ternary Dividers

14.11 Quantum-DNA Ternary Divider

The ternary divider is constructed for 1-qutrit and 2-qutrit dividend and divisor for the quantum system in Chapter 11, and the 1-bit and 2-bit sequences dividend and divisor in the DNA system in Chapter 13. Now, in this section, the architecture of the ternary Quantum-DNA cross-platform environment will be constructed.

The general block diagrams of the Quantum-DNA ternary dividers are shown in Figure 14.19. From the Figure 14.19, it is seen that the inputs of the divider will be in the form of qutrits but the outputs will be in the form of DNA ternary bit sequences. As usual, the first part of the circuit will contain the quantum ternary system, then quantum ternary cache memory will store the qutrit output values generated from the quantum ternary system. Then the NMR relaxation process will convert the qutrits into the equivalent DNA sequences. And the rest of the operation will perform in the DNA ternary system which will produce the final outputs.

14.11.1 The Construction of the Quantum-DNA Ternary Divider

Figure 14.20 shows the operational circuit diagram of the Quantum-DNA ternary 1-qutrit divider where a part of the division operations is performed in the quantum system, then the quantum ternary cache memory is used to store the qutrits temporarily. NMR relaxation circuits at zero Kelvin temperature are used as data conversion units to convert the qutrits values into the equivalent DNA bit sequences. And then, the DNA ternary system performs the rest of the operations and produces the desired outputs.

The upgraded version of that circuit is shown below (figure 14.21) where the heat transfer is also included.

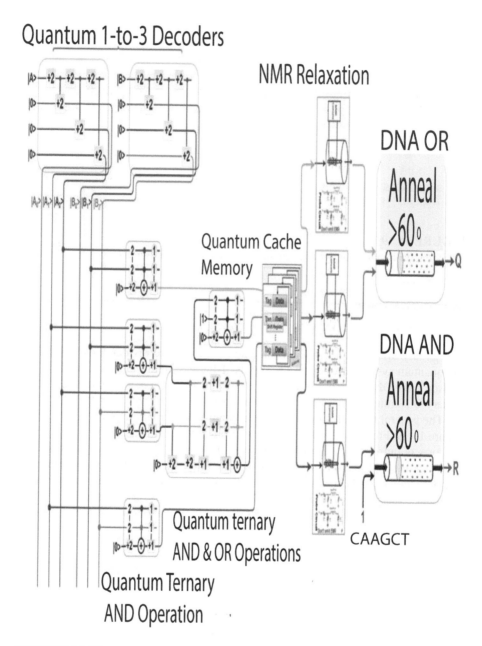

FIGURE 14.20
The Circuit Architecture of a Quantum-DNA Ternary 1-Qutrit Divider

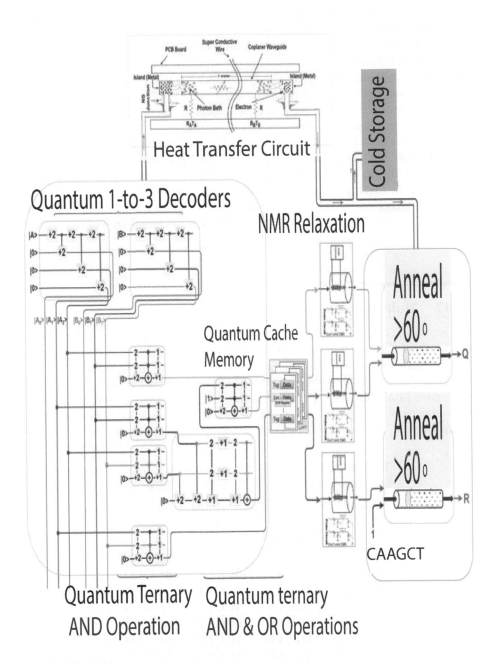

FIGURE 14.21
The Circuit Architecture of a Quantum-DNA Ternary 1-Qutrit Divider Including
Heat Conductance Circuit

The generated excessive heat in the quantum system will be passing through the heat conductance circuit to cold storage. And the division operations are performed according to the previously described working mechanism of Quantum and DNA ternary divider. Similarly, it is possible to construct the general architecture for the Quantum-DNA ternary 2-qutrit divider. Figure 14.22 depicts the general architecture of the Quantum-DNA ternary 2-qutrit divider.

In the above architecture, the heat conductance circuit dispels the massive heat produced by the quantum system.

The values of |A1>, |A2> will be found from the decoded output of the value |A>. Similarly, the values |B1> and |B2> will be found from the decoded output of the value |B>. In this case, the other output values from the decoder (|A0>, and |B0>) won't be required to perform the operations. In a similar way, it is possible to design the algorithm for the 2-qutrit Quantum-DNA ternary divider and notice the DNA system, two DNA ternary operations will take place simultaneously.

14.12 Multiple-Valued Quantum-DNA Comparator

A comparator is a circuit that compares two input signals and gives output as which input is larger or smaller or equal. Here, in Quantum-DNA circuit, the input of |A> and |B> is Quantum qubit DNA sequence and is the output of X, Y, and Z is DNA sequence. Table 14.5 shows the truth table of Quantum-DNA 1-bit comparator.

Figure 14.23 displays the block diagram of Quantum-DNA 1-bit comparator. Here nine multi-valued DNA AND operations and six multi-valued DNA OR operations are used. The output will be DNA sequences. In the input section, two 1-to-3

TABLE 14.5

Quantum-DNA 1-Bit Comparator Truth Table

Inputs		Outputs		
|A>	|B>	X	Y	Z
		A < B	A= B	A > B
|0>	|0>	ACCTAG	TGGATC	ACCTAG
|0>	|1>	TGGATC	ACCTAG	ACCTAG
|0>	|2>	TGGATC	ACCTAG	ACCTAG
|1>	|0>	ACCTAG	ACCTAG	TGGATC
|1>	|1>	ACCTAG	TGGATC	ACCTAG
|1>	|2>	TGGATC	ACCTAG	ACCTAG
|2>	|0>	ACCTAG	ACCTAG	TGGATC
|2>	|1>	ACCTAG	ACCTAG	TGGATC
|2>	|2>	ACCTAG	TGGATC	ACCTAG

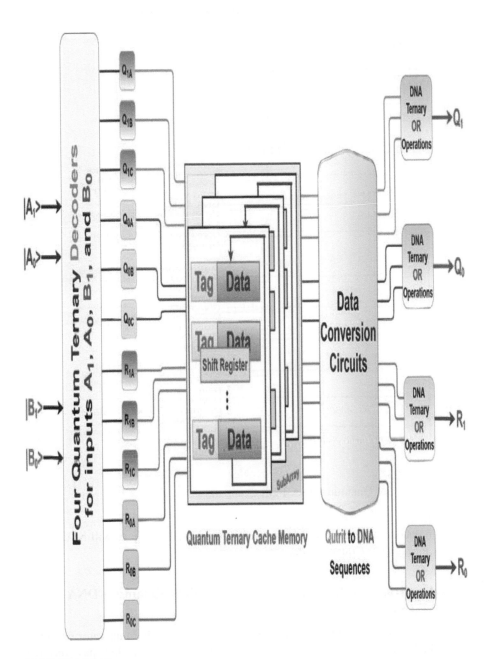

FIGURE 14.22
The Architecture of a Quantum-DNA Ternary 2-Qutrit Divider

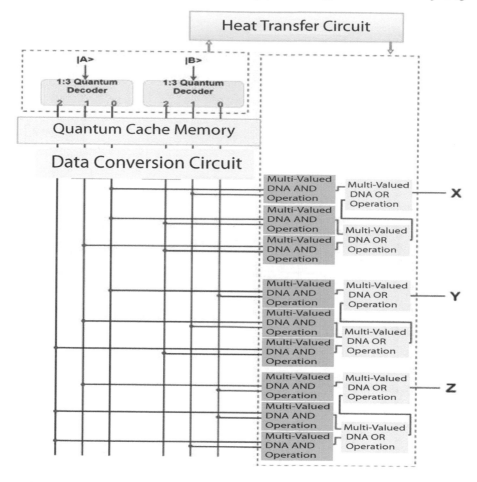

FIGURE 14.23
Block Diagram of Quantum-DNA 1-Bit Comparator

quantum decoders are used to decode the input qubits. Two inputs will be decoded into six through these quantum decoders.

14.12.1 Circuit Architecture of Multiple-Valued Quantum-DNA Comparator

The construction of the Multi-Valued Quantum-DNA 1-qubit comparator circuit is shown in Fig. 14.24. In this circuit, the input, |A> and |B>, at first, pass through two 1-to-3 Decoders separately so that inserted qubit, i.e., |0>, |1>, or |2> can be identified using the output of 1-to-3 decoders.

Nine DNA AND operation is used in Quantum-DNA 1-qubit comparator where each input of these AND operations come from the various combinations of the out-

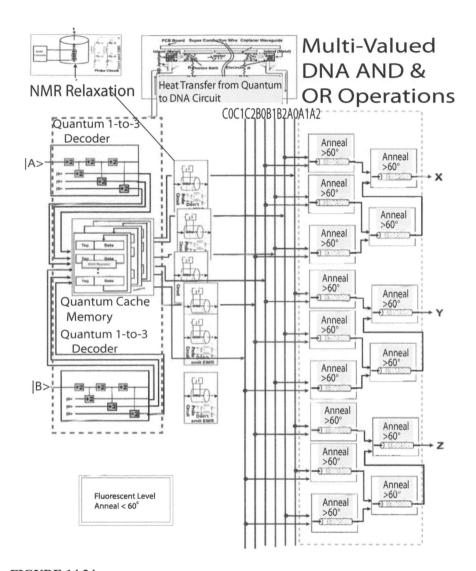

FIGURE 14.24
Circuit Architecture of Multi-Valued Quantum-DNA Comparator

put of 1-to-3 decoder. However, as the output of Quantum operation will act as an input of DNA operation, thus each qubit is converted to its corresponding DNA molecular sequence by using NMR relaxation technique. Before the conversion Quantum outputs are stored in a Quantum Cache memory shortly to match the speed to operations of different kinds.

Finally, based on the value of |A> and |B>, only one line of each 1-to-3 decoder generates sequence **TGGATC** along the line whereas others provide **ACCTAG**. Thus, only one DNA AND operation gives the result **TGGATC**, where other outputs remain inactive. i.e., **ACCTAG**.

14.12.2   Working Principle of Multiple-Valued Quantum-DNA Comparator

1.   For input sequences |A>, |B> = **|0>, |0>**, the multi-valued 1-to-3 decoder will perform (|A0> && |B0>) and the 4th AND line will be open. As a result, the output sequence of Y equal **TGGATC** and remaining lines, X and Z will remain closed**, i.e. ACCTAG**.

2.   For input sequences |A>, |B> = **|0>, |1>**, the multi-valued 1-to-3 decoder will perform (|A0> && |B1>) and the 1st AND line will be open. As a result, the output sequence of X equal **TGGATC** and remaining lines, Y and Z will remain closed**, i.e. ACCTAG.**

3.   For input sequences |A>, |B> = **|0>, |2>**, the multi-valued 1-to-3 decoder will perform (|A0> && |B2>) and the 2nd AND line will be open. As a result, the output sequence of X equal **TGGATC** and remaining lines, Y and Z will remain closed**, i.e. ACCTAG**

4.   For input sequences |A>, |B> = **|1>, |0>**, the multi-valued 1-to-3 decoder will perform (|A1> && |B0>) and the 7th AND line will be open. As a result, the output sequence of Z equal **TGGATC** and remaining lines, X and Y will remain closed**, i.e. ACCTAG**.

5.   For input sequences |A>, |B> = **|1>, |1>** the multi-valued 1-to-3 decoder will perform (|A1> && |B1>) and the 5th AND line will be open. As a result, the output sequence of Y equal **TGGATC** and remaining lines, X and Z will remain closed**, i.e. ACCTAG**.

6.   For input sequences |A>, |B> = **|1>, |2>**, the multi-valued 1-to-3 decoder will perform (|A1> && |B2>) and the 3rd AND line will be open. As a result, the output sequence of X equal **TGGATC** and remaining lines, Y and Z will remain closed**, i.e. ACCTAG.**

7.   For input sequences A, B= **|2>, |0>**, the multi-valued 1-to-3 decoder will perform (|A2> && |B0>) and the 8th AND line will be open. As a result, the output sequence of Z equal **TGGATC** and remaining lines, X and Y will remain closed**, i.e. ACCTAG**.

8.   For input sequences |A>, |B> = **|2>, |1>**, the multi-valued 1-to-3 decoder will perform (|A2> && |B1>) and the 9th AND line will be open. As a result, the output sequence of Z equal **TGGATC** and remaining lines, X and Y will remain closed**, i.e. ACCTAG**.

9.   For input sequences |A>, |B> = **|2>, |2>**, the multi-valued 1-to-3 decoder will perform (|A2> && |B2>) and the 6th AND line will be open. As a result, the output sequence of Y equal **TGGATC** and remaining lines, X and Z will remain closed**, i.e. ACCTAG**.

14.13 Summary

Multi-Valued arithmetic circuits in quantum-DNA circuits are presented in this chapter. The combination of a multi-valued quantum circuit and a multi-valued DNA circuit is actually quantum molecular biology. Here one form of quantum molecular biology which is called quantum-DNA computing is described for multi-valued arithmetic circuits. These circuits are discussed already in the quantum and DNA part of this book in previous chapters. All construction procedures and working principles are described here in easy language. Necessary figures are also shown in this chapter. This combination process will bring a revolutionary change in modern science in the computing world.

Bibliography

[1] Hallworth, R. P., & Heath, F. G. (1962). Semiconductor circuits for ternary logic. Proceedings of the IEE-Part C: Monographs, 109(15), 219-225.

[2] Dhande, A. P., & Ingole, V. T. (2005, March). Design and implementation of 2 bit ternary ALU slice. In Proc. Int. Conf. IEEE-Sci. Electron., Technol. Inf. Telecommun (Vol. 17).

[3] Mandal, S. B., Chakrabarti, A., & Sur-Kolay, S. (2011, May). Synthesis techniques for ternary quantum logic. In 2011 41st IEEE International Symposium on Multiple-Valued Logic (pp. 218-223). IEEE.

[4] Li, T., Gao, H., Wang, C., Cheng, Z., Xue, J., Zhang, Z., ... & Cao, J. (2022). Oil utilization degree at various pore sizes via different displacement methods. Journal of Petroleum Exploration and Production Technology, 1-17.

[5] Giri, S., & Saraswathi, M. N. (2012). Implementation of combinational circuits using ternary multiplexer. International Journal of Computational Engineering Research, 2(2), 457-463.

15

Multiple-Valued Arithmetic Operations in DNA-Quantum Computing

15.1 Introduction

Molecular biology is the field of science concerned with studying the chemical structures and processes of biological phenomena that involve the basic units of life and molecules. The field of molecular biology is focused especially on nucleic acids (e.g., DNA and RNA) and proteins, these molecules interact and behave within cells. Molecular biology emerged in the 1930s.

When quantum computing combines with this DNA computing to form a new computing process, it is quantum molecular biology. In this chapter, another form of quantum molecular biology will be discussed with multi-valued logic. In the previous chapter, multi-valued arithmetic operations in quantum-DNA computing are discussed. This chapter will describe multi-valued arithmetic circuits in DNA-quantum circuits. The same operations which were discussed in the previous chapter will be discussed here.

15.2 Establishing DNA-Quantum Cross-Platform for the Multi-Valued Logic System

The system was divided into four parts. The first part contained the DNA system, then a DNA cache memory where the DNA sequences were stored, then the data conversion unit to convert the DNA information into qubits, and the last part was the quantum system.

In the multiple-valued logic system, the same thing will be done. The general organization of the DNA-Quantum cross-platform for the multiple-valued logic system is depicted in Figure 15.1. Every ternary operation in the DNA-Quantum system will contain –

1. Multiple-valued DNA System,

DOI: 10.1201/9781003381938-15

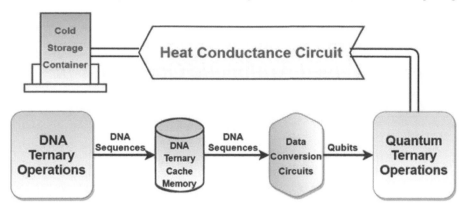

FIGURE 15.1

The General Organization Multiple-Valued DNA-Quantum Computing with Data Conversion Circuits, DNA Ternary Cache Memory, and Heat Conductance Circuit

2. Multiple-valued DNA Cache Memory,

3. Data Conversion Unit, and

4. Multiple-valued Quantum System.

Among them, the multiple-valued quantum system, multiple-valued DNA system, and the data conversion unit are learned which will remain the same. That means the NMR process can be used, and the trap ion also can be used to convert the DNA information (DNA strands) into the equivalent quantum ternary data (qutrits). Here, to convert DNA sequences to the corresponding qutrits, either the NMR process or trap ions is used.

And again, to complete the organization, it is needed to use a heat conductance circuit through which the tremendous heat can be transferred from the quantum system to a cooler. Now, one thing needs to be considered, as DNA system operations will be performed in the first place, it is not possible to provide the required heat to the DNA system from the quantum system, because the heat is generated in the quantum system only when quantum operations are performed. Therefore, it is not possible to provide the required heat to the DNA system from the quantum system. Rather, this heat can be provided from the outside.

Another thing to consider now is – the DNA ternary cache memory to store the DNA information. The DNA cache memory is designed already. But that cannot store the DNA sequence for the ternary DNA sequence data unless it is made in a way to allow storing and retrieving the DNA sequences for the ternary logic. So, a DNA cache memory for the ternary system to store the ternary DNA sequences and retrieve the ternary sequences from the DNA ternary cache memory for further process.

TABLE 15.1

Truth Table of Multi-Valued DNA-Quantum
Half Adder

Inputs		Outputs	
A	B	|Carry>	|Sum>
ACCTAG	ACCTAG	|0>	|0>
ACCTAG	CAAGCT	|0>	|1>
ACCTAG	TGGATC	|0>	|2>
CAAGCT	ACCTAG	|0>	|1>
CAAGCT	CAAGCT	|0>	|2>
CAAGCT	TGGATC	|1>	|0>
TGGATC	ACCTAG	|0>	|2>
TGGATC	CAAGCT	|1>	|0>
TGGATC	TGGATC	|1>	|1>

15.3 Multiple-Valued DNA-Quantum Half Adder

A Multi-Valued half adder is a type of adder, an electronic circuit that performs the addition of ternary numbers. Multi-Valued Half Adder is constructed in chapter 11 and chapter 13 in details. Now here to design a half adder in which DNA sequences act as the input however the outputs of this half adder will produce in Quantum qubits. Table 15.1 shows the truth Table of DNA-Quantum Half adder. From the truth table, the equations for the sum and carry are

$$\textbf{Sum} = A^0.B^2 + A^1.B^1 + A^2.B^0 + 1. (A^0.B^1 + A^1.B^0 + A^2.B^2)$$

$$\textbf{Carry} = 1.(A^1.B^2 + A^2.B^1 + A^2 B^2)$$

To design a DNA-Quantum ternary half adder, DNA decoders are used so that it is possible to decode the input sequence into the corresponding bits. Except for the two DNA decoders at the input level, other operations will be conducted using Quantum operations.

15.3.1 Circuit Architecture of Multiple-Valued DNA-Quantum Half Adder

As shown in Figure 15.1, the multi-valued DNA decoder produces 3 outputs line for each input value. If 1-to-3 decoders are not used, it will not be possible to access all combinations of these two inputs. Decoder for input A produces A0, A1, A2, for input B produce B0, B1, B2. Only one output line is true (**TGGATC**) and the other remains false (**ACCTAG**) for an input. These nine DNA sequence outputs of the DNA decoder are temporarily stored in DNA cache memory. Then they convert into Quantum qubits with the help of trap ions. After the conversion, they act as the inputs of the remaining Quantum operations

From the truth table 15.2, the sum of the half adder produces three |2> qubits and three |1> qubits as true values. For these total six true values, six quantum AND operations are designed and each of the outputs of the quantum AND operation will go through quantum OR operation to produce the true output when one of the outputs of the quantum AND operation is true. But as the output |1> is needed in three conditions, a quantum AND operation is executed with a fixed input|1> before the final quantum OR operation which is used to generate the maximum output among |2>, |1> and |0>.

In the case of carry of the half adder, three input combinations produce qubit |1> as true values. After minimization, the carry function produces three 1 values. Thus, three quantum AND operations and two quantum OR operations are used. Again. For the purpose of generating maximum qubit at most |1> a quantum AND operation is added at last to get Carry bit.

Figure 15.2 shows the circuit architecture of multi-valued DNA-quantum half adder.

15.3.2 Working Principle of Multiple-Valued DNA-Quantum Half Adder

A 2-input ternary DNA-Quantum half-adder contains 2 DNA sequence input as input and 2 output Quantum one for Sum and another for Carry.

1. When the input sequences A, B = ACCTAG (0), DNA decoders generate A0, A1, A2 = (2, 0, 0) and B0, B1, B2 = (2, 0, 0). None of the AND operations (AND1-AND9) are connected to the |A0> and |B0> line. As a result, none of the AND gate values will be true. So the output of Sum and Carry both are |0> (0).

2. When the input sequences A = ACCTAG (0) and B = CAAGCT (1), DNA decoders generate A0, A1, A2 = (2, 0, 0) and B0, B1, B2 = (0, 2, 0). AND4 is connected to the A0 and B1 line. As a result, the AND4 value (|2>) is also AND with |1> to produces minimum sum |1>. Here, Carry is |0>

3. When the input sequences A = ACCTAG (0) and B = TGGATC (2), DNA decoders generate A0, A1, A2 = (2, 0, 0) and B0, B1, B2 = (0, 0, 2). AND1 is connected to the A0 and B2 line. AND1 operation values will be true (|2>) and produces minimum as sum |2> and minimum as carry |0>

4. When the input sequences A = CAAGCT (1) and B = ACCTAG (0), DNA decoders generate A0, A1, A2 = (0, 2, 0) and B0, B1, B2 = (2, 0, 0). AND5 operation is connected to A1 and B0 lines. As a result, AND5 output will be true (|2>). So AND5 value (|2>) is also AND with |1> and produces minimum as sum |1> and carry is |0>.

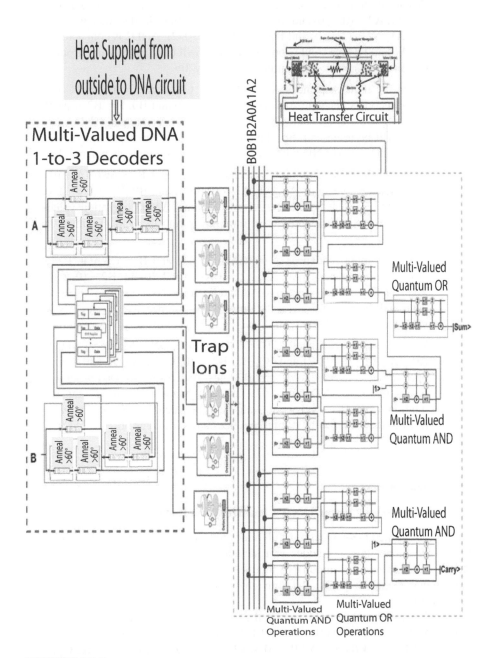

FIGURE 15.2
Circuit Architecture of Multi-Valued DNA-Quantum Half Adder

5. When the input sequences A, B = CAAGCT (1), DNA decoders generate A0, A1, A2 = (0, 2, 0) and B0, B1, B2 = (0, 2, 0). DNA decoders generate A0, A1, A2 = (2, 0, 0) and B0, B1, B2 = (0, 0, 2). AND2 is connected to the A1 and B1 line. AND2 operation value will be true ($|2>$) and produce minimum as sum $|2>$ and minimum as carry $|0>$.

6. When the input sequences A = CAAGCT (1), and B = TGGATC (2), DNA decoders generate A0, A1, A2 = (0, 2, 0) and B0, B1, B2 = (0, 0, 2). AND7 gate is connected to A1 and B2 lines. As a result, AND7 gate value will be true ($|2>$). So, the output of differences is $|0>$ and AND7 value ($|2>$) is also AND with $|1>$ and produces minimum as borrow $|1>$.

7. When the input sequences A = TGGATC (2), and B = ACCTAG (0), DNA decoders generate A0, A1, A2 = (0, 0, 2) and B0, B1, B2 = (2, 0, 0). AND3 is connected to the A2 and B0 line. AND3 operation values will be true ($|2>$) and produce minimum as sum $|2>$ and minimum as carry $|0>$.

8. When the input sequences A = TGGATC (2), and B = CAAGCT (1), DNA decoders generate A0, A1, A2 = (0, 0, 2) and B0, B1, B2 = (0, 2, 0). The AND7 gate is connected to A2 and B1 lines. As a result, AND8 gate value will be true ($|2>$). So, the output of differences is $|0>$ and AND8 value ($|2>$) is also AND with $|1>$ and produces minimum as borrow $|1>$.

9. When the input sequences A, B = TGGATC (2), DNA decoders generate A0, A1, A2 = (0, 0, 2) and B0, B1, B2 = (0, 0, 2). AND6 and AND9 operation is connected to A2 and B2 lines. As a result, AND6 output will be true ($|2>$). So AND6 value ($|2>$) is also AND with $|1>$ and produces a minimum sum $|1>$. AND9 gate value will be true ($|2>$). So AND9 value ($|2>$) is also AND with $|1>$ and produces a minimum as borrow $|1>$.

―――――――――

15.4   Multiple-Valued DNA-Quantum Full Adder

A Multiple-Valued DNA-Quantum full adder removes the drawback of a Multiple-Valued DNA-Quantum half adder as it includes carry bit in the addition operation. Hence, there are a total of three DNA inputs A, B, and C and this circuit will generate the output in Quantum qubits. The truth table of DNA-Quantum Full Adder is shown in Table 15.2.

As discussed in chapters 11 and 13, the following equations can be obtained.

Sum = A2.B0.C0 +A1.B0.C1 + A1. B1.C0+ A0.B1.C1 +A0.B2.C0 + A2.B2.C1 + 1. (A1.B0.C0 + A0.B0.C1 + A0.B1.C0 + A2.B1.C1 + A2.B2.C0 + A1.B2.C1)

Carry = 1 (A0. B2. C0+A2.B1. C0+A2.B2.C0 +A1.B1. C1+B2.C1+A2.C1)

     = 1. ((A0. B2+A2.B1+A2B2).C0 + (A1. B1+ B2+A2).C1)

TABLE 15.2

Multi-Valued DNA-Quantum Full Adder Truth Table

Inputs			Outputs	
A	**B**	**C**	**|Carry>**	**|Sum>**
ACCTAG	ACCTAG	ACCTAG	|0>	|0>
ACCTAG	CAAGCT	ACCTAG	|0>	|1>
ACCTAG	TGGATC	ACCTAG	|0>	|2>
CAAGCT	ACCTAG	ACCTAG	|0>	|1>
CAAGCT	CAAGCT	ACCTAG	|0>	|2>
CAAGCT	TGGATC	ACCTAG	|1>	|0>
TGGATC	ACCTAG	ACCTAG	|0>	|2>
TGGATC	CAAGCT	ACCTAG	|1>	|0>
TGGATC	TGGATC	ACCTAG	|1>	|1>
ACCTAG	ACCTAG	CAAGCT	|0>	|1>
ACCTAG	CAAGCT	CAAGCT	|0>	|2>
ACCTAG	TGGATC	CAAGCT	|1>	|0>
CAAGCT	ACCTAG	CAAGCT	|0>	|2>
CAAGCT	CAAGCT	CAAGCT	|1>	|0>
CAAGCT	TGGATC	CAAGCT	|1>	|1>
TGGATC	ACCTAG	CAAGCT	|1>	|0>
TGGATC	CAAGCT	CAAGCT	|1>	|1>
TGGATC	TGGATC	CAAGCT	|1>	|2>

15.4.1 Circuit Architecture of Multiple-Valued DNA-Quantum Full Adder

At first, the multi-valued DNA decoder initially produces three output lines for each input value. It will be impossible to access all combinations of these two inputs if the 1-to-3 decoders are not used. For input A, the decoder produces A0, A1, and A2, while for input B, the decoder produces B0, B1, and B2 and finally for input C, the decoder produces C0 and C1 (Figure 15.3). For an input, only one output line is true (**TGGATC**), while the other remains false (**ACCTAG**). The DNA decoder's eight DNA sequence outputs are temporarily stored in DNA cache memory. Then, with the help of trap ions, they are converted into quantum qubits.

The sum of the full adder produces six |2> and six |1> values. Six multi-valued quantum AND operations are used for these six |2> values, and each of the multi-valued quantum AND operations output will conduct a multi-valued quantum OR operation to produce the true output when one of the multi-valued quantum AND operation outputs is true. In addition, there are six |1> values in total. Six multi-valued quantum AND operations are designed for these six |1> values and each of the

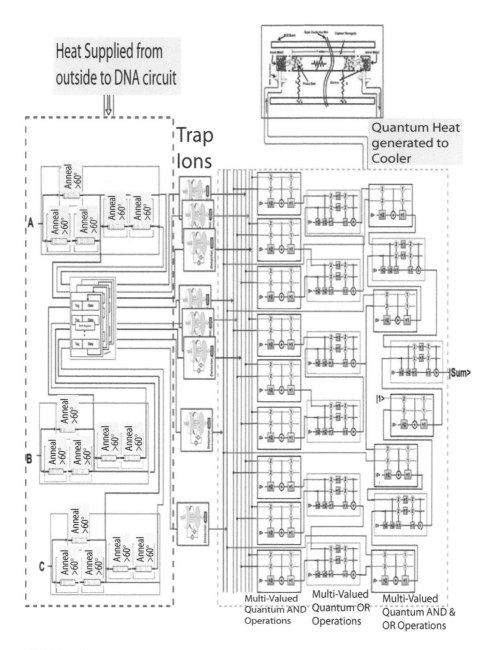

FIGURE 15.3
Circuit Architecture of Multi-Valued DNA-Quantum Full Adder (Sum)

multi-valued quantum AND operation outputs will perform a multi-valued quantum OR operation to produce the true ($|2>$) output when one of the multi-valued quantum AND operation outputs is true. However, because the output $|1>$ is required, it is needed to perform the multi-valued quantum AND operation with the output of the multi-valued quantum OR operation as input along with $|1>$. Finally, multi-valued quantum AND operation produce the minimum value. So, the output of the multi-valued quantum AND operation will always $|1>$. At last, a multi-valued quantum OR operation is used to generate the maximum output among $|2>$, $|0>$ and $|0>$, $|1>$ as shown in Figure 15.3.

As for carry of the full adder produce nine $|1>$ values. Nine multi-valued quantum AND operations are designed and each of the multi-valued quantum AND operation output will perform multi-valued quantum AND operations to produce the true ($|2>$) output when one of the outputs of the multi-valued quantum AND operation is true. But as the output $|1>$ is required, it is needed to perform multi-valued quantum AND operation taking the multi-valued quantum OR operation output as input along with $|1>$.

So, the output of the multi-valued quantum AND operation will always 1 for one of the true inputs among nine multi-valued quantum AND operations. The output of this last multi-valued quantum AND operation gives the result of Carry qubit as shown in Figure 15.4.

Figure 15.3 shows the circuit architecture of Multi-Valued DNA-Quantum Full Adder (Sum) and Figure 15.4 shows the circuit architecture of Multi-Valued DNA-Quantum Full Adder (Carry).

15.4.2 Working Principle of Multiple-Valued DNA-Quantum Full Adder

A DNA-Quantum Full-Adder contains 3 DNA input sequences as input and 2 output Quantum qubit, one for sum and the other for carry. To understand the working principle of DNA-Quantum Full Adder, let's consider the following case.

When the input sequence A, B = ACCTAG (0), and C = CAAGCT (1), DNA decoder generate A0, A1, A2 = (2, 0, 0), B0, B1, B2 = (2, 0, 0) and C0, C1, C2 = (0, 2, 0). The 13th AND gate is connected to the A0 and B0 and C1 line. As a result, the 13th AND gate value, that evaluates A0.B0.C1, will be true ($|2>$). It is seen that the 13th AND gate is connected to the OR gate, so the output of the OR gate will be true ($|2>$). This true value ($|2>$) is also AND with $|1>$ and produces minimum $|1>$ as output. This output will go through the OR gate and generate $|1>$ as sum.

In the full-adder carry circuit, it is seen that none of the AND operations and OR operations are connected to A0, B0 and C1 lines. As a result, none of the AND operations and OR operations value will be true. Thus, they produce $|0>$. This truth value will TAND with $|1>$, and produce a minimum $|0>$ as output. This output will go through the OR gate and generate $|0>$ as carry.

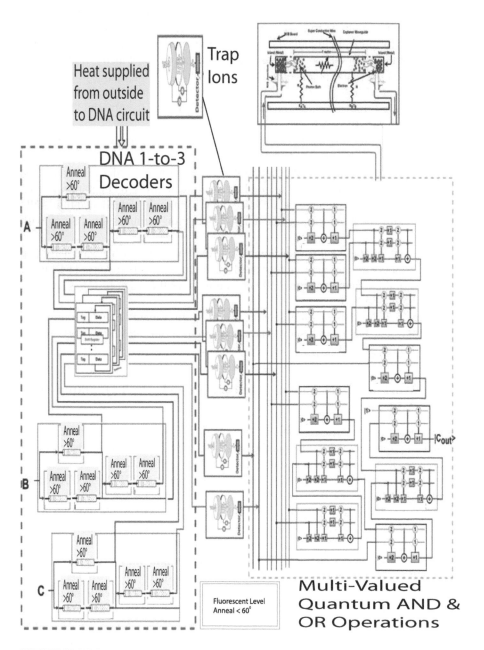

FIGURE 15.4
Circuit Architecture of Multi-Valued DNA-Quantum Full Adder (Carry)

TABLE 15.3

Truth Table of DNA-Quantum Ternary Half Subtractor

Inputs		Outputs	
A	**B**	**|Borrow>**	**|D>**
ACCTAG	ACCTAG	|0>	|0>
ACCTAG	CAAGCT	|1>	|2>
ACCTAG	TGGATC	|1>	|1>
CAAGCT	ACCTAG	|0>	|1>
CAAGCT	CAAGCT	|0>	|0>
CAAGCT	TGGATC	|1>	|2>
TGGATC	ACCTAG	|0>	|2>
TGGATC	CAAGCT	|0>	|1>
TGGATC	TGGATC	|0>	|0>

15.5 Multiple-Valued DNA-Quantum Half Subtractor

A ternary half subtractor is a type of subtractor, an electronic circuit that performs the subtractions of ternary numbers. The DNA-Quantum half subtractor can subtract two single ternary digits which is provided as DNA sequence and provide the output, difference and a borrow value in Quantum qubit representation. The truth table below (Table 15.3) shows the outputs of a DNA-Quantum ternary half subtractor.

From the truth table, the equations for the difference and borrow can be found as

Difference $= A^0.B^1 + A^1.B^2 + A^2.B^0 + 1.(A^0.B^2 + A^1.B^0 + A^2.B^1)$

Borrow $= 1.(A^0.B^1 + A^0.B^2 + A^1.B^2)$

The general design of a ternary half subtractor is already explained in chapters 11 and 13. However, the major difference in this DNA-Quantum half subtractor is that the DNA Decoder is used at the starting as the inputs as DNA sequences and except that the other operations are designed using Quantum gates as the final output will be in Quantum qubit.

15.5.1 Circuit Architecture of Multiple-Valued DNA-Quantum Half Subtractor

Initially, as Figure 15.5 depicts, multi-valued DNA decoder produces 3 outputs line for each input value. If the 1-to-3 decoders are not used, it will not be possible to access all combinations of these two inputs. Decoder for input A produces A0, A1, A2, for input B produce B0, B1, B2. Only one output line is true (**TGGATC**) and other remains false (**ACCTAG**) for an input. These six DNA sequence outputs of the DNA decoder are temporarily stored in DNA cache memory. Then they convert into

Quantum qubits with the help of trap ions. After the conversion, they act as the inputs of the remaining Quantum operations.

According to truth table 15.3, the Difference (D) of the half subtractor yields three $|2>$ qubits and three $|1>$ qubits as true values. For these total six true values, six quantum AND operations are used, and each of the quantum AND operation's outputs will go through a quantum OR operation to produce the true output when one of the quantum AND operation's outputs is true. However, because the output $|1>$ is required in three conditions, a quantum AND operation with a fixed input $|1>$ is performed before the final quantum OR operation, which is used to generate the maximum output among $|2>$, $|1>$, and $|0>$. In the case of half subtractor borrow (B_{out}) three input combinations produce qubit $|1>$ as true values. The circuit architecture of Multi-Valued DNA-Quantum Half Subtractor is shown in Figure 15.5.

The carry function produces three 1 values. As a result, three quantum AND operations and two quantum OR operations are employed. Again. A quantum AND operation is added at the end to obtain the borrow qubit in order to generate the maximum qubit at most $|1>$.

15.5.2   Working Principle of Multiple-Valued DNA-Quantum Half Subtractor

A 2-input multi-valued DNA-Quantum half-subtractor contains 2 DNA sequence input as input and 2 Quantum outputs one for difference and another for borrow.

1. When the input sequences A, B = ACCTAG (0), DNA decoders generate A0, A1, A2 = (2, 0, 0) and B0, B1, B2 = (2, 0, 0). None of the AND gates are connected to the A0 and B0 line. As a result, none of the AND gate values will be true. So the output of differences and borrow both are $|0>$.

2. When the input sequences A = ACCTAG (0) and B = CAAGCT (1), DNA decoders generate A0, A1, A2 = (2, 0, 0) and B0, B1, B2 = (0, 2, 0). D0 and D6 AND gates are connected to the A0 and B1 line. As a result, D0 and D6 AND gates value will be true ($|2>$). So, the output of differences is $|2>$ and D6 value ($|2>$) is also AND with $|1>$ and produce minimum as borrow $|1>$.

3. When the input sequences A = ACCTAG (0) and B = TGGATC (2), DNA decoders generate A0, A1, A2 = (2, 0, 0) and B0, B1, B2 = (0, 0, 2). D3 and D7 AND gates are connected to the A0 and B2 line. As a result, D3 and D7 AND gates value will be true ($|2>$). So D3 value ($|2>$) is also AND with $|1>$ and produce minimum as difference $|1>$ and D7 value ($|2>$) is also AND with $|1>$ and produce minimum as borrow $|1>$.

4. When the input sequences A = CAAGCT (1) and B = ACCTAG (0), DNA decoders generate A0, A1, A2 = (0, 2, 0) and B0, B1, B2 = (2, 0, 0). D4 gate is connected to A1 and B0 line. As a result, D4 AND gates value will be true ($|2>$). So D4 value ($|2>$) is also AND with $|1>$ and produce minimum as difference $|1>$ and borrow is $|0>$.

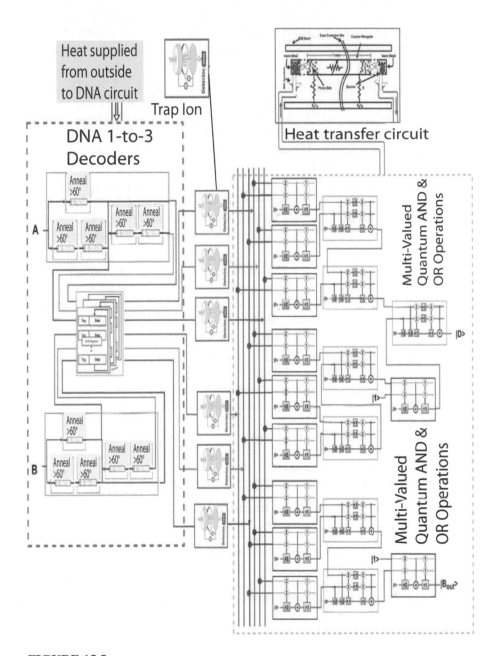

FIGURE 15.5
Circuit Architecture of Multi-Valued DNA-Quantum Half Subtractor

5. When the input sequences A, B = CAAGCT (1), DNA decoders generate A0, A1, A2 = (0, 2, 0) and B0, B1, B2 = (0, 2, 0). None of the AND gates (D0-D8) are connected to A1 and B1 lines. As a result, none of the AND gate value will be true. So, the output of differences and borrow both are |0>.

6. When the input sequences A = CAAGCT (1), and B = TGGATC (2), DNA decoders generate A0, A1, A2 = (0, 2, 0) and B0, B1, B2 = (0, 0, 2). D1 and D8 AND gates are connected to A1 and B2 lines. As a result, D1 and D8 AND gates value will be true (|2>). So, the output of differences is |2> and D8 value (|2>) is also AND with |1> and produce minimum as borrow |1>.

7. When the input sequences A = TGGATC (2), and B = ACCTAG (0), DNA decoders generate A0, A1, A2 = (0, 0, 2) and B0, B1, B2 = (2, 0, 0). D2 gate is connected to A1 and B0 line. As a result, the D2 AND gate value will be true (TGGATC). So, the output of differences is |2> and borrow is |0>.

8. When the input sequences A = TGGATC (2), and B = CAAGCT (1), DNA decoders generate A0, A1, A2 = (0, 0, 2) and B0, B1, B2 = (0, 2, 0). D5 gate is connected to the A2 and B1 line. As a result, D5 AND gates value will be true (|2>). So D5 value (|2>) is also AND with |1> and produce minimum as difference |1> and borrow is |0>.

9. When the input sequences A, B = TGGATC (2), DNA decoders generate A0, A1, A2 = (0, 0, 2) and B0, B1, B2 = (0, 0, 2). None of the AND gates (D0-D8) are connected to the A2 and B2 lines. As a result, none of the AND gate values will be true. So, the output of differences and borrow both are |0>.

15.6 Multiple-Valued DNA-Quantum Full Subtractor

Multi-Valued full subtractor performs subtraction operation taking borrow qubit in consideration, thus it solves the problem that half subtractor cannot overcome. In a Multi-Valued full subtractor, there are a total of three inputs A, B, and C. The truth Table of Multi-Valued DNA-Quantum full subtractor is given in Table 15.4.

The difference and borrow as –

$D = A0.B1.C0 + A1.B2.C0 + A2.B0.C0 + A0.B0.C1 + A1.B1.C1 + A2.B2.C1.$ 1. $(A0.B2.C0 + A1.B0.C0 + A2.B1.C0 + A0.B1.C1 + A1.B2.C1 + A2.B0.C1)$
$= (A0.B1 + A1.B2 + A2.B0).C0 + (A0.B0 + A1.B1 + A2.B2).C1.$ 1. $((A0.B2 + A1.B0 + A2.B1).C0 + (A0.B1 + A1.B2 + A2.B0).C1)$

TABLE 15.4

Truth Table of DNA-Quantum Ternary Full Subtractor

Inputs			Outputs	
A	**B**	**C**	**|D>**	**|Bout>**
ACCTAG	ACCTAG	ACCTAG	|0>	|0>
ACCTAG	CAAGCT	ACCTAG	|2>	|1>
ACCTAG	TGGATC	ACCTAG	|1>	|1>
CAAGCT	ACCTAG	ACCTAG	|1>	|0>
CAAGCT	CAAGCT	ACCTAG	|0>	|0>
CAAGCT	TGGATC	ACCTAG	|2>	|1>
TGGATC	ACCTAG	ACCTAG	|2>	|0>
TGGATC	CAAGCT	ACCTAG	|1>	|0>
TGGATC	TGGATC	ACCTAG	|0>	|0>
ACCTAG	ACCTAG	CAAGCT	|2>	|1>
ACCTAG	CAAGCT	CAAGCT	|1>	|1>
ACCTAG	TGGATC	CAAGCT	|0>	|1>
CAAGCT	ACCTAG	CAAGCT	|0>	|0>
CAAGCT	CAAGCT	CAAGCT	|2>	|1>
CAAGCT	TGGATC	CAAGCT	|1>	|1>
TGGATC	ACCTAG	CAAGCT	|1>	|0>
TGGATC	CAAGCT	CAAGCT	|0>	|0>
TGGATC	TGGATC	CAAGCT	|2>	|1>

B_{out} = 1. (A0. B1. C0+A0.B2.C0 + A1.B2.C0 + A0.C1 +B2.C1+A1.B1.C1)

= 1. ((A0. B1+A0.B2+A1.B2).C0 + (A0+B2+A1.B1).C1)

In chapters 11 and 13, the general design of a ternary full subtractor is already explained. The main difference in this DNA-Quantum full subtractor is that a DNA Decoder is used at the beginning because the inputs will be in DNA sequences, and the other operations will be designed using Quantum gates because the final output will be in a Quantum qubit.

15.6.1 Architecture of Multiple-Valued DNA-Quantum Full Subtractor

At first, the multi-valued DNA decoder initially produces three output lines for each input value. It will be impossible to access all combinations of these two inputs if the 1-to-3 decoders are not used. For input A, the decoder produces A0, A1, and A2, while for input B, the decoder produces B0, B1, and B2 and finally for input C, the decoder produces C0 and C1 (Figure 15.6). For an input, only one output line is true (**TGGATC**), while the other remains false (**ACCTAG**). The DNA decoder's eight DNA sequence outputs are temporarily stored in DNA cache memory. Then, with the help of trap ions, they are converted into quantum qubits. Following the conversion, they serve as inputs to the remaining Quantum operations.

According to the truth table, the difference of the full Subtractor produces six |2> and six |1> values. Six multi-valued quantum AND operations are designed for these six |2> values, and each of the multi-valued quantum AND operations output will conduct a multi-valued quantum OR operation to produce the true output when one of the multi-valued quantum AND operation outputs is true. In addition, there are six |1> values in total. Six multi-valued quantum AND operations are designed for these six |1> values and each of the multi-valued quantum AND operation outputs will perform a multi-valued quantum OR operation to produce the true (|2>) output when one of the multi-valued quantum AND operation outputs is true. However, because the output |1> is required, the multi-valued quantum AND operation is performed with the output of the multi-valued quantum OR operation as input along with |1>. Finally, the multi-valued quantum AND operation yield the smallest value. As a result, the result of the multi-valued quantum AND operation is always |1>. Finally, as shown in fig. 5.4.6, a multi-valued quantum OR operation is used to generate the maximum output among |2>, |0>, and |0>, |1>. (a).

The borrow of the full Subtractor yields nine |1> values. When one of the multi-valued quantum AND operation outputs is true, the multi-valued quantum AND operation is performed to produce the true (|2>) output. However, because the output |1> is required, the multi-valued quantum AND operation is performed with the output of the multi-valued quantum OR operation as input along with |1>. The minimum value is produced by the multi-valued quantum AND operation. As a result, the output of the multi-valued quantum AND operation will always be 1 for one of the nine true inputs. The result of this final multi-valued quantum AND operation is the Borrow qubit, as shown in Figure 15.7.

Figure 15.6 shows the circuit architecture of multi-valued DNA-Quantum Full Subtractor (Difference) and Figure 15.7 shows the circuit architecture of multi-valued DNA-Quantum Full Subtractor (Borrow)

15.6.2   Working Principle of Multiple-Valued DNA-Quantum Full Subtractor

A ternary DNA-Quantum Full-Subtractor contains 3 DNA input sequences as input and 2 output Quantum qubits, one for sum and another for carry. To understand the working principle of DNA-Quantum Full Subtractor, let's consider the following case:

When the input sequence A, B = ACCTAG (0), and C = CAAGCT (1), DNA decoder generate A0, A1, A2 = (2, 0, 0), B0, B1, B2 = (2, 0, 0) and C0, C1, C2 = (0, 2, 0). The 10th AND gate is connected to the A0 and B0 and C1 line. As a result, the 10th AND gate value, that evaluates A0.B0.C1, will be true (|2>). It is seen that the 13th AND gate is connected to the OR gate, so the output of the OR gate will be true (|2>). This output will go through the last OR gate and generate |2> as difference (D).

In the full-subtractor borrow circuit, it is seen that none of the AND operations and OR operations are connected to A0, B0 and C1 lines. As a result, none of the

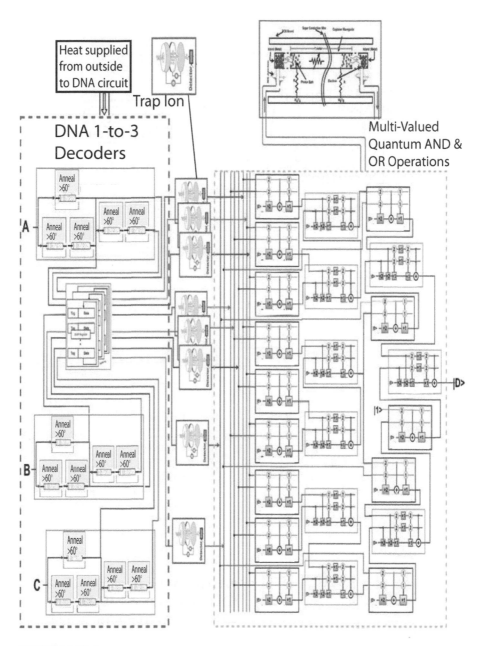

FIGURE 15.6
Circuit Architecture of Multi-Valued DNA-Quantum Full Subtractor (Difference)

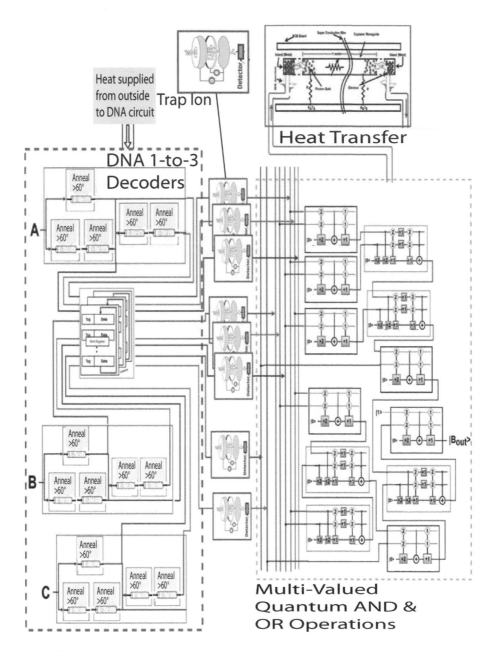

FIGURE 15.7
Circuit Architecture of Multi-Valued DNA-Quantum Full Subtractor (Borrow)

AND operations and OR operation value will be true. Thus, they produce |0>. This truth value will TAND with |1>, and produce minimum |0> as output. This output will go through the OR gate and generate |0> as borrow (Bout).

15.7 DNA-Quantum Ternary Parallel Adder

Therefore, now it is known how to construct the ternary parallel adder in cross-platform. So, it will be much easier now. From here, the opposite procedures of everything will be followed as what is done earlier. That means all the quantum ternary systems will be replaced by the corresponding DNA ternary systems and all the DNA ternary systems will be replaced by the corresponding quantum systems. And DNA ternary cache memory is used instead of quantum ternary cache memory.

So, let's see how to construct the ternary parallel adder in the DNA-Quantum cross-platform.

15.7.1 The Architecture of Quantum-DNA Ternary Parallel Adder

The N-bit ternary parallel adder is divided into two distinct segments to keep the architecture simple. So, the architecture will contain N ternary full-adders. It can be divided into two ways – i) N-1 ternary full adders will be in the DNA system and the last ternary full-adder will be in the DNA system, and, ii) N-X ternary full-adders will be in the quantum system and the X ternary full-adders will be in the DNA system. The 1st approach will be much simpler than the other one.

15.7.2 The General Organizations of the DNA-Quantum Ternary Parallel Adder

Using the first approach, the general organizations of DNA-Quantum 4-bit ternary parallel adder can be constructed which is shown in Figure 15.8. The Figure shows, that three ternary full-adder operations are performed in the DNA system and the last ternary full-adder operation is performed in the quantum system. A DNA ternary cache memory is used to store the bit sequence output and the input sequences for the last full-adder operation.

The sequences are converted into quantum ternary bits (qutrits) using the data conversion circuits. Besides, a heat transfer circuit is used to transfer heat from the quantum system to a cooling system. And now it is not possible to take the required heat in the DNA system from the quantum system because the DNA operations will be executed first. That is why the required heat is taken in the DNA system from an external heat source.

Figure 15.9 shows the general organizations of the DNA-Quantum ternary N-bit parallel adder.

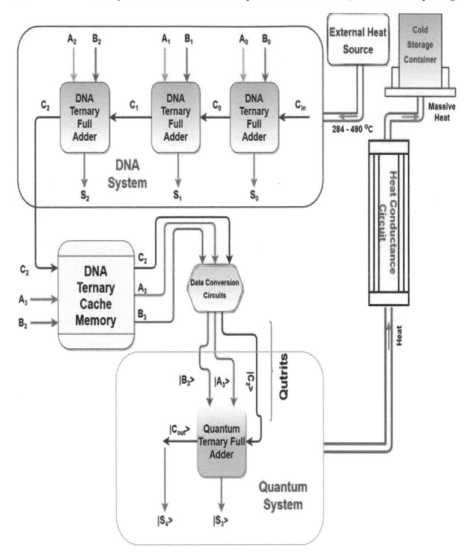

FIGURE 15.8
The General Organizations of the DNA-Quantum Ternary 4-Bit Parallel Adder

The Figure shows, N-1 ternary full-adder operations are performed in the DNA system and the last ternary full-adder operation is performed in the quantum system. A DNA ternary cache memory is used to store the bit sequence output and the input sequences for the last full-adder operation. The DNA sequences are converted into quantum ternary bits using the data conversion circuits. Besides, a heat transfer circuit is used to transfer heat from the quantum system to a cooling system.

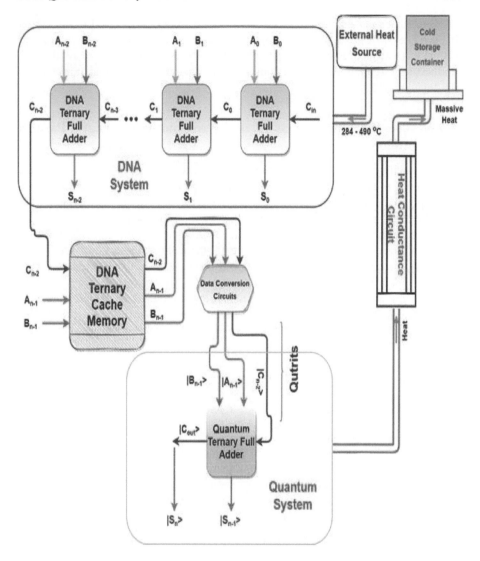

FIGURE 15.9
The General Organizations of DNA-Quantum Ternary N-Bit Parallel Adder

15.7.3 The Circuit Design of the DNA-Quantum Ternary Parallel Adder

The general organization of the DNA-Quantum ternary 4-bit and N-bit parallel adder is shown in Figures 15.8 and 15.9. And the architecture contains DNA ternary full-adders, DNA ternary cache memory, data conversion circuits, quantum ternary full adders, an external heat source, and the heat conductance circuit. Every component was constructed earlier. The DNA ternary cache memory is designed in the earlier chapter. The NMR process or trap ions can be used as a data conversion circuit to convert the DNA sequences into the equivalent quantum ternary bits. And lastly the heat conductance circuit, connect each component to design the DNA-Quantum ternary parallel adder.

15.7.4 The Working Principles of DNA-Quantum Ternary Parallel Adder

There is nothing new to understand the working procedure. All it needs to combine two systems. In the 1st three DNA ternary full-adders in the DNA system, the input and output mapping will be according to the DNA ternary parallel adder operations. Then the NMR process or trap ions will receive the DNA sequences from a DNA ternary cache memory. It will convert the DNA sequences to the equivalent qutrits. Then the converted qutrits will be the inputs to the quantum ternary full-adder. And eventually, the quantum ternary full-adder will generate the output of the last ternary full-adder operation.

Let us test an example. Let the addend and augend qutrits are A and B, respectively. Then, for a 4-bit input the input A will be $A_3A_2A_1A_0$ and the input B will be $B_3B_2B_1B_0$. Let's take some value for those inputs and observe the behavior of the circuit.

Suppose A = 1020 and B = 1100. Therefore, the values of A_0 = ACCTAG, A_1 = TGGATC, A2 = ACCTAG, and A_3 = CAAGCT. And the values of B_0 = ACCTAG, B1 = ACCTAG, B2 = CAAGCT, and B_3 = CAAGCT.

1. For the inputs A_0, and B_0 both are ACCTAG. And the carry input qubit C_{in} is also ACCTAG - The first DNA ternary full-adder will get all the inputs as ACCTAG. Therefore, it will produce output ACCTAG as the sum and also ACCTAG as the carry output sequence. This carry will be working as the carry input of the next DNA ternary full-adder.

2. The next input sequences to the second DNA full-adder are A_1 = TGGATC, B_1 = ACCTAG, and the carry input C_0 is also ACCTAG. Therefore, the second DNA ternary full-adder will generate TGGATC as the sum output and again ACCTAG as the carry output of C_1. The value of C_1 will be delivered to the next DNA ternary full-adder as a carry input.

3. The input to the third DNA ternary full-adder are A_2 = ACCTAG, B_1 = CAAGCT, and the carry input C_1 is again ACCTAG. Therefore, the third DNA ternary full-adder will generate CAAGCT as the sum output and also again ACCTAG as the carry output of $|C_2>$.

4. Now the DNA part is done. The carry output of the very last DNA ternary full-adder operation will be stored in a DNA ternary cache memory. And the next addend and augend input sequences are also stored in the DNA ternary cache memory.

5. From cache memory, the values $A_3 = $ CAAGCT, $B_3 = $ CAAGCT, and C_2 $= $ ACCTAG will be delivered to the NMR process or trap ions. From here the DNA sequences values will be converted into the equivalent qutrits. The converted values will be $|A_3> = |1>$, $|B_3> = |1>$, and $|C_2> = |0>$. These values will be the inputs to the last ternary full-adder which is in the quantum system.

6. The final ternary full-adder is in the quantum system. It will receive the value as $|A_3> = |1>$, $|B_3> = |1>$, and $|C_2> = |0>$. Therefore, it will generate the output as $|2>$ for the sum, and $|0>$ for the last carry output which will be the most significant bit of the addition result eventually.

Therefore, it is possible to get the expected result from the designed architecture for the DNA-Quantum ternary parallel adder.

15.8   DNA-Quantum Ternary Carry-Lookahead Adder

To develop DNA-Quantum carry-lookahead adder by dividing the whole system into two segments – one for the DNA system and the other for the quantum system, as before, this can be done in two ways. Dividing the system for N-bit into (N-1) bit in the DNA system and the last one bit in the quantum system. Or (N-X) bit in the DNA system and X bit in the quantum system. Again, let us see how to construct it in the 1st way which is much simpler.

15.8.1   The General Organizations of DNA-Quantum Ternary Carry-Lookahead Adder

Figure 15.10 shows the general organization of a 4-bit DNA-Quantum ternary carry-lookahead adder. Here, among 4-bit input addition operations, the first three operations are occurring in the DNA system, and the last bit addition operation is occurring in the quantum system. The DNA ternary cache memory store the DNA sequence values of the most significant bit of the inputs and the last DNA system's carry value obtained by the combinational DNA operations. The heat transfer circuit transfers the heat from the quantum system, and the DNA system gets the necessary heat from an external system.

15.8.2   The Circuit Architecture of Quantum-DNA Ternary Carry-Lookahead Adder

Figure 15.11 shows the circuit diagram of a DNA-Quantum 3-bit ternary carry-lookahead adder. And the required heat to perform the operations in the DNA system is provided from an external heat source. The working mechanism of all those

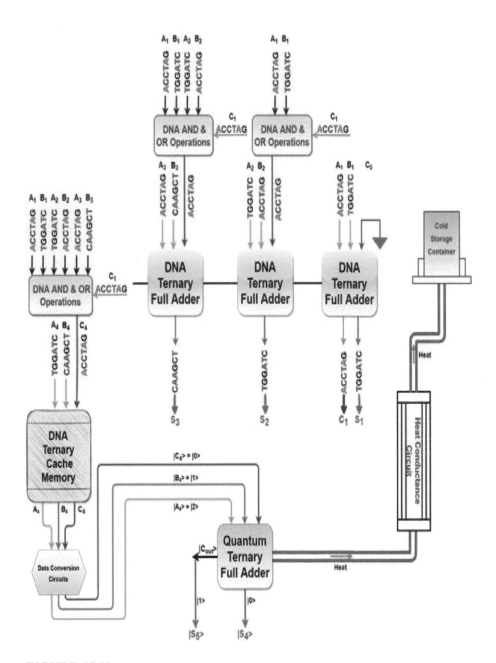

FIGURE 15.10
The General Organizations of DNA-Quantum Ternary 4-Bit Carry-Lookahead Adder

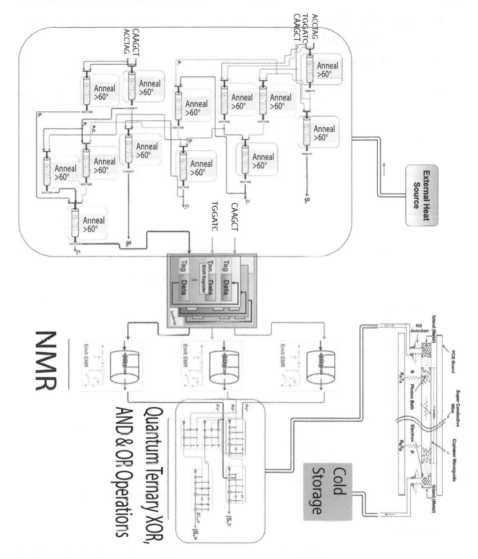

FIGURE 15.11
The Circuit Diagram of DNA-Quantum Ternary 3-Bit Carry-Lookahead Adder Operation

systems was explained in previous sections and chapters. Figure 15.11 depicted to understand the working procedures of the 3-bit DNA-Quantum ternary carry-lookahead adder.

15.8.3 The Working Principles of DNA-Quantum Ternary Carry-Lookahead Adder

The working procedures of the first part of the DNA-Quantum system which is the DNA ternary system are the same as it is discussed in the working procedures of the DNA ternary carry-lookahead adder. As the circuit diagram shows, 1st 2-bit addition and the carry input sequence for the last addition operation is occurring in the DNA system. Then carry bit sequence along with the other two input sequences are stored in the DNA ternary cache memory. From the cache memory, they are passed through the NMR processes to become the input of the quantum ternary full-adder. Then the quantum system will perform the quantum ternary full-adder operation and will generate the final sum outputs.

15.9 DNA-Quantum Ternary Carry-Skip Adder

In the ternary carry-skip adder, a block of ternary full-adder is created, and then the block propagates are determined before operating the actual ternary full-adder operations. If the block propagates BP is equal to 1 or 2 (for DNA computing in ternary logic), then the output carry will be the same as the carry input given to the block.

Here, block propagate,

$BP_i = P_0 . P_1 . P_2 \dots . P_i$

Where, $P_i = (A_i \textbf{ DTXOR } B_i)$

This is how to reduce the time delay to propagate the carry from one block to another. The ternary carry-skip adder is exactly the same as the ternary parallel adder when the BP is not equal to 1 or 2. Therefore, n number of ternary full-adders are designed of an n-bit block. And if BP is equal to 1 or 2 then it is needed to bypass the value of the carry input bit of the block to the carry output using a 3-to-1 ternary multiplexer.

15.9.1 The General Organizations of DNA-Quantum Ternary Carry-Skip Adder

The general overview of the circuit diagram for the DNA-Quantum ternary carry-skip adder is depicted in Figure 15.12.

From the Figure 15.12, it is easy to understand that the DNA-Quantum ternary carry-skip adder is going to be built by dividing it into two parts – one part is in the DNA system which is consists of connected DNA ternary parallel adders, and the second part is in the quantum system which is consist of a 3x1 quantum ternary multiplexer that will bypass the carry output of the block.

And a heat transfer circuit is connected to the quantum system to transfer the massive produced heat into the cooler. Moreover, the necessary heat can be transferred

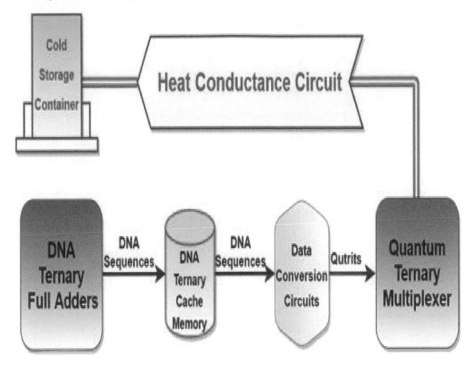

FIGURE 15.12
The Front-View of the DNA-Quantum Ternary Carry-Skip Adder

to the DNA system to operate the DNA system. Figure 15.13 shows the front-view of the DNA-Quantum Ternary 4-Bit Carry-Skip Adder.

In Figure 15.13, the DNA system Includes the four DNA ternary full-adders are shown. The first carry input to the block and carry output of the block are stored in a DNA ternary cache memory so that they can be passed to the next operations.

The data conversion unit converts the data and delivers the quantum ternary bits to the quantum system as the inputs which is a quantum 3-to-1 ternary multiplexer. And the quantum multiplexer does the rest of the operations. It bypasses the carry output qutrit based on the selection input. And as it is already known, it passes the carry input from the block to the next block only if the value of the BP (block propagate) is either $|1>$ or $|2>$. Otherwise, the ternary 3-to-1 multiplexer will bypass the last ternary full adder's carry output as the block's carry output.

15.9.2 The Circuit Architecture of the DNA-Quantum Ternary Carry-Skip Adder

Figure 15.14 shows an inside view of the DNA-Quantum ternary 4-bit carry-skip adder operation. The value of P_i is determined first, and the block propagate is determined by the DNA ternary AND operations accordingly.

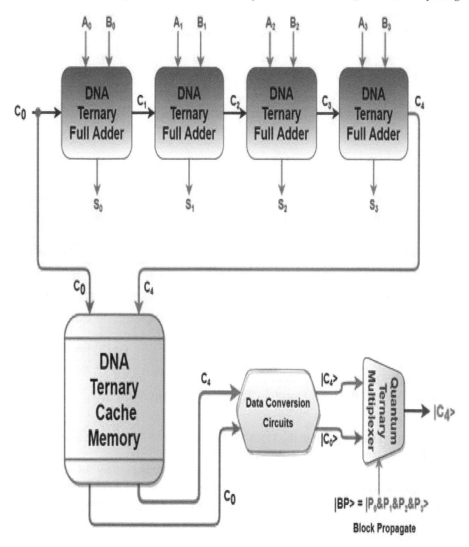

FIGURE 15.13

The Front-View of the DNA-Quantum Ternary 4-Bit Carry-Skip Adder

The value of block propagates BP, and the two carry input sequences C_0 and C_4 (as the Figure shows) will be stored in the DNA ternary cache memory. And the data conversion unit will convert them accordingly. The converted values will be the input of the quantum ternary 3-to-1 multiplexer and will generate one qutrit as the carry output. Note that, the block propagate BP is working as the select input of the quantum ternary multiplexer. If BP is $|1>$ or $|2>$, then the multiplexer will bypass

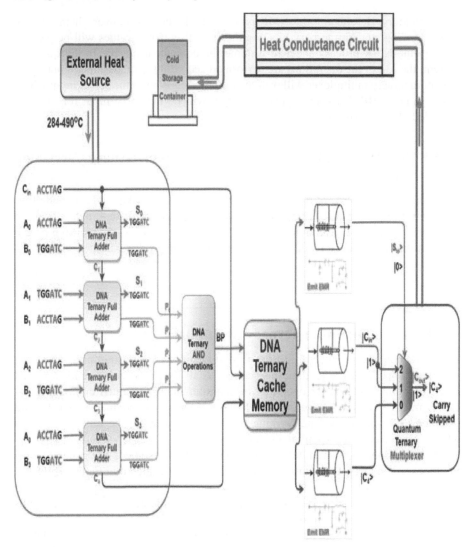

FIGURE 15.14

Circuit Architecture of the DNA-Quantum Ternary 4-bit Carry-Skip Adder

the value of $|C_0>$ (1st carry input to the block) as output. Otherwise, the value of $|C_4>$ will be passed through the quantum ternary 3-to-1 multiplexer.

It is easy to design the circuit diagram by connecting the components shown in Figure 15.14 accordingly. The DNA system will contain four DNA ternary full-adders, the four DNA ternary XOR operations to determine the value of P_i. It will also have three DNA ternary AND operations to get the output of BP. The three DNA sequence values (values of C_0, C_4, and BP) will be stored in the DNA ternary cache memory. Here again, the NMR process can be used to convert them into the corresponding quantum ternary bits. So, the stored sequence values will be passing through the NMR process and will become the equivalent qutrits. In the quantum system, the rest of the operations are performed according to quantum computing. A Quantum ternary multiplexer will pass the output according to the select input which is the value of BP. Thus, the bypassed output will be the carry input for the next cascaded block if existed.

15.9.3   The Working Principles of DNA-Quantum Ternary Carry-Skip Adder

The working procedures of the DNA-quantum ternary carry-skip adder include four main units which are discussed already. They are Quantum system, data storing to cache memory, data conversion, and the DNA system. Besides, the working procedures of the circuit is discussed from the very beginning of this parent section.

15.10   DNA-Quantum Ternary Multiplier

In this section, the architecture of the DNA-Quantum ternary multiplier operation will be develped. Before reading this section, as always, it is recommended to read again the ternary multiplier operation in quantum multiple-valued computing and DNA multiple-valued computing which have been explained in Chapter 11 and Chapter 13, respectively. The working mechanism is exactly the same and that is why it is not needed to explain it again in this chapter.

A 2x2 bit DNA-Quantum ternary multiplier will be designed. The general working block of the DNA-Quantum ternary multiplier can be shown in Figure 15.15.

The DNA 1-bit sequence multiplier is constructed to implement the ternary multiplication rules. Using the 1-bit multiplier it is possible to design the larger bit sequence size ternary multiplication. Figure 15.16 shows the general working block of a DNA-Quantum ternary 2×2 multiplication operation where two input sequences A and B are two-bit values in size, and they produce five qutrits product values.

Figure 15.16 shows that two input sequences $A = A_1 A_0$ and $B = B_1 B_0$ are multiplied by a DNA-Quantum system and produce the production values in the quantum system where $|P> = |P_4 P_3 P_2 P_1 P_0>$.

15.10.1   The Circuit Architecture of the DNA-Quantum Ternary 2×2 Multiplier

Figure 15.17 shows an architecture of the DNA-Quantum ternary 2×2 multiplier. The architecture divides the multiplier into two systems – the DNA system and the quantum system. DNA system, which includes four 1-bit DNA ternary multiplier

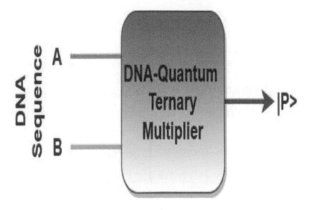

FIGURE 15.15
The General Working Block of a DNA-Quantum Ternary Multiplier

operations, and four DNA ternary half-adders. The outputs from this DNA system are stored in the DNA cache memory. And here, trap ions are used as the data conversion units. After converting the DNA sequences into the equivalent qutrits, the qutrits are entered into the quantum system. Where the quantum system includes one ternary quantum half-adder and two ternary quantum full-adders. Finally, the outputs of the product in the form of quantum ternary bits are gained.

Figure 15.18 shows the complete architecture of a DNA-Quantum ternary 2×2 multiplier. The heat conductance circuit is added and through the heat conductance circuit the generated massive heat from the quantum system will be transferred to cold storage. And notice, this time that the NMR process is used to convert the DNA sequences into the corresponding qutrits. To be mentioned, here the NMR process will be performed at zero kelvin temperature. And all the other operations were also described earlier.

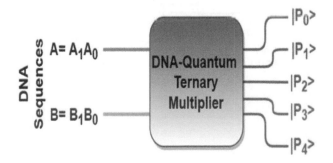

FIGURE 15.16
The General Working Block of a DNA-Quantum Ternary 2×2 Multiplier

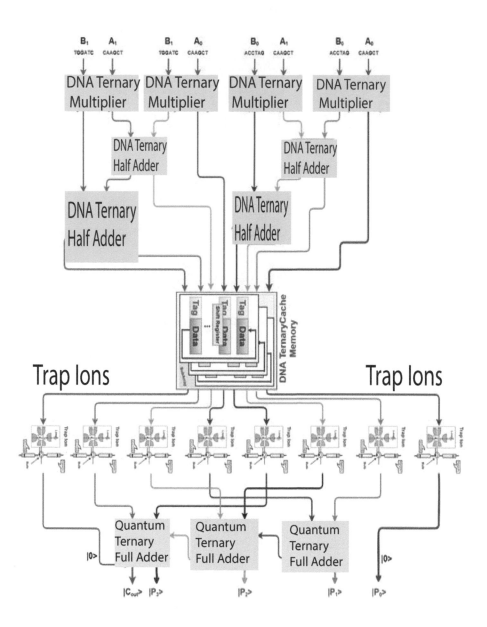

FIGURE 15.17

Circuit Architecture of a DNA-Quantum Ternary 2 × 2 Multiplier

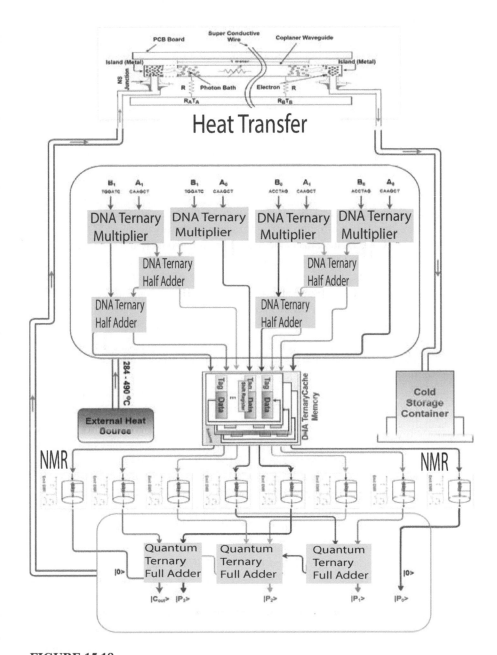

FIGURE 15.18
The Complete Architecture of a DNA-Quantum Ternary 2 × 2 Multiplier

15.10.2   The Working Procedure of the DNA-Quantum Ternary 2×2 Multiplier

The above Figures have already shown the working procedures of performing multiplication operations of two inputs. Observe the operations on each component.

Let's see the operational behavior of the designed architecture for the DNA-Quantum ternary 2×2 multiplier, shown in Figure 15.18.

Suppose, it is needed to multiply two ternary input values in the DNA-Quantum multiplier, where A = 02, and B = 10. After multiplication operations, these inputs will produce $P = P_4 P_3 P_2 P_1 P_0$, where these product values are in the form of qutrits. Here, the value of A_0 = TGGATC, and A_1 = ACCTAG. And the value of B_0 = ACCTAG, and B_1 = CAAGCT.

1. In the DNA system, the 1st DNA ternary 1-bit multiplier will get the input values of A_0 and B_0. Here, A_0 = TGGATC, and B_0 = ACCTAG. Therefore, this multiplier will generate two outputs m_0 = ACCTAG and c_0 = ACCTAG. Here, m_0 is the multiplication product, and c_0 is the carry value. The value of m_0 will be the least significant qutrit of the final product. Therefore, it is needed to pass the value of m_0 to the DNA ternary cache memory. And the value of c_0 will be work as an input to the 1st DNA ternary half-adder shown in Figure 15.18 (from the right side of the Figure).

2. The 2nd DNA ternary 1-bit multiplier will get the input values of A_1 and B_0. Here, A_1 = ACCTAG, and B_0 = ACCTAG. Therefore, this multiplier will generate two outputs m_1 = ACCTAG and c_1 = ACCTAG. The value of m_1 will be work as the 2nd input value to the 1st DNA ternary half-adder. And the value of c_1 will be work as an input to the 3rd DNA ternary half-adder.

3. The 3rd DNA ternary 1-bit multiplier will get the input values of A_0 and B_1. Here, A_0 = TGGATC, and B_1 = CAAGCT. And, this multiplier will generate two outputs m_2 = TGGATC and c_2 = ACCTAG. The value of m_2 will be stored to the DNA ternary cache memory which will work as an input to the 1st quantum ternary half adder eventually. And the value of c_2 will work as an input to the 2nd DNA ternary half-adder.

4. The 4th DNA ternary 1-bit multiplier will get the input values of A_1 and B_1. Here, A_1 = ACCTAG, and B_1 = CAAGCT. And, this multiplier will generate two outputs m_3 = ACCTAG and c_3 = ACCTAG. The value of m_3 will be work as another input value to the 2nd DNA ternary half-adder. And the value of c_3 will work as an input to the 4th DNA ternary half-adder.

5. Now, the 1st DNA ternary half-adder will get inputs c_0 = ACCTAG (from step 1), and m_1 = ACCTAG (from step 2). So, this half-adder will produce two outputs. Here, from the 1st half adder, S_0 = ACCTAG, and the carry output will also be ACCTAG. The value of S_0 will be stored in the DNA ternary cache memory and it will work as another input of the 1st quantum

ternary half-adder further. And the carry value from this half-adder will work as another input value of the 3rd DNA ternary half-adder (according to Figure 15.18).

6. The 2nd DNA ternary half-adder will get inputs c_2 = ACCTAG (from step 3), and m_3 = ACCTAG (from step 4). So, this half-adder will produce two outputs, S_1 = ACCTAG, and the carry output will also be ACCTAG (see the truth Table of DNA ternary half-adder). The value of S_1 will be stored in the cache memory and it will work as an input of the 1st quantum ternary full-adder eventually. And the carry value from this half-adder will work as another input value of the 4th DNA ternary half-adder (see Figure 15.18).

7. The 3rd DNA ternary half-adder will get inputs c_1 = ACCTAG (from step 2), and also ACCTAG from the carry output of the 1st half-adder (from step 5). So, this half-adder will produce two outputs, S_2 = ACCTAG, and the carry output will also be ACCTAG (see the truth Table of ternary half-adder). The value of S_2 will be stored in the cache memory and it will work as an input of the 1st quantum ternary full-adder. And the carry value from this half-adder will also be stored in the cache memory and will work as an input value of the 2nd quantum ternary full-adder (according to Figure 15.18).

8. The 4th DNA ternary half-adder will get inputs c_3 = ACCTAG (from step 4), and also ACCTAG from the carry output of the 2nd half-adder (from step 6). So, this half-adder will produce two outputs, S_3 = ACCTAG, and the carry output will also be ACCTAG (see the truth Table of ternary half-adder). The value of S_3 will be stored in the cache memory and will work as an input of the 2nd quantum ternary full-adder. And the carry value from this half-adder produces the most significant qutrit value of the final product value, and thus it will be stored in the cache memory as well.

9. The DNA system operations are done. All the outputs are stored from the DNA system to the DNA ternary cache memory. Now, from the cache memory, all the values will be passed through the NMR process to convert them into their equivalent quantum ternary bits.

10. In the quantum system, The 1st quantum ternary half-adder will get inputs $|S_0>$ = $|0>$ (from the converted value from step 5), and also $|m_2>$ = $|2>$ from step 3. So, this half-adder will produce the value of $|P_1>$ = $|2>$. And a carry out value $|0>$ which will work as a carry input to the 1st quantum full-adder.

11. Now, the 1st quantum ternary full-adder operations can be performed. It gets three inputs as, $|S_1>$ = $|0>$, $|S_2>$ = $|0>$, and also $|0>$ (from step 9). Therefore the sum of the addition will be $|0>$ which is the value of $|P_2>$. And the carry will also be $|0>$ which will be the carry input to the 2nd quantum ternary full-adder.

12. The last operation is the 2nd quantum ternary full-adder. It gets the three inputs as - $|S_3> = |0>$, $|0>$ from the converted value of step 7, and also $|0>$ from step 10. Therefore it will also produce sum $|0>$ and carry output as $|0>$. The sum output will be the value of $|P_3>$ and the carry output will be the value of $|P_4>$. And therefore, both the value as $|0>$ is obtained.

13. And it is possible to get the value of $|P_0>$ from the converted qutrit value of $m_0 = ACCTAG$ from step 1. Therefore, the value of $|P_0>$ as $|0>$ is obtained.

So, finally, here all the product values – $|P_0> = |0>$, $|P_1> = |2>$, $|P_2> = |0>$, $|P_3> = |0>$, and $|P_4> = |0>$. Therefore, there is the product of the ternary multiplication operation of $|P> = |00020>$ in the DNA-Quantum ternary multiplication system, which is absolutely correct for the given inputs.

15.11   DNA-Quantum Ternary Divider

In the DNA-Quantum ternary divider, the first part of the operations will be performed in the DNA system and the last part of the operations will be performed in the quantum system. That implies, the input will be the DNA sequences and the generated output will be the quantum ternary bits. The DNA-Quantum ternary divider block diagrams are shown in Figure 15.19.

The above figure shows that for an n-bit DNA-Quantum ternary divider, the dividend and the divisor will be the n-bit DNA sequences, but the outputs (quotient and remainder) will be in the form of n-qutrit.

15.11.1   The Construction of the DNA-Quantum Ternary 2 × 2 Multiplier

As the divider blocks show, a ternary divider circuit is needed to construct, where the components will be the DNA system followed by the DNA ternary cache memory, which will again be followed by the data conversion units to convert the DNA base sequences into the quantum ternary bits. And finally, the converted qutrits will generate the outputs performing the rest of the operations.

Figure 15.20 shows the circuit architecture of the DNA-quantum ternary 1-bit divider. The extra part is the DNA ternary cache memory which is used to store the DNA ternary information and the data conversion units which will convert the DNA information into the equivalent quantum ternary bits.

It is known that the working mechanism of the 2-bit DNA ternary divider and 2-qutrit quantum ternary divider. Therefore, it is possible to construct the general architecture of the DNA-Quantum ternary 2-bit divider. Figure 15.21 shows the general architecture of the 2-bit DNA-Quantum ternary divider operation.

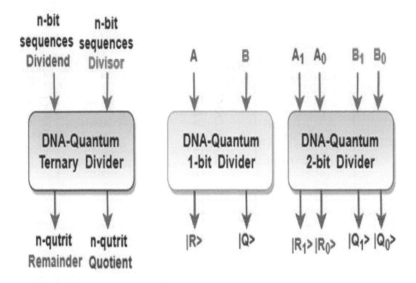

FIGURE 15.19
The General Block Diagrams of DNA-Quantum Ternary Dividers

15.12 Multiple-Valued DNA-Quantum Comparator

As discussed in the previous chapter, the comparator is a circuit that compares two input signals and gives output as which input is larger or smaller, or equal. Here, in the DNA-Quantum circuit, the input of A and B is the DNA sequence and the Quantum qubit is the output of $|Y>$ and $|Z>$ where $|X>$. Table 15.5 shows the truth table of the DNA-Quantum 1-bit comparator.

Figure 15.22 displays the block diagram of the DNA-Quantum 1-bit comparator. Here two DNA decoders are used in the DNA part and in the quantum part, nine multi-valued quantum AND and six multi-valued quantum OR operations are used.

15.12.1 Circuit Architecture of Multiple-Valued DNA-Quantum Comparator

The construction of the Multi-Valued DNA-Quantum 1-qubit comparator circuit is shown in Fig.15.23. In this circuit, the input, A and B, at first, pass through two 1-to-3 Decoders separately so that inserted DNA molecular sequence, i.e. **ACCTAG, CAAGCT** or **TGGATC,** which represent "1", "2" and "3", respectively, can be identified using the output of 1 to 3 decoders. Nine Quantum AND operation is used in DNA-Quantum 1-qubit comparator where each input of these AND operations come from the various combinations of the output of 1-to-3 decoder. However, as the output of DNA operation will act as an input of Quantum operation, thus each DNA molec-

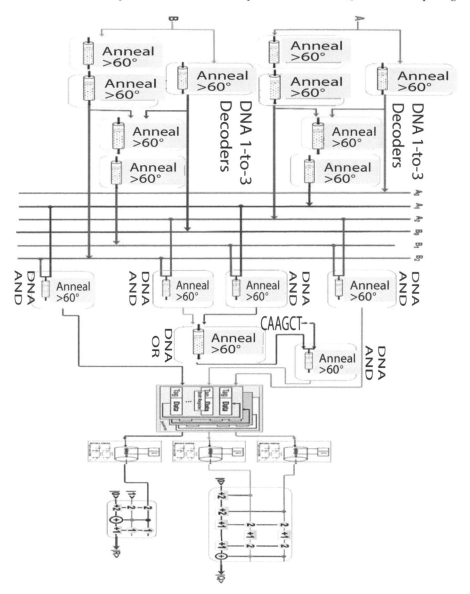

FIGURE 15.20
Circuit Architecture of the DNA-Quantum Ternary 1-Bit Divider

ular sequence is converted to its corresponding qubit by using the trap ion technique. Before the conversion DNA outputs are stored in a DNA Cache memory shortly to match the speed to different operations. Finally, based on the value of A and B, only one line of each 1-to-3 decoder generates qubit $|2>$ along the line whereas others

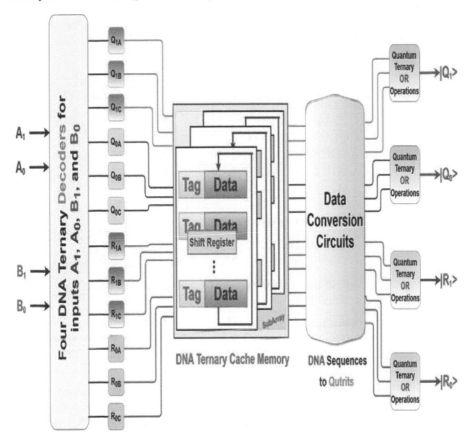

FIGURE 15.21
The General Architecture of the DNA-Quantum Ternary 2-Bit Divider

provide |0>. Thus only one Quantum OR operation gives the result |2>, where other outputs, remain inactive. i.e. |0>.

15.12.2 Working Principle of Multiple-Valued DNA-Quantum Comparator

1. For input sequences, A, B =**ACCTAG, ACCTAG**, the multi-valued 1-to-3 decoder will perform (A0 && B0) and the 4th line will be open. As a result, the output qubit of |Y> equal |2> and remaining lines |X> to |Z> will remain closed |0>.

2. For input sequences, A, B =**ACCTAG, CAAGCT**, the multi-valued 1-to-3 decoder will perform (A0 && B1) and the 1st line will be open. As a

TABLE 15.5

Truth Table of DNA-Quantum1-Bit Comparator

Inputs		Outputs		
A	B	|X>	|Y>	|X>
		A < B	A= B	A > B
ACCTAG	ACCTAG	|0>	|2>	|0>
ACCTAG	CAAGCT	|2>	|0>	|0>
ACCTAG	TGGATC	|2>	|0>	|0>
CAAGCT	ACCTAG	|0>	|0>	|2>
CAAGCT	CAAGCT	|0>	|2>	|0>
CAAGCT	TGGATC	|2>	|0>	|0>
TGGATC	ACCTAG	|0>	|0>	|2>
TGGATC	CAAGCT	|0>	|0>	|2>
TGGATC	TGGATC	|0>	|2>	|0>

result, the output qubit of |X> equal **|2>** and remaining lines |Y> to |Z> will remain closed **|0>**.

3. For input sequences, A, B =**ACCTAG, TGGATC** , the multi-valued 1-to-3 decoder will perform (A0 && B2) and the 2nd line will be open. As a result, the output qubit of |X> equal **|2>** and remaining lines |Y> to |Z> will remain closed **|0>**.

4. For input sequences, A, B =**CAAGCT, ACCTAG**, the multi-valued 1-to-3 decoder will perform (A1 && B0) and the 7th line will be open. As a result, the output qubit of |Z> equal **|2>** and remaining lines |X> to |Y> will remain closed **|0>**.

5. For input sequences, A, B =**CAAGCT, CAAGCT**, the multi-valued 1-to-3 decoder will perform (A1 && B1) and the 5th line will be open. As a result, the output qubit of |Y> equal **|2>** and remaining lines |X> to |Z> will remain closed **|0>**.

6. For input sequences, A, B =**CAAGCT, TGGATC** , the multi-valued 1-to-3 decoder will perform (A1 && B2) and the 3rd line will be open. As a result, the output qubit of |X> equal **|2>** and remaining lines |Y> to |Z> will remain closed **|0>**.

7. For input sequences, A, B =**TGGATC** , **ACCTAG**, the multi-valued 1-to-3 decoder will perform (A2 && B0) and the 8th line will be open. As a result, the output qubit of |Z> equal **|2>** and remaining lines |X> to |Y> will remain closed **|0>**.

8. For input sequences, A, B =**TGGATC** , **CAAGCT**, the multi-valued 1-to-3 decoder will perform (A2 && B1) and the 9th line will be open. As a result, the output qubit of |Z> equal **|2>** and remaining lines |X> to |Y> will remain closed **|0>**.

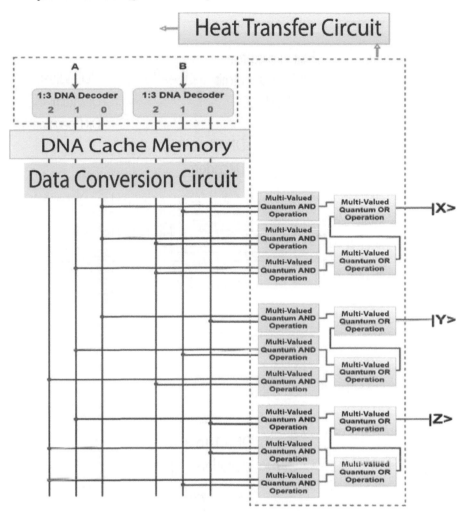

FIGURE 15.22
Block Diagram of Multi-Valued DNA-Quantum 1-Bit Comparator

9. For input sequences, A, B =**TGGATC, TGGATC**, the multi-valued 1-to-3 decoder will perform (A2 && B2) and the 6th line will be open. As a result, the output qubit of |Y> equal **|2>** and remaining lines |X> to |Z> will remain closed **|0>**.

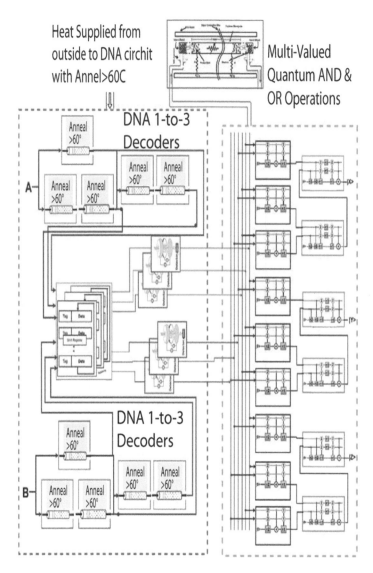

FIGURE 15.23
Circuit Architecture of Multi-Valued DNA-Quantum Comparator

15.13   Summary

The cross-platform for the ternary logic operation in Quantum and DNA computing can be formed in two ways namely, Quantum-DNA Ternary system and

DNA-Quantum Ternary system, called quantum molecular biology. In the Ternary Quantum-DNA system, the system is divided into two parts where the first part contains the Quantum ternary system and the second part contains the DNA ternary system. In the DNA-Quantum system, the system is also divided into two parts where the first part contains the DNA ternary system and the second part contains the Quantum ternary system. To store the quantum ternary data also called qutrits, Quantum Ternary Cache Memory is used. And DNA information for ternary data can be stored in the DNA Ternary Cache Memory. The required time to perform the arithmetic operations on the Quantum-DNA or DNA-Quantum system depends on the DNA system only as the DNA system takes a much longer time to perform operations.

Bibliography

[1] Britannica, T. (2020). Editors of Encyclopaedia. Argon. Encyclopedia Britannica.

[2] Hallworth, R. P., & Heath, F. G. (1962). Semiconductor circuits for ternary logic. Proceedings of the IEE-Part C: Monographs, 109(15), 219-225.

[3] Dhande, A. P., & Ingole, V. T. (2005, March). Design and implementation of 2 bit ternary ALU slice. In Proc. Int. Conf. IEEE-Sci. Electron., Technol. Inf. Telecommun (Vol. 17).

[4] Mandal, S. B., Chakrabarti, A., & Sur-Kolay, S. (2011, May). Synthesis techniques for ternary quantum logic. In 2011 41st IEEE International Symposium on Multiple-Valued Logic (pp. 218-223). IEEE.

[5] Li, T., Gao, H., Wang, C., Cheng, Z., Xue, J., Zhang, Z., ... & Cao, J. (2022). Oil utilization degree at various pore sizes via different displacement methods. Journal of Petroleum Exploration and Production Technology, 1-17.

Part IV

Multiple-Valued Combinational Circuits in Quantum Molecular Biology

Overview

Multi-Valued Quantum computing is a type of computing that performs calculations using the collective properties of quantum states, such as superposition, interference, and entanglement. Quantum computers are machines that are capable of performing quantum computations. Quantum computing makes use of quantum mechanics to make a huge leap forward in computation in order to tackle certain problems. The design, simulation, and development of multi-valued combinational circuits using quantum technology are discussed in the first chapter of this part. The general concept of quantum computing is discussed at the beginning of the book in order to gain a better grasp of it. Quantum gates and basic quantum processes are the foundational elements of quantum combinational Circuits. DNA computing is another type of computing where molecular biology is involved. DNA molecules are used to represent information. A huge amount of data can be stored in multi-valued DNA computing. DNA computing is a new branch of computing that uses DNA, biochemistry, and molecular biology hardware to replace traditional electronic computing. DNA computing's theory, experiments, and applications are all part of the field research and development. A brief explanation is required before the actual designs. The main focus of this part is on the design of multi-valued computing for various combinational circuits in quantum molecular biology. This part documents everything from the architectures of basic multi-valued quantum and DNA combinational circuit components to the design of each circuit, as well as the main architecture of quantum and DNA combinational circuits and their working principle. Multi-Valued multiplexer, multi-valued demultiplexer, multi-valued encoder, and multi-valued decoder will be discussed in this part in quantum, DNA computing and quantum molecular biology. The first two chapters will be about quantum computing and DNA computing in a multi-valued logic system. The last two chapters of this part are about the cross-platform of these two computers. That means quantum molecular biology. Quantum molecular biology consists of two forms, one is quantum-DNA computing and the other is DNA-quantum computing.

16

Multiple-Valued Quantum Combinational Circuits

16.1 Introduction

The qutrit is the information unit of a three-valued quantum system (ternary quantum system). The ternary quantum system, which has the basis states |0>, |1>, and |2>, is a form of the three-dimensional quantum system. These basic states are referred to as qutrit states, and they are represented by three one-dimensional vectors (Equation 16.1).

$$|0> = \begin{bmatrix} 1 \\ 0 \\ 0 \end{bmatrix}, |1> = \begin{bmatrix} 0 \\ 1 \\ 0 \end{bmatrix}, |2> = \begin{bmatrix} 0 \\ 0 \\ 1 \end{bmatrix} \tag{16.1}$$

In a ternary quantum system, a qutrit can be defined as a linear superposition of the above-mentioned basis states with the following equation:

$\psi = \alpha |0> + \text{ß} |1> + \gamma |2>$

Where α, ß and γ are the complex quantities to represent the probability amplitudes of the basis states and ψ is the wave function. This chapter will discuss about the multi-valued quantum multiplexer, demultiplexer, encoder and decoder.

16.2 Multiple-Valued Quantum Multiplexer

One of the most vital block designs in a binary digital system is the multiplexer. It is a device that allows only one input signal from a group of input signals, and the input signal selected by the multiplexer is transmitted into a single medium. Multiplexers, in fact, help to improve the efficiency of the communication system. It enables the transmission of data such as audio, video, and so on from various channels via cables.

This circuit is useful for creating multi-valued arithmetic circuits. An n-to-1 multiplexer is composed of three parts. There are three n inputs, one output, and one selector line. A multi-valued Quantum n-to-1 multiplexer is depicted in Figure 16.1.

DOI: 10.1201/9781003381938-16

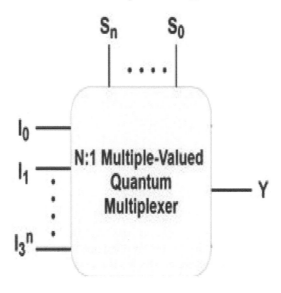

FIGURE 16.1
Multi-Valued Quantum n-to-1 Multiplexer

Table 16.1 and Figure 16.2, respectively, show the truth table and associative block diagram of Multi-Valued Quantum 3-to-1 Multiplexer.

For, Multi-Valued Quantum 3-to-1 Multiplexer, one 1-to-3 Quantum decoder is needed to get the output of selection line S. Additionally, three multi-valued quantum AND and one multi-valued quantum OR operation are required as shown in Figure 16.2.

When total inputs are 9 and the selection lines iare 2, as shown in Table 16.2, then it is multi-valued quantum 9-to-1 Multiplexer. Here, among the quantum AND operations, only one AND will provide its input to the output line via quantum OR operation. Table 16.2 shows the 9-to-1 Quantum Multiplexer Truth Table.

Figure 16.3 shows the block diagram of multi-valued quantum 9-to-1 multiplexer. For 9-to-1 multiplexer, two quantum decoders and one quantum ternary OR are needed with nine quantum ternary AND operations. The output of nine quantum

TABLE 16.1
Truth Table of 3-to-1 Quantum Multiplexer

Inputs	Outputs
$|S>$	$|Y>$
$|0>$	$|I0>$
$|1>$	$|I1>$
$|2>$	$|I2>$

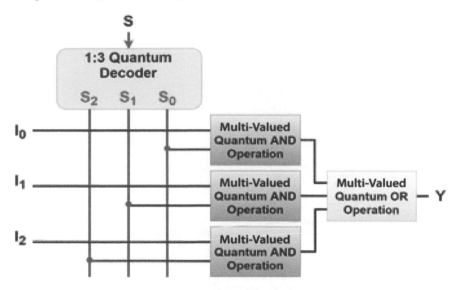

FIGURE 16.2
Block Diagram of Multi-Valued Quantum 3-to-1 Multiplexer

ternary AND operation will be the input for multi-valued quantum OR operation which will produce the final output.

Additionally, to design a Multi-Valued Quantum *n-to-*1 Multiplexer (Figure 16.4), there will be a 3^n input line that will be connected to 3^n Quantum AND operation. Then a total of n selection lines will decide which input will pass through the output of Quantum OR operation. The block diagram of multi-valued quantum *n-to-*1 multiplexer is shown in Figure 16.4.

TABLE 16.2
Truth Table of 9-to-1 Quantum Multiplexer

Inputs		Outputs
$|S1>$	$|S0>$	$|Y>$
$|0>$	$|0>$	$|I0>$
$|0>$	$|1>$	$|I1>$
$|0>$	$|2>$	$|I2>$
$|1>$	$|0>$	$|I3>$
$|1>$	$|1>$	$|I4>$
$|1>$	$|2>$	$|I5>$
$|2>$	$|0>$	$|I6>$
$|2>$	$|1>$	$|I7>$
$|2>$	$|2>$	$|I8>$

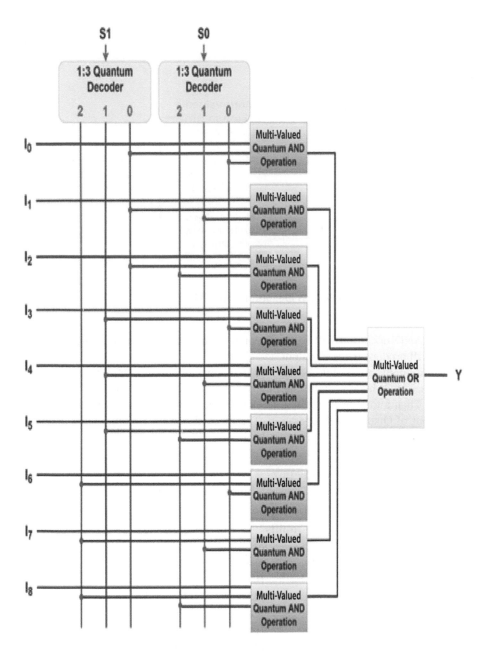

FIGURE 16.3
Block Diagram of Multi-Valued Quantum 9-to-1 Multiplexer

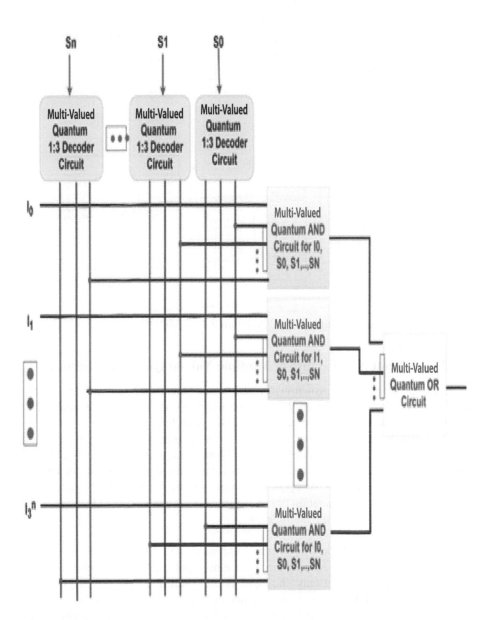

FIGURE 16.4
Block Diagram of Multi-Valued Quantum *n-to*-1 Multiplexer

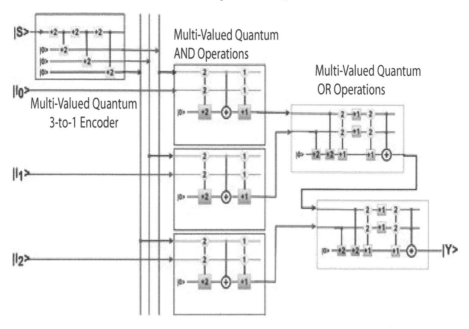

FIGURE 16.5
Circuit Architecture of Multi-Valued Quantum 3-to-1 Multiplexer

16.2.1 Circuit Architecture of Multiple-Valued Quantum Multiplexer

Figure 16.5 shows the construction of multi-valued Quantum 3-to-1 multiplexer circuit. In this circuit, all the inputs |I0> - |I2> connect to three different Quantum AND gates in which another input comes from the Selection line |S>.

To select one particular input, qubit |0>, |1> or |2> is providing through |S> based on which one of the three output lines of the Quantum 1-to-3 Decoder gets activated and passes qubit |2> through it. Other output lines of the decoder provide qubit |0>. Thus only one Quantum AND operation is open to take the input to the output. Finally, the Quantum OR operation passes the input that comes via the Quantum AND operation to the output of the multiplexer.

The architecture of the Multi-Valued Quantum 9-to-1 Multiplexer is also the same (Figure 16.6). However, as there are nine inputs, |I0> - |I8>, two selection lines are required to select only one input as the result of output |Y>. Figure 16.6 shows the circuit architecture of multi-valued quantum 9-to-1 multiplexer

The whole circuit contains two quantum 1-to-3 decoders, eighteen multi-valued quantum AND operations and eight multi-valued quantum OR operations which will produce the final output.

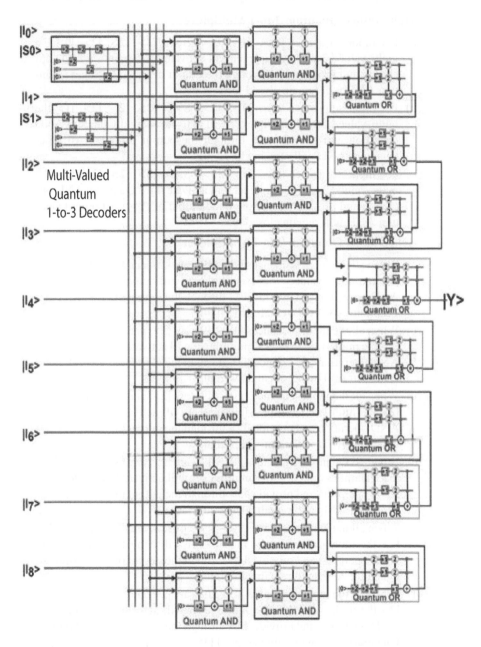

FIGURE 16.6
Circuit Architecture of Multi-Valued Quantum 9-to-1 Multiplexer

16.2.2   Working Principle of Multiple-Valued Quantum Multiplexer

1. **Multi-Valued Quantum 3-to-1 Multiplexer**

    (a) For input sequences |I0>, |I1>, |I2>, |S> = |2>, |2>, |2>, |0> the multi-valued 3-to-1 multiplexer will activate first Quantum AND operation and provide |I0> through the output of |Y> . All other Quantum AND will remain closed |0>.

    (b) For input sequences |I0>, |I1>, |I2>, |S> = |2>, |2>, |2>, |1> the multi-valued 3-to-1 multiplexer will activate second Quantum AND operation and provide |I1> through the output of |Y> . All other Quantum AND will remain closed |0>.

    (c) For input sequences |I0>, |I1>, |I2>, |S> = |2>, |2>, |2>, |2> the multi-valued 3-to-1 multiplexer will activate third Quantum AND operation and provide |I2> through the output of |Y> . All other Quantum AND will remain closed |0>.

2. **Multi-Valued Quantum 9-to-1 Multiplexer**

    (a) When all the input sequences |I0>- |I9> are |2>, and |S0>,|S1> = |0>, |0> the multi-valued 9-to-1 multiplexer will activate first Quantum AND operation and provide |I0> through the output of |Y> . All other Quantum AND will remain closed |0>.

    (b) When all the input sequences |I0>- |I9> are |2>, and |S0>,|S1> = |0>, |1> the multi-valued 9-to-1 multiplexer will activate second Quantum AND operation and provide |I1> through the output of |Y> . All other Quantum AND will remain closed |0>.

    (c) When all the input sequences |I0>- |I9> are |2>, and |S0>,|S1> = |0>, |2> the multi-valued 9-to-1 multiplexer will activate third Quantum AND operation and provide |I2> through the output of |Y> . All other Quantum AND will remain closed |0>.

    (d) When all the input sequences |I0>- |I9> are |2>, and |S0>,|S1> = |1>, |0> the multi-valued 9-to-1 multiplexer will activate fourth Quantum AND operation and provide |I3> through the output of |Y> . All other Quantum AND will remain closed |0>.

    (e) When all the input sequences |I0>- |I9> are |2>, and |S0>,|S1> = |1>, |1> the multi-valued 9-to-1 multiplexer will activate fifth Quantum AND operation and provide |I4> through the output of |Y> . All other Quantum AND will remain closed |0>.

    (f) When all the input sequences |I0>- |I9> are |2>, and |S0>,|S1> = |1>, |2> the multi-valued 9-to-1 multiplexer will activate sixth Quantum AND operation and provide |I5> through the output of |Y> . All other Quantum AND will remain closed |0>.

    (g) When all the input sequences |I0>- |I9> are |2>, and |S0>,|S1> = |2>, |0> the multi-valued 9-to-1 multiplexer will activate seventh Quantum AND operation and provide |I6> through the output of |Y> . All other Quantum AND will remain closed |0>.

(h) When all the input sequences |I0>- |I9> are |2>, and |S0>,|S1> = |2>, |1> the multi-valued 9-to-1 multiplexer will activate eighth Quantum AND operation and provide |I7> through the output of |Y> . All other Quantum AND will remain closed |0>.

(i) When all the input sequences |I0>- |I9> are |2>, and |S0>,|S1> = |2>, |2> the multi-valued 9-to-1 multiplexer will activate ninth Quantum AND operation and provide |I8> through the output of |Y> . All other Quantum AND will remain closed |0>.

16.3 Multiple-Valued Quantum Demultiplexer

Demultiplexer circuits are among the most important circuits in the design of complex hardware. It maps a single input to multiple outputs. Thereby, a 1-to-n multi-valued quantum demultiplexer consists of three main parts: one input, n selection line and 3^N outputs as shown in Figure 16.7.

The truth table of 1-to-3 demultiplexer is shown in Table: 16.3. Here input is I, outputs are $|Y_0>, |Y_1>, |Y_2>$ and selection lines are $S_0, S_1, S_2.$ The selected output equals the input I and all other outputs are 0.

Figure 16.8 depicts the 1-to-3 demultiplexer's block diagram. Depending on the combination chosen, one of the decoder outputs becomes 2 while the others remain 0.

One 1-to-3 quantum decoder is needed first, then three multi-valued quantum AND operations are needed to produce the outputs.

Table 16.4 shows the truth table of 1-to-9 Demultiplexer. Here 2 selection lines S_1, S_2 are needed to select one output line among nine.

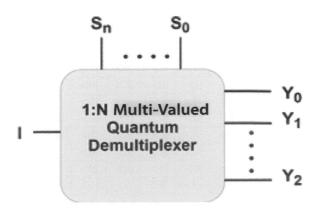

FIGURE 16.7
1-to-n Multi-Valued Quantum Demultiplexer

TABLE 16.3

Truth Table of 1-to-3 Quantum
Demultiplexer

Inputs	Outputs		
|S>	|Y_2>	|Y_1>	|Y_0>
|0>	|0>	|0>	|I>
|1>	|0>	|I>	|0>
|2>	|I>	|0>	|0>

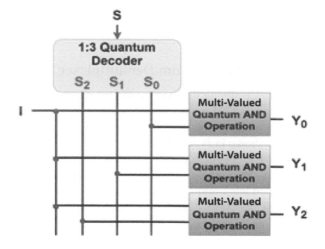

FIGURE 16.8

Block Diagram of Multi-Valued Quantum 1-to-3 Demultiplexer

TABLE 16.4

Truth Table of 1-to-9 Quantum Demultiplexer

Inputs		Outputs								
|S1>	|S0>	|Y0>	|Y1>	|Y2>	|Y3>	|Y4>	|Y5>	|Y6>	|Y7>	|Y8>
|0>	|0>	|I>	|0>	|0>	|0>	|0>	|0>	|0>	|0>	|0>
|0>	|1>	|0>	|I>	|0>	|0>	|0>	|0>	|0>	|0>	|0>
|0>	|2>	|0>	|0>	|I>	|0>	|0>	|0>	|0>	|0>	|0>
|1>	|0>	|0>	|0>	|0>	|I>	|0>	|0>	|0>	|0>	|0>
|1>	|1>	|0>	|0>	|0>	|0>	|I>	|0>	|0>	|0>	|0>
|1>	|2>	|0>	|0>	|0>	|0>	|0>	|I>	|0>	|0>	|0>
|2>	|0>	|0>	|0>	|0>	|0>	|0>	|0>	|I>	|0>	|0>
|2>	|1>	|0>	|0>	|0>	|0>	|0>	|0>	|0>	|I>	|0>
|2>	|2>	|0>	|0>	|0>	|0>	|0>	|0>	|0>	|0>	|I>

Figure 16.9 shows the block diagram of multi-valued Quantum 1-to-9 Demultiplexer. Here the block diagram contains two 1-to-3 quantum decoders and nine multi-valued quantum AND operations to produce final outputs.

16.4 Circuit Architecture of Multiple-Valued Quantum Demultiplexer

The construction of the Multi-Valued Quantum 1-to-3 Demultiplexer circuit is shown in Figure 16.10. In this circuit, the input |I> is directly connected to all the Quantum AND operations as one input. The various outputs of selection line |S> contribute as another input to these AND operations. The result of |S> decides which output line must be activated.

To select one particular output, qubit |0>, |1> or |2> is provided through |S> based on which one of the three output lines of the Quantum 1-to-3 Decoder gets activated and passes qubit |2> through it. Other output lines of the decoder provide qubit |0>. Thus only one Quantum AND operation is open to select one output among |Y0> to |Y2>.

The architecture of the Multi-Valued Quantum 1-to-9 Demultiplexer also applies the same procedure shown in Figure 16.11. However, as there are nine outputs, |Y0> - |Y8>, two selection lines are required to active only one output among |Y0> - |Y8>, where other outputs remain inactive. i.e. |0>.

In the above circuit, inputs are inserted into the quantum 1-to-3 decoders, then they have passed through nine multi-valued quantum AND oprerations and entered into another nine multi-valued quantum AND operations which produced the final outputs.

16.4.1 Working Principle of Multiple-Valued Quantum Demultiplexer

1. **Multi-Valued Quantum 1-to-3 Demultiplexer**

   (a) For input sequences |I>, |S> = **|2>, |0>** the multi-valued 1-to-3 Demultiplexer will activate first Quantum AND and provide |2> through the output of |Y0> to activate it. All other outputs |Y0> - |Y2> will remain close as remaining Quantum AND operation will generate **|0>**.

   (b) For input sequences |I>, |S> = **|2>, |1>** the multi-valued 1-to-3 Demultiplexer will activate second Quantum AND and provide |2> through the output of |Y1> to activate it. All other outputs |Y0> - |Y2> will remain close as remaining Quantum AND operation will generate **|0>**.

   (c) For input sequences |I>, |S> = **|2>, |2>** the multi-valued 1-to-3 Demultiplexer will activate third Quantum AND and provide |2>

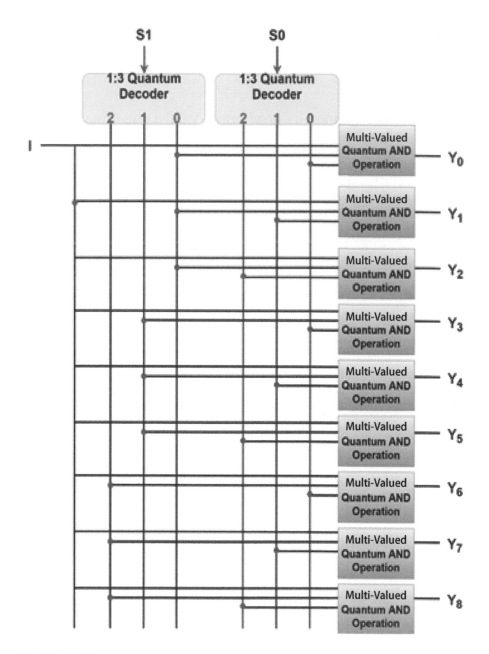

FIGURE 16.9
Block Diagram of Multi-Valued Quantum 1-to-9 Demultiplexer

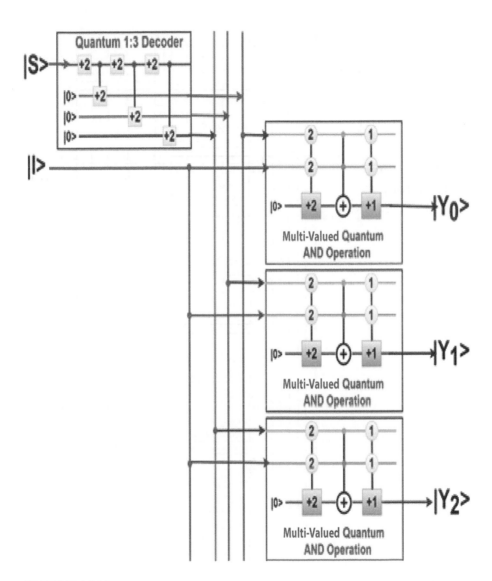

FIGURE 16.10
Circuit Architecture of Multi-Valued Quantum 1-to-3 Demultiplexer

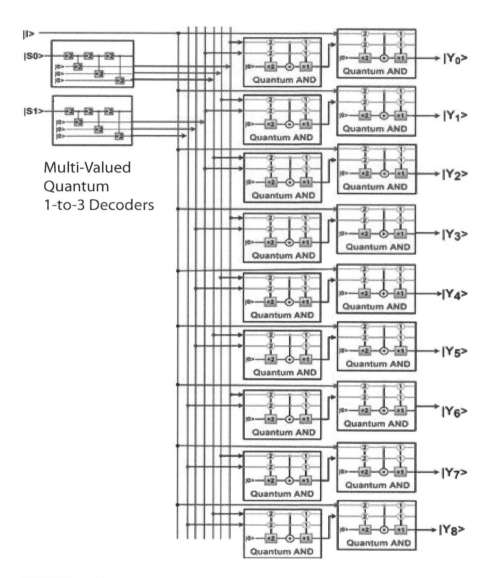

FIGURE 16.11
Circuit Architecture of Multi-Valued Quantum 1-to-9 Demultiplexer

through the output of |Y2> to activate it. All other outputs |Y0> - |Y2> will remain close as remaining Quantum AND operation will generate **|0>**.

2. **Multi-Valued Quantum 1-to-9 Demultiplexer**

(a) When the input qubit |I> is |2>, and |S0>,|S1> = **|0>, |0>** the multi-valued 1-to-3 Demultiplexer will activate first Quantum AND and provide |2> through the output of |Y0> to activate it . All other outputs |Y0> - |Y9> will remain close as remaining Quantum AND operation will generate **|0>**.

(b) When the input qubit |I> is |2>, and |S0>,|S1> = **|0>, |1>** the multi-valued 1-to-3 Demultiplexer will activate second Quantum AND and provide |2> through the output of |Y1> to activate it . All other outputs |Y0> - |Y9> will remain close as remaining Quantum AND operation will generate **|0>**.

(c) When the input qubit |I> is |2>, and |S0>,|S1> = **|0>, |2>** the multi-valued 1-to-3 Demultiplexer will activate third Quantum AND and provide |2> through the output of |Y2> to activate it . All other outputs |Y0> - |Y9> will remain close as remaining Quantum AND operation will generate **|0>**.

(d) When the input qubit |I> is |2>, and |S0>,|S1> = **|1>, |0>** the multi-valued 1-to-3 Demultiplexer will activate fourth Quantum AND and provide |2> through the output of |Y3> to activate it . All other outputs |Y0> - |Y9> will remain close as remaining Quantum AND operation will generate **|0>**.

(e) When the input qubit |I> is |2>, and |S0>,|S1> = **|1>, |1>** the multi-valued 1-to-3 Demultiplexer will activate fifth Quantum AND and provide |2> through the output of |Y4> to activate it . All other outputs |Y0> - |Y9> will remain close as remaining Quantum AND operation will generate **|0>**.

(f) When the input qubit |I> is |2>, and |S0>,|S1> = **|1>, |2>** the multi-valued 1-to-3 Demultiplexer will activate sixth Quantum AND and provide |2> through the output of |Y5> to activate it . All other outputs |Y0> - |Y9> will remain close as remaining Quantum AND operation will generate **|0>**.

(g) When the input qubit |I> is |2>, and |S0>,|S1> = **|2>, |0>** the multi-valued 1-to-3 Demultiplexer will activate seventh Quantum AND and provide |2> through the output of |Y6> to activate it . All other outputs |Y0> - |Y9> will remain close as remaining Quantum AND operation will generate **|0>**.

(h) When the input qubit |I> is |2>, and |S0>,|S1> = **|2>, |1>** the multi-valued 1-to-3 Demultiplexer will activate eighth Quantum AND and provide |2> through the output of |Y7> to activate it . All other outputs |Y0> - |Y9> will remain close as remaining Quantum AND operation will generate **|0>**.

TABLE 16.5

Truth Table of 3-to-1 Quantum
Encoder

Inputs			Outputs
$|D2>$	$|D1>$	$|D0>$	$|A>$
$|0>$	$|0>$	$|2>$	$|0>$
$|0>$	$|2>$	$|0>$	$|1>$
$|2>$	$|0>$	$|0>$	$|2>$

(i) When the input qubit $|I>$ is $|2>$, and $|S0>,|S1> = \mathbf{|2>, |2>}$ the multi-valued 1-to-3 Demultiplexer will activate ninth Quantum AND and provide $|2>$ through the output of $|Y8>$ to activate it . All other outputs $|Y0> - |Y9>$ will remain close as remaining Quantum AND operation will generate $\mathbf{|0>}$.

16.5   Multiple-Valued Quantum Encoder

Multi-Valued Quantum Encoder is a combinational logic circuit which acts as a multi inputs and multi outputs device as it has 3^n input lines and n output lines (Figure 16.12). The encoder is an important part of a digital computer. Multi-Valued Quantum Encoder encodes a decimal number to its equivalent ternary-qubit number. For example, if an input of Multi-Valued Quantum Encoder is decimal value '6', then the encoder will provide qubit $|2>|0>$ as a result.

Table 16.5 represents the truth table of 3-to-1 Multi-Valued Quantum Encoder where $|D0>$, $|D1>$, and $|D2>$ are the inputs lines and $|A>$ is the output line.

Figure 16.13 gives the idea of how a 3-to-1 Multi-Valued Quantum Encoder is designed. As given in the diagram, 3-to-1 Multi-Valued Quantum Encoder needs two

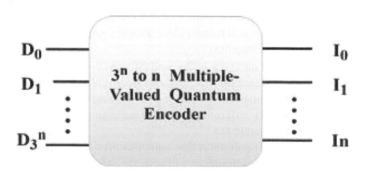

FIGURE 16.12
Multi-Valued Quantum 3^n-to-n Encoder

TABLE 16.6

Truth Table of 9-to-2 Quantum Encoder

Inputs									Outputs	
|D8>	|D7>	|D6>	|D5>	|D4>	|D3>	|D2>	|D1>	|D0>	|A1>	|A0>
|0>	|0>	|0>	|0>	|0>	|0>	|0>	|0>	|2>	|0>	|0>
|0>	|0>	|0>	|0>	|0>	|0>	|0>	|2>	|0>	|0>	|1>
|0>	|0>	|0>	|0>	|0>	|0>	|2>	|0>	|0>	|0>	|2>
|0>	|0>	|0>	|0>	|0>	|2>	|0>	|0>	|0>	|1>	|0>
|0>	|0>	|0>	|0>	|2>	|0>	|0>	|0>	|0>	|1>	|1>
|0>	|0>	|0>	|2>	|0>	|0>	|0>	|0>	|0>	|1>	|2>
|0>	|0>	|2>	|0>	|0>	|0>	|0>	|0>	|0>	|2>	|0>
|0>	|2>	|0>	|0>	|0>	|0>	|0>	|0>	|0>	|2>	|1>
|2>	|0>	|0>	|0>	|0>	|0>	|0>	|0>	|0>	|2>	|2>

multi-valued quantum AND, one multi-valued quantum NOT and one multi-valued quantum OR operation.

For, Multi-Valued Quantum 9-to-2 Encoder, Table 16.6 shows the truth table where |D0> to |D8> are the nine inputs lines and |A0> and |A1> are the outputs. The block diagram of 9-to-2 Multi-Valued Quantum Encoder is displayed in Figure 16.14.

In Figure 16.14, six multi-valued quantum OR operations and two multi-valued quantum AND operations are used. Here are nine inputs and two outputs.

16.5.1 Circuit Architecture of Multiple-Valued Quantum Encoder

Figure 16.15 depicts the circuit architecture of multi-valued Quantum 3-to-1 Encoder. When an input line is on, it will generate qubit |2> whereas others will be closed at |0>. Hence, each input line passes through some Quantum operations before conducting the Quantum OR operation. First, input line |D0> pass through a Quantum NOT operation to perform a Standard Ternary Inversion (STI). i.e. qubit |0> will become |2> or vice versa and |D1> conduct a DNA AND operation with qubit |1> as the other input, so that it can generate the out of |A> to |1> whenever |D1> is on. Additionally, |D2> is directly connected to the Quantum OR operation as it produces output |2>.

Figure 16.16 shows the circuit architecture of multi-valued quantum 9-to-2 encoder. For a Multi-Valued Quantum 9-to-2 Encoder (Figure 16.16), the input of |D1>, |D2>, |D4>, |D5>, |D7>, and |D8> are used to get the result of |A0> after performing five Quantum OR and one Quantum AND operation. Whereas |D3>, |D4>, |D5>, |D6>, |D7>, and |D8> is needed to get the outcome of |A1> after conducting another five Quantum OR and one Quantum AND operation.

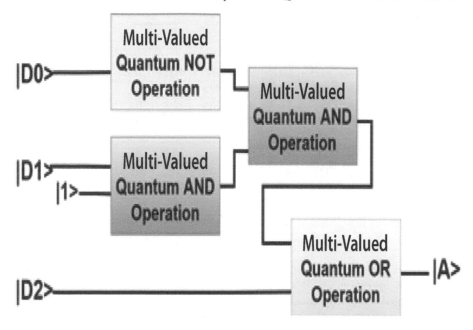

FIGURE 16.13
Block Diagram of Multi-Valued Quantum 3-to-1 Encoder

16.5.2 Working Principle of Multiple-Valued Quantum Encoder

1. **Multi-Valued Quantum 3-to-1 Encoder**

    (a) To get Output |A> = |0>, the multi-valued 3-to-1 Encoder |D0> will
        be activated, whereas other input line remain closed, i.e. |0>

    (b) To get Output |A> = |1>, the multi-valued 3-to-1 Encoder |D1> will
        be activated, whereas other input line remain closed, i.e. |0>

    (c) To get Output |A> = |2>, the multi-valued 3-to-1 Encoder |D2> will
        be activated, whereas other input line remain closed, i.e. |0>

2. **Multi-Valued Quantum 9-to-1 Multiplexer**

    (a) To get Output |A1>, |A0> = |0>, |0>, the multi-valued 9-to-2 En-
        coder |D0> will be activated, whereas other input line remain closed,
        i.e. |0>

    (b) To get Output |A1>, |A0> = |0>, |1>, the multi-valued 9-to-2 En-
        coder |D1> will be activated, whereas other input line remain closed,
        i.e. |0>

    (c) To get Output |A1>, |A0> = |0>, |2>, the multi-valued 9-to-2 En-
        coder |D2> will be activated, whereas other input line remain closed,
        i.e. |0>

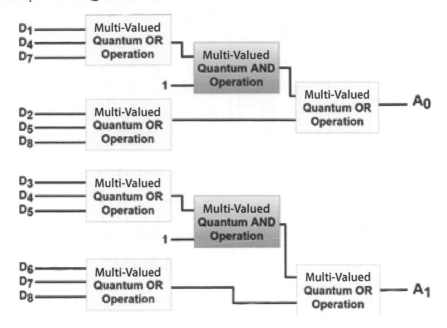

FIGURE 16.14
Block Diagram of Multi-Valued Quantum 9-to-2 Encoder

    (d) To get Output $|A1\rangle$, $|A0\rangle$ = $|1\rangle$, $|0\rangle$, the multi-valued 9-to-2 Encoder $|D3\rangle$ will be activated, whereas other input line remain closed, i.e. $|0\rangle$

    (e) To get Output $|A1\rangle$, $|A0\rangle$ = $|1\rangle$, $|1\rangle$, the multi-valued 9-to-2 Encoder $|D4\rangle$ will be activated, whereas other input line remain closed, i.e. $|0\rangle$

    (f) To get Output $|A1\rangle$, $|A0\rangle$ = $|1\rangle$, $|2\rangle$, the multi-valued 9-to-2 Encoder $|D5\rangle$ will be activated, whereas other input line remain closed, i.e. $|0\rangle$

    (g) To get Output $|A1\rangle$, $|A0\rangle$ = $|2\rangle$, $|0\rangle$, the multi-valued 9-to-2 Encoder $|D6\rangle$ will be activated, whereas other input line remain closed, i.e. $|0\rangle$

    (h) To get Output $|A1\rangle$, $|A0\rangle$ = $|2\rangle$, $|1\rangle$, the multi-valued 9-to-2 Encoder $|D7\rangle$ will be activated, whereas other input line remain closed, i.e. $|0\rangle$

    (i) To get Output $|A1\rangle$, $|A0\rangle$ = $|2\rangle$, $|2\rangle$, the multi-valued 9-to-2 Encoder $|D8\rangle$ will be activated, whereas other input line remain closed, i.e. $|0\rangle$

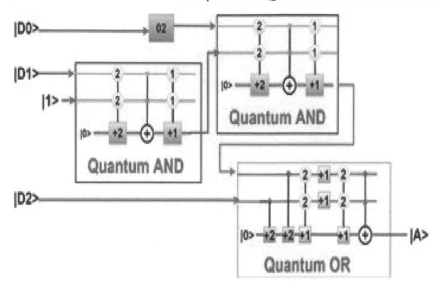

FIGURE 16.15
Circuit Architecture of Multi-Valued Quantum 3-to-1 Encoder

16.6 Multiple-Valued Quantum Decoder

Multi-Valued Quantum Decoder is a combinational logic circuit that is a multi-input and multi outputs circuit as it has n input line and 3^n output lines. The multi-valued quantum n-to-3^n decoder is shown in Figure 16.17. Basically, it decodes ternary qubits to their corresponding decimal number.

Two Multi-Valued 1-to-3 Quantum Decoders and nine Multi-Valued Quantum AND (D0-D8) operations are used to create a 2-to-9 Quantum Decoder (Figure 16.18). The truth table for 2-to-9 Quantum Decoders is given in Table 16.7.

Figure 16.18 shows the block diagram multi-valued quantum 2-to-9 decoder. Here inputs |A> and |B> are decoded by 1-to-3 quantum decoders. After decoding, three inputs are produced from each input. Single inputs are now multi-valued inputs. So, two inputs are decoded into six inputs.

In the case of n number of input Quantum decoder, a total n number of 1-to-3 decoder is needed to design an n-to-3^n Quantum decoder. Then |D0> to |D3^n> can be obtained (Figure 16.19). Figure 16.19 shows the block diagram multi-valued quantum n-to-3^n decoder.

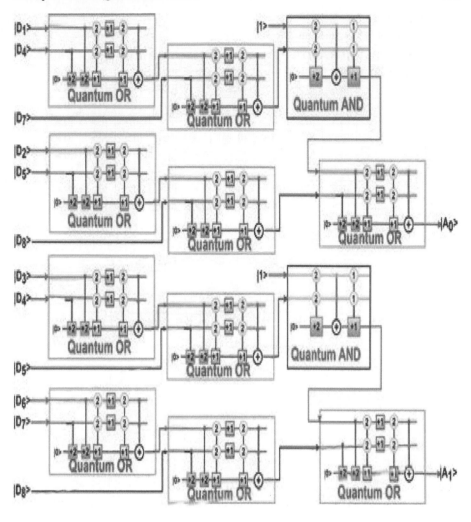

FIGURE 16.16
Circuit Architecture of Multi-Valued Quantum 9-to-2 Encoder

16.6.1 Circuit Architecture of Multiple-Valued Quantum Decoder

The construction of the Multi-Valued Quantum 2-to-9 decoder circuit is shown in Figure 16.20. In this circuit, the input, $|A>$ and $|B>$, at first, pass through two 1-to-3 Decoders separately so that inserted qubit, i.e. $|0>$, $|1>$, or $|2>$ can be identified using the output of 1-to-3 decoders. Nine Quantum AND operation is used in 2-to-9 Quantum Decoder where each input of these AND operations come from the various combinations of the output of 1-to-3 decoder.

Based on the value of $|A>$ and $|B>$ only one line of each 1-to-3 decoder generates qubit $|2>$ along the line whereas others provide $|0>$. Thus only one quantum AND

TABLE 16.7

Truth Table of Quantum 2-to-9 Decoder

|A>	|B>	|D8>	|D7>	|D6>	|D5>	|D4>	|D3>	|D2>	|D1>	|D0>
|0>	|0>	|0>	|0>	|0>	|0>	|0>	|0>	|0>	|0>	|2>
|0>	|1>	|0>	|0>	|0>	|0>	|0>	|0>	|0>	|2>	|0>
|0>	|2>	|0>	|0>	|0>	|0>	|0>	|0>	|2>	|0>	|0>
|1>	|0>	|0>	|0>	|0>	|0>	|0>	|2>	|0>	|0>	|0>
|1>	|1>	|0>	|0>	|0>	|0>	|2>	|0>	|0>	|0>	|0>
|1>	|2>	|0>	|0>	|0>	|2>	|0>	|0>	|0>	|0>	|0>
|2>	|0>	|0>	|0>	|2>	|0>	|0>	|0>	|0>	|0>	|0>
|2>	|1>	|0>	|2>	|0>	|0>	|0>	|0>	|0>	|0>	|0>
|2>	|2>	|2>	|0>	|0>	|0>	|0>	|0>	|0>	|0>	|0>

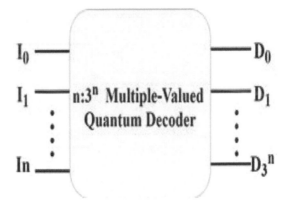

FIGURE 16.17

Multi-Valued Quantum n-to-3^n Decoder

operation among |D0> to |D8>, gives the result |2>, where other outputs, |D0>-|D8> remain inactive. i.e. **|0>**.

16.6.2 Working Principle of Multiple-Valued Quantum Decoder

1. For input sequences |A>, |B> = **|0>**, **|0>**, the multi-valued 1-to-3 decoder will perform (|A0> && |B0>) and the |D0> line will be open. As a result, the output qubit of |D0> equal **|2>** and remaining lines |D1> to |D8> will remain closed **|0>**.

2. For input sequences |A>, |B> = **|0>**, **|1>**, the multi-valued 1-to-3 decoder will perform (|A0> && |B1>) and the |D1> line will be open. As a result, the output qubit of |D1> equal **|2>** and remaining lines, |D0>, |D2> to |D8> will remain closed**|0>**.

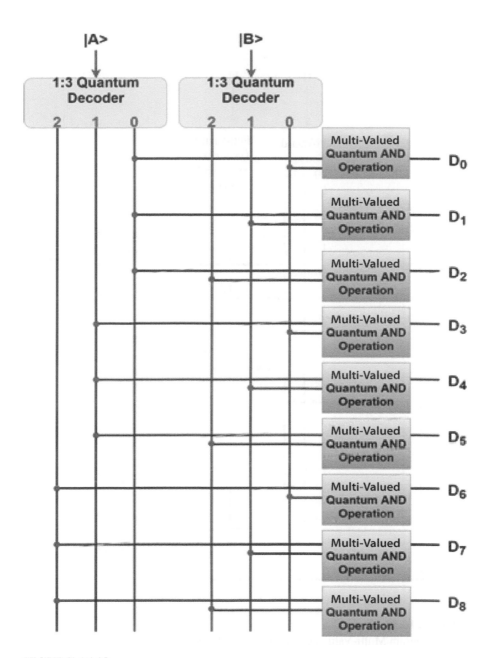

FIGURE 16.18
Block Diagram Multi-Valued Quantum 2-to-9 Decoder

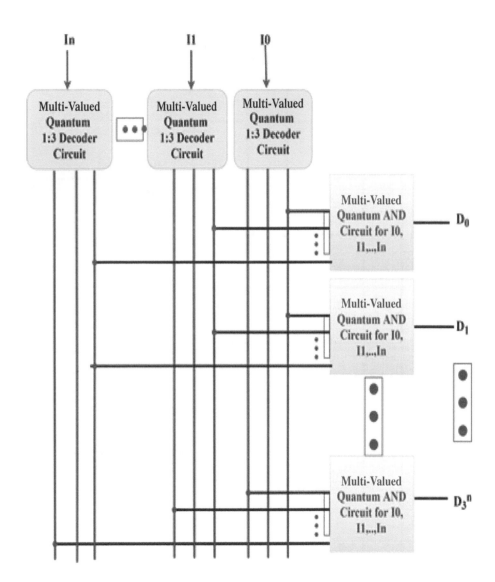

FIGURE 16.19
Block Diagram Multi-Valued Quantum n-to-3^n Decoder

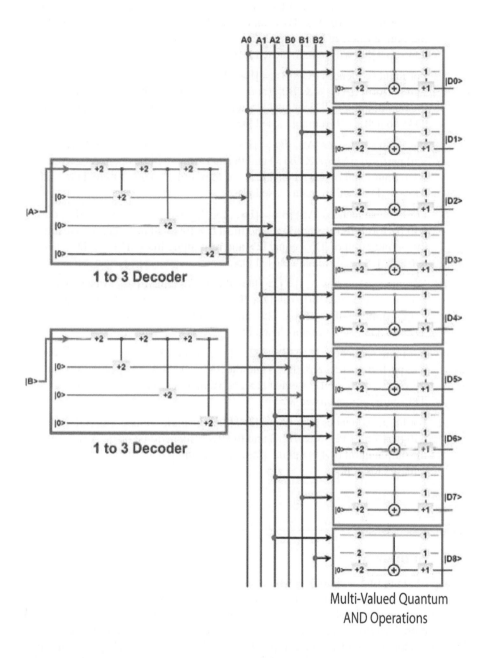

FIGURE 16.20
Circuit Architecture of Multi-Valued Quantum 2-to-9 Decoder

3. For input sequences |A>, |B> = **|0>, |2>**, the multi-valued 1-to-3 decoder will perform (|A0> && |B2>) and the |D2> line will be open. As a result, the output qubit of |D2> equal **|2>** and remaining lines, |D0>, |D1> and |D3> to |D8> will remain closed **|0>**.

4. For input sequences |A>, |B> = **|1>, |0>**, the multi-valued 1-to-3 decoder will perform (|A1> && |B0>) and the |D3> line will be open. As a result, the output qubit of |D3> equal **|2>** and remaining lines, |D0> to |D2> and |D4> to |D8> will remain closed **|0>**.

5. For input sequences |A>, |B> = **|1>, |1>** the multi-valued 1-to-3 decoder will perform (|A1> && |B1>) and the |D4> line will be open. As a result, the output qubit of |D4> equal **|2>** and remaining lines, |D0> to |D3> and |D5> to |D8> will remain closed **|0>**.

6. For input sequences |A>, |B> = **|1>, |2>**, the multi-valued 1-to-3 decoder will perform (|A1> && |B2>) and the |D5> line will be open. As a result, the output qubit of |D5> equal **|2>** and remaining lines, |D0> to |D4> and |D6> to |D8> will remain closed **|0>**.

7. For input sequences A, B= **|2>, |0>**, the multi-valued 1-to-3 decoder will perform (|A2> && |B0>) and the |D6> line will be open. As a result, the output qubit of |D6> equal **|2>** and remaining lines, |D0> to |D5> and |D7> to |D8> will remain closed **|0>**.

8. For input sequences |A>, |B> = **|2>, |1>**, the multi-valued 1-to-3 decoder will perform (|A2> && |B1>) and the |D7> line will be open.As a result, the output qubit of |D7> equal **|2>** and remaining lines, |D0> to |D6> and |D8> will remain closed **|0>**.

9. For input sequences |A>, |B> = **|2>, |2>**, the multi-valued 1-to-3 decoder will perform (|A2> && |B2>) and the |D8> line will be open. As a result, the output qubit of |D8> equal **|2>** and remaining lines, |D0> to |D7> will remain closed **|0>**.

16.7 Summary

Quantum computing technology with many values is gaining attraction in today's fast-paced environment. An attempt is made to give architectural approaches for designing various multi-valued quantum combinational circuits in this chapter. Each combinational circuit's algorithms, such as adder, subtractor, multiplexer, demultiplexer, encoder, decoder, and comparator, are displayed. This chapter looks into specific architecture concepts for multi-valued quantum combinational circuits with great performance. Some combinational circuits are also shown for 2-qubit and N-qubit, together with their design techniques and functioning principles. Multi-Valued Quantum computers, as it is known, generate a lot of heat, which causes disorder among qubits. As a result, the amount of heat created by these quantum combinational circuits is displayed, giving a clear picture of their efficiency. A speed calculation has also been performed to indicate how much faster they will be if this multi-valued quantum combinational circuit is implemented perfectly. All of these articles have been hypothetically implemented.

Bibliography

[1] Dhande, A. P., & Ingole, V. T. (2005, March). Design and implementation of 2 bit ternary ALU slice. In Proc. Int. Conf. IEEE-Sci. Electron., Technol. Inf. Telecommun (Vol. 17).

[2] Lin, S., Kim, Y. B., & Lombardi, F. (2009). CNTFET-based design of ternary logic gates and arithmetic circuits. IEEE Transactions on Nanotechnology, 10(2), 217-225.

[3] Heung, A., & Mouftah, H. T. (1985). Depletion/enhancement CMOS for a lower power family of three-valued logic circuits. IEEE Journal of Solid-State Circuits, 20(2), 609 616.

[4] Monfared, A. T., & Haghparast, M. (2017). Design of novel quantum/reversible ternary adder circuits. International Journal of Electronics Letters, 5(2), 149-157.

[5] Zadeh, R. P., & Haghparast, M. (2011). A new reversible/quantum ternary comparator. Australian Journal of Basic and Applied Sciences, 5(12), 2348-2355.

17

Multiple-Valued DNA Combinational Circuits

DOI: 10.1201/9781003381938-17

17.1 Introduction

A ternary or three-valued logic function has two inputs that can take one of three states (0, 1, or 2) and produces one output signal that can take one of these three states. Two DNA sequences are utilized as inputs and one DNA sequence is used as output in ternary DNA computing. When computing ternary DNA, the sequence ACCTAG is regarded as "0," the sequence CAAGCT strands as "1" and the sequence TGGATC as "2".

The fluorescence level is utilized to detect the DNA sequence in Ternary DNA computing. Fluorescence is defined as fluorescent molecules temporarily absorbing electromagnetic wavelengths from the visible light spectrum and then emitting light at a lower energy level. When it appears in a living creature, it's considered bio-fluorescence. The light that is emitted is a different color than the light that is absorbed as a result of this. An electron is excited by stimulating light, causing its energy to rise to an unstable level.

17.2 Multiple-Valued DNA Multiplexer

A balanced multiplexer is an important component of a digital system that converts multiple inputs to a single output. This circuit is useful for creating multi-valued arithmetic circuits. An n-to-1 multiplexer is composed of three parts. There are three N inputs, one output, and one selector line. A block diagram of multi-valued DNA n-to-1 Multiplexer is depicted in Figure 17.1.

Table 17.1 and Figure 17.2, respectively, show the truth table and associative block diagram of Multi-Valued DNA 3-to-1 Multiplexer. Using the truth table (Table 17.1), the block diagram is designed where a 1-to-3 Decoder is used to activate the selection line.

In Figure 17.2, the block diagram of multi-valued DNA 3-to-1 multiplexer consists of three multi-valued DNA AND operations and one OR operation. Three inputs and one selection line is decoded into three selection lines by 1-to-3 DNA decoder.

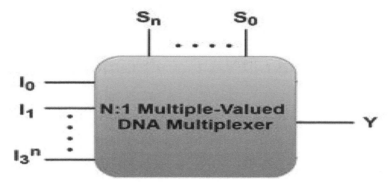

FIGURE 17.1
Multi-Valued n-to-1 DNA Multiplexer

TABLE 17.1
Truth Table of Multi-Valued DNA 3-to-1
Multiplexer

Inputs	Outputs
S	Y
ACCTAG	I0
CAAGCT	I1
TGGATC	I2

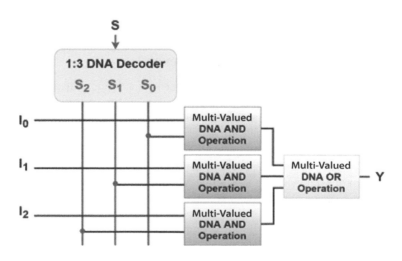

FIGURE 17.2
Block Diagram of Multi-Valued DNA 3-to-1 Multiplexer

TABLE 17.2

Truth Table of Multi-Valued DNA
9-to-1 Multiplexer

Inputs		Outputs
S1	**S0**	**Y**
ACCTAG	ACCTAG	I0
ACCTAG	CAAGCT	I1
ACCTAG	TGGATC	I2
CAAGCT	ACCTAG	I3
CAAGCT	CAAGCT	I4
CAAGCT	TGGATC	I5
TGGATC	ACCTAG	I6
TGGATC	CAAGCT	I7
TGGATC	TGGATC	I8

When total inputs are 9 and the selection line is 2, Table 17.2 provides a truth table of Multi-Valued DNA 9-to-1 Multiplexer. Here, among the DNA AND operations, only one DNA AND operation will provide its input to the output line via DNA OR operation.

Figure 17.3 gives the block diagram of the Multi-Valued DNA 9-to-1 Multiplexer. Additionally, a block diagram of Multi-Valued DNA n-to-1 Multiplexer is shown in Figure 17.4, there will be a 3^n input line which will be connected to 3^n DNA AND operation. Then a total of n selection lines will decide which input will pass through the output of DNA OR operation.

17.2.1 Circuit Architecture of Multiple-Valued DNA Multiplexer

Figure 17.5 shows the construction of Multi-Valued DNA 3-to-1 multiplexer circuit. In this circuit, all the inputs I0 - I2 connect to three different DNA AND gates in which another input comes from the Selection line S.

To select one particular input, molecular sequence **ACCTAG**, **CAAGCT** or **TGGATC** is provided through S to represent 1, 2, or 3, respectively. Based on this output of the selection line one of the three output lines of the DNA 1-to-3 Decoder gets activated and passes molecular sequence **TGGATC** through it. Other output lines of the decoder provide molecular sequence **ACCTAG**. Thus only one DNA AND operation is open to take the input to the output. Finally, the DNA OR operation passes the input that comes via the DNA AND operation to the output of the multiplexer.

The circuit architecture of the Multi-Valued DNA 9-to-1 Multiplexer is also the same which is shown in Figure 17.6. However, as there are nine inputs, I0 - I8, two selection lines are required to select only one input as the result of output Y.

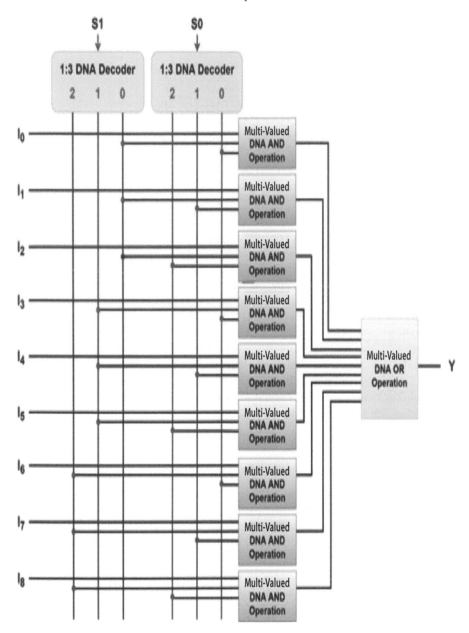

FIGURE 17.3
Block Diagram of Multi-Valued DNA 9-to-1 Multiplexer

17.2.2 Working Principle of Multiple-Valued DNA Multiplexer

The working principle of Multiple-Valued DNA multiplexers are explained in this section.

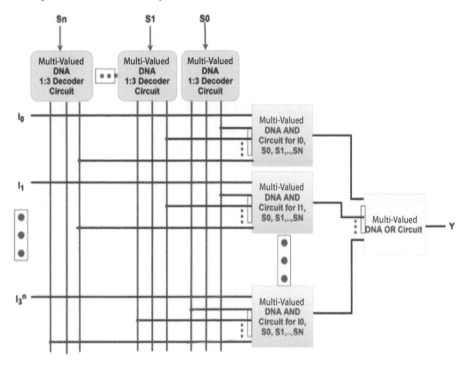

FIGURE 17.4
Block Diagram of Multi-Valued DNA n-to-1 Multiplexer

1. **Multi-Valued DNA 3-to-1 Multiplexer**

    (a) For input sequences I0, I1, I2, S = **TGGATC, TGGATC, TGGATC,** and **ACCTAG** which represent 2, 2, 2, and 0, the multi-valued 3-to-1 multiplexer will activate first DNA AND and provide I0 through the output of Y . All other DNA AND will remain closed i.e. **ACCTAG**.

    (b) For input sequences I0, I1, I2, S = **TGGATC, TGGATC, TGGATC,** and **CAAGCT** which represent 2, 2, 2, and 1, the multi-valued 3-to-1 multiplexer will activate first DNA AND and provide I1 through the output of Y . All other DNA AND will remain closed i.e. **ACCTAG**.

    (c) For input sequences I0, I1, I2, S = **TGGATC, TGGATC, TGGATC,** and **TGGATC** which represent 2, 2, 2, and 2, the multi-valued 3-to-1 multiplexer will activate first DNA AND and provide I2 through the output of Y . All other DNA AND will remain closed i.e. **ACCTAG**.

2. **Multi-Valued DNA 9-to-1 Multiplexer**

    (a) When all the input sequences I0-I9 are **TGGATC**, and S0, S1 = **ACCTAG** , **ACCTAG** the multi-valued 9-to-1 multiplexer will activate first DNA AND and provide I0 through the output of Y . All other DNA AND will remain closed i.e. **ACCTAG**. .

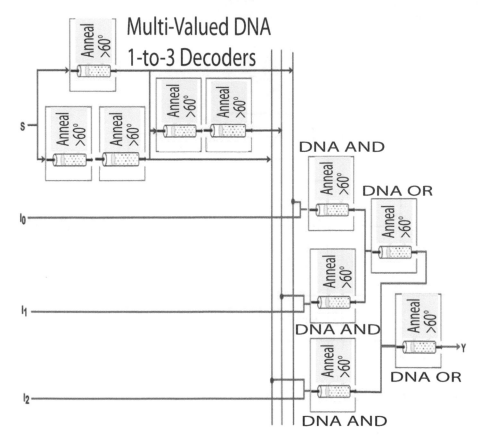

FIGURE 17.5
Circuit Architecture of Multi-Valued DNA 3-to-1 Multiplexer

- (b) When all the input sequences I0-I9 are **TGGATC**, and S0, S1 = **AC-CTAG** , **CAAGCT** the multi-valued 9-to-1 multiplexer will activate second DNA AND and provide I1 through the output of Y . All other DNA AND will remain closed i.e. **ACCTAG**.

- (c) When all the input sequences I0-I9 are **TGGATC**, and S0, S1 = **AC-CTAG** , **TGGATC** the multi-valued 9-to-1 multiplexer will activate a third DNA AND and provide I2 through the output of Y . All other DNA AND will remain closed i.e. **ACCTAG**.

- (d) When all the input sequences I0-I9 are **TGGATC**, and S0, S1 = **CAAGCT** , **ACCTAG** the multi-valued 9-to-1 multiplexer will activate fourth DNA AND and provide I3 through the output of Y . All other DNA AND will remain closed i.e. **ACCTAG**. .

- (e) When all the input sequences I0-I9 are **TGGATC**, and S0, S1 = **CAAGCT** , **CAAGCT** the multi-valued 9-to-1 multiplexer will

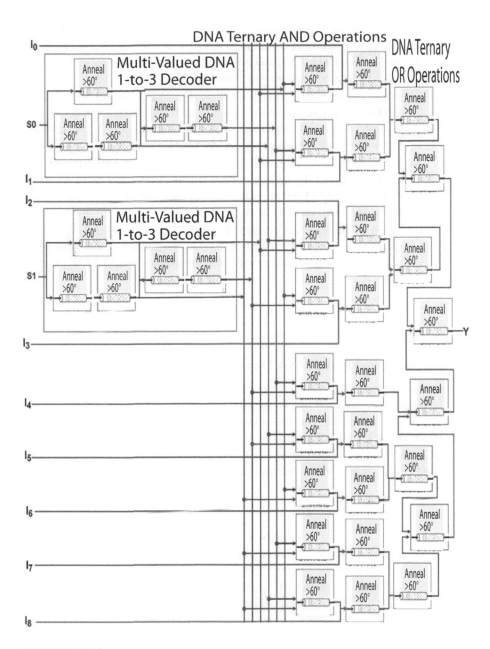

FIGURE 17.6
Circuit Architecture of Multi-Valued DNA 9-to-1 Multiplexer

activate fifth DNA AND and provide I4 through the output of Y . All other DNA AND will remain closed i.e. **ACCTAG**.

(f) When all the input sequences I0-I9 are **TGGATC**, and S0, S1 = **CAAGCT** , **TGGATC** the multi-valued 9-to-1 multiplexer will activate a sixth DNA AND and provide I5 through the output of Y . All other DNA AND will remain closed i.e. **ACCTAG**.

(g) When all the input sequences I0-I9 are **TGGATC**, and S0, S1 = **TGGATC, ACCTAG** the multi-valued 9-to-1 multiplexer will activate seventh DNA AND and provide I6 through the output of Y . All other DNA AND will remain closed i.e. **ACCTAG**. .

(h) When all the input sequences I0-9 are **TGGATC**, and S0, S1 = **TGGATC, CAAGCT** the multi-valued 9-to-1 multiplexer will activate eighth DNA AND and provide I7 through the output of Y . All other DNA AND will remain closed i.e. **ACCTAG**.

(i) When all the input sequences I0-I9 are **TGGATC**, and S0, S1 = **TGGATC, TGGATC** the multi-valued 9-to-1 multiplexer will activate the ninth DNA AND and provide I8 through the output of Y. All other DNA AND will remain closed i.e. **ACCTAG**.

17.3   Multiple-Valued DNA Demultiplexer

Demultiplexer circuits are among the most important circuits in the design of complex hardware. It convets a single input to multiple outputs. Thereby, a 1-to-n Multi-Valued DNA Demultiplexer consists of three main parts: one input, N selection line and 3^N outputs as shown in Figure 17.7.

The truth table of Multi-Valued DNA 1-to-3 Demultiplexer is shown in Table 17.3 where only the selected output equals the input I and all other outputs are 0. To

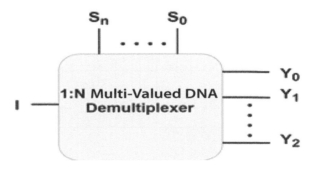

FIGURE 17.7
Multi-Valued 1-to-n DNA Demultiplexer

TABLE 17.3

Truth Table of 1-to-3 Demultiplexer

Inputs	Outputs		
S	Y_2	Y_1	Y_0
ACCTAG	ACCTAG	ACCTAG	I
CAAGCT	ACCTAG	I	ACCTAG
TGGATC	I	ACCTAG	ACCTAG

represent 0, 1, and 2 in the molecular sequence, **ACCTAG**, **CAAGCT** and **TGGATC** are used, respectively.

Figure 17.8 depicts the 1-to-3 demultiplexer's block diagram. Depending on the combination chosen, one of the decoder outputs becomes 2 while the others remain 0.

Table 17.4 shows the truth table of multi-valued DNA 1-to-9 Demultiplexer, where 2 selection lines are needed to select one output line among nine. Figure 17.9 gives the block diagram of Multi-Valued DNA 1-to-9 Demultiplexer.

17.3.1 Circuit Architecture of Multiple-Valued DNA Demultiplexer

The construction of the Multi-Valued DNA 1-to-3 Demultiplexer circuit is shown in Figure 17.10. In this circuit, the input I is directly connected to all the DNA AND operations as one input. The various outputs of selection line S contribute as another input of these AND operations. The result of S decides which output line must be activated.

To select one particular input, molecular sequence **ACCTAG**, **CAAGCT** or **TGGATC** is provided through S to represent 1, 2, or 3, respectively. Based on this output of the selection line one of the three output lines of the DNA 1-to-3 Decoder

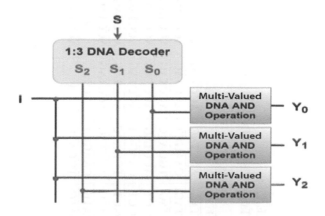

FIGURE 17.8

Block Diagram of Multi-Valued DNA 1-to-3 Demultiplexer

TABLE 17.4
Truth Table of 1-to-9 Demultiplexer

Inputs		Outputs								
S1	S0	Y_8	Y_7	Y_6	Y_5	Y_4	Y_3	Y_2	Y_1	Y_0
ACCTAG	ACCTAG	ACCTAG	ACCTAG	ACCTAG	ACCTAG	ACCTAG	ACCTAG	ACCTAG	ACCTAG	I
ACCTAG	CAAGCT	ACCTAG	ACCTAG	ACCTAG	ACCTAG	ACCTAG	ACCTAG	ACCTAG	I	ACCTAG
ACCTAG	TGGATC	ACCTAG	ACCTAG	ACCTAG	ACCTAG	ACCTAG	ACCTAG	I	ACCTAG	ACCTAG
CAAGCT	ACCTAG	ACCTAG	ACCTAG	ACCTAG	ACCTAG	ACCTAG	I	ACCTAG	ACCTAG	ACCTAG
CAAGCT	CAAGCT	ACCTAG	ACCTAG	ACCTAG	ACCTAG	I	ACCTAG	ACCTAG	ACCTAG	ACCTAG
CAAGCT	TGGATC	ACCTAG	ACCTAG	ACCTAG	I	ACCTAG	ACCTAG	ACCTAG	ACCTAG	ACCTAG
TGGATC	ACCTAG	ACCTAG	ACCTAG	I	ACCTAG	ACCTAG	ACCTAG	ACCTAG	ACCTAG	ACCTAG
TGGATC	CAAGCT	ACCTAG	I	ACCTAG	ACCTAG	ACCTAG	ACCTAG	ACCTAG	ACCTAG	ACCTAG
TGGATC	TGGATC	I	ACCTAG	ACCTAG	ACCTAG	ACCTAG	ACCTAG	ACCTAG	ACCTAG	ACCTAG

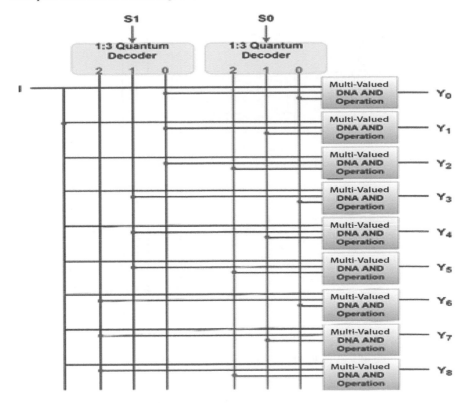

FIGURE 17.9
Block Diagram of Multi-Valued DNA 1-to-9 Demultiplexer

gets
activated and passes molecular sequence **TGGATC** through it. Other output lines
of the decoder provide molecular sequence **ACCTAG**. Thus only one DNA AND
operation is open to select one output among Y0 to Y2.

The architecture of the Multi-Valued DNA 1-to-9 Demultiplexer also applies the
same procedure (Figure 17.11). However, as there are nine outputs, Y0 - Y8, two
selection lines are required to activate only one output among Y0 - Y8, where other
outputs remain inactive. i.e. **ACCTAG**.

17.3.2 Working Principle of Multiple-Valued DNA Demultiplexer

The working principle of Multiple-Valued DNA Demultiplexers are explained in this
section.

1. **Multi-Valued DNA 1-to-3 Demultiplexer**

    (a) For input sequences I, S = **TGGATC**, **ACCTAG** which represent 2,
        and 0, the multi-valued 1-to-3 Demultiplexer will activate first DNA

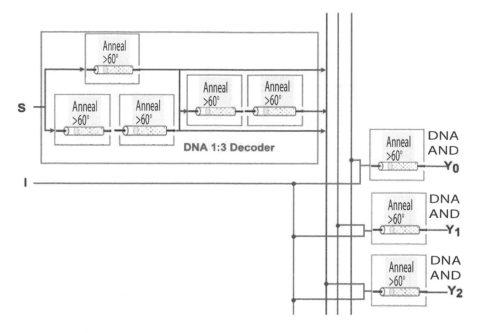

FIGURE 17.10
Circuit Architecture of Multi-Valued DNA 1-to-3 Demultiplexer

AND and provide "2", **TGGATC**, through the output of Y0 . All other outputs Y0 - Y2 will remain close as remaining DNA AND operation will generate "0", **ACCTAG**.

(b) For input sequences I, S = **TGGATC**, **CAAGCT** which represent 2, and 0, the multi-valued 1-to-3 Demultiplexer will activate a second DNA AND and provide "2", **TGGATC**, through the output of Y1 . All other outputs Y0 - Y2 will remain close as remaining DNA AND operation will generate "0", **ACCTAG**.

(c) For input sequences I, S = **TGGATC**, **TGGATC** which represent 2, and 0, the multi-valued 1-to-3 Demultiplexer will activate third DNA AND and provide "2", **TGGATC**, through the output of Y2. All other outputs Y0 - Y2 will remain close as remaining DNA AND operation will generate "0", **ACCTAG**.

2. **Multi-Valued DNA 1-to-9 Demultiplexer**

(a) When the input qubit, I = **TGGATC**, and S0, S1 = **ACCTAG, ACC-TAG** ; the multi-valued 1-to-9 Demultiplexer will activate first DNA AND and provide "2", **TGGATC**, through the output of Y0 . All other outputs Y0 - Y8 will remain close as remaining DNA AND operation will generate "0", **ACCTAG**.

(b) When the input qubit, I = **TGGATC**, and S0, S1 = **ACCTAG, CAAGCT** ; the multi-valued 1-to-9 Demultiplexer will activate

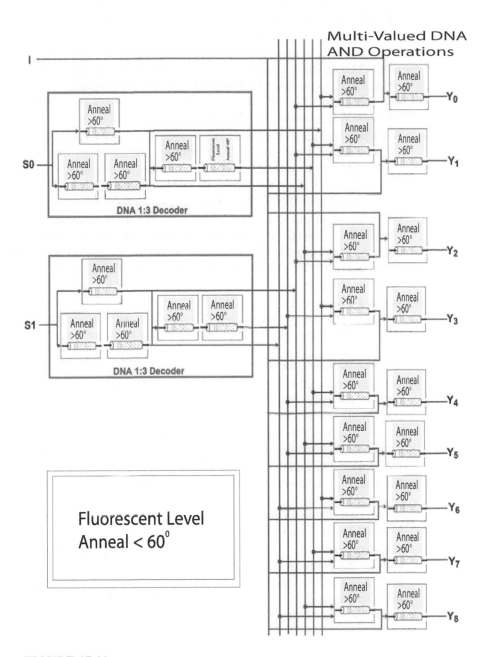

FIGURE 17.11
Circuit Architecture of Multi-Valued DNA 1-to-9 Demultiplexer

second DNA AND and provide "2", **TGGATC**, through the output of Y1 . All other outputs Y0 - Y8 will remain close as remaining DNA AND operation will generate "0", **ACCTAG**.

(c) When the input qubit, I = **TGGATC**, and S0, S1 = **ACCTAG, TGGATC;** the multi-valued 1-to-9 Demultiplexer will activate third DNA AND and provide "2", **TGGATC**, through the output of Y2 . All other outputs Y0 - Y8 will remain close as remaining DNA AND operation will generate "0", **ACCTAG**.

(d) When the input qubit, I = **TGGATC**, and S0, S1 = **CAAGCT , AC-CTAG** ; the multi-valued 1-to-9 Demultiplexer will activate fourth DNA AND and provide "2", **TGGATC**, through the output of Y3 . All other outputs Y0 - Y8 will remain close as remaining DNA AND operation will generate "0", **ACCTAG**.

(e) When the input qubit, I = **TGGATC**, and S0, S1 = **CAAGCT, CAAGCT** ; the multi-valued 1-to-9 Demultiplexer will activate fifth DNA AND and provide "2", **TGGATC**, through the output of Y4 . All other outputs Y0 - Y8 will remain close as remaining DNA AND operation will generate "0", **ACCTAG**.

(f) When the input qubit, I = **TGGATC**, and S0, S1 = **CAAGCT, TGGATC;** the multi-valued 1-to-9 Demultiplexer will activate sixth DNA AND and provide "2", **TGGATC**, through the output of Y5 . All other outputs Y0 - Y8 will remain close as remaining DNA AND operation will generate "0", **ACCTAG**.

(g) When the input qubit, I = **TGGATC**, and S0, S1 = **TGGATC, AC-CTAG** ; the multi-valued 1-to-9 Demultiplexer will activate seventh DNA AND and provide "2", **TGGATC**, through the output of Y6 . All other outputs Y0 - Y8 will remain close as remaining DNA AND operation will generate "0", **ACCTAG**.

(h) When the input qubit, I = **TGGATC**, and S0, S1 = **TGGATC, CAAGCT** ; the multi-valued 1-to-9 Demultiplexer will activate eighth DNA AND and provide "2", **TGGATC**, through the output of Y7 . All other outputs Y0 - Y8 will remain close as remaining DNA AND operation will generate "0", **ACCTAG**.

(i) When the input qubit, I = **TGGATC**, and S0, S1 = **TGGATC, TG-GATC;** the multi-valued 1-to-9 Demultiplexer will activate ninth DNA AND and provide "2", **TGGATC**, through the output of Y8 . All other outputs Y0 - Y8 will remain close as remaining DNA AND operation will generate "0", **ACCTAG**.

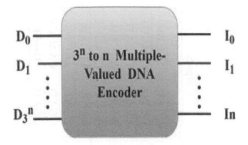

FIGURE 17.12
Multi-Valued DNA 3^n-to-n Encoder

17.4 Multiple-Valued DNA Encoder

Multi-Valued DNA Encoder is a combinational logic circuit which acts as a multi-inputs and multi outputs device as it has 3^n input lines and n output lines (Figure 17.12). Encoder is an important part of a digital computer. Multi-Valued DNA Encoder encodes a decimal number to its equivalent ternary-molecular sequence number. For example, if an input of a Multi-Valued DNA Encoder is decimal value '6', then the encoder will provide the molecular sequence '**TGGATC**' '**ACCTAG**' as a result.

Table 17.5 represents the truth table of 3-to-1 Multi-Valued DNA Encoder where D0, D1, and D2 are the input lines and A is the output line. Figure 17.13 gives the idea of how a 3-to-1 Multi-Valued DNA Encoder is designed. As given in the diagram, 3-to-1 Multi-Valued DNA Encoder needs two DNA AND, one DNA NOT and one DNA OR operation.

Table 17.6 shows the truth table for multi-valued DNA 9-to-2 Encoder where D0 to D8 are the nine input lines and A0 and A1 are the outputs. The block diagram of 9-to-2 Multi-Valued DNA Encoder is displayed in Figure 17.14.

TABLE 17.5
Truth Table of Multi-Valued 3-to-1 DNA Encoder

Inputs			Outputs
D2	D1	D0	A
ACCTAG	ACCTAG	TGGATC	ACCTAG
ACCTAG	TGGATC	ACCTAG	CAAGCT
TGGATC	ACCTAG	ACCTAG	TGGATC

TABLE 17.6
Truth Table of Multi-Valued 9-to-2 DNA Encoder

Inputs									Outputs	
D8	D7	D6	D5	D4	D3	D2	D1	D0	A1	A0
ACCTAG	ACCTAG	ACCTAG	ACCTAG	ACCTAG	ACCTAG	ACCTAG	ACCTAG	TGGATC	ACCTAG	ACCTAG
ACCTAG	ACCTAG	ACCTAG	ACCTAG	ACCTAG	ACCTAG	ACCTAG	TGGATC	ACCTAG	ACCTAG	CAAGCT
ACCTAG	ACCTAG	ACCTAG	ACCTAG	ACCTAG	ACCTAG	TGGATC	ACCTAG	ACCTAG	ACCTAG	TGGATC
ACCTAG	ACCTAG	ACCTAG	ACCTAG	ACCTAG	TGGATC	ACCTAG	ACCTAG	ACCTAG	CAAGCT	ACCTAG
ACCTAG	ACCTAG	ACCTAG	ACCTAG	TGGATC	ACCTAG	ACCTAG	ACCTAG	ACCTAG	CAAGCT	CAAGCT
ACCTAG	ACCTAG	ACCTAG	TGGATC	ACCTAG	ACCTAG	ACCTAG	ACCTAG	ACCTAG	CAAGCT	TGGATC
ACCTAG	ACCTAG	TGGATC	ACCTAG	ACCTAG	ACCTAG	ACCTAG	ACCTAG	ACCTAG	TGGATC	ACCTAG
ACCTAG	TGGATC	ACCTAG	ACCTAG	ACCTAG	ACCTAG	ACCTAG	ACCTAG	ACCTAG	TGGATC	CAAGCT
TGGATC	ACCTAG	ACCTAG	ACCTAG	ACCTAG	ACCTAG	ACCTAG	ACCTAG	ACCTAG	TGGATC	TGGATC

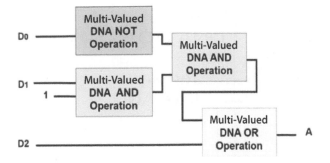

FIGURE 17.13
Block Diagram of Multi-Valued DNA 3-to-1 Encoder

17.4.1 Circuit Architecture of Multiple-Valued DNA Encoder

Figure 17.15 depicts the architecture of Multi-Valued DNA 3-to-1 Encoder. As when an input line is on, it will generate a molecular sequence '**TGGATC**', which represents '2', whereas others will be closed at '**ACCTAG**', as it represents '0'. Hence, each input line passes through some DNA operations before conducting the DNA OR operation. First, input line D0 passes through a DNA NOT operation to

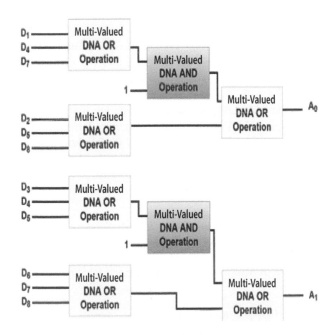

FIGURE 17.14
Block Diagram of Multi-Valued 9-to-2 DNA Encoder

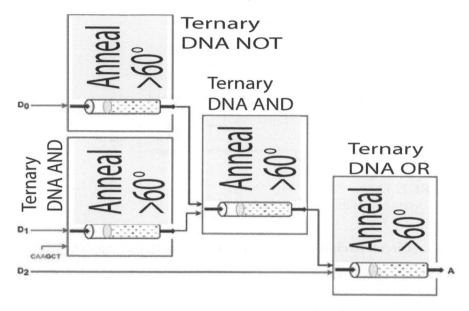

FIGURE 17.15
Circuit Architecture of Multi-Valued DNA 3-to-1 Encoder

perform a Standard Ternary Inversion (STI). That is, sequence '**ACCTAG**' will become '**TGGATC**' or vice versa and D1 conduct a DNA AND operation with the molecular sequence '**CAAGCT**', which represent '1', as the other input, so that it can generate the out of A to '**CAAGCT**' whenever D1 is on. Additionally, D2 directly connected to the DNA OR operation as it produces output '**TGGATC**'.

For a Multi-Valued DNA 9-to-2 Encoder (Figure 17.16), the input of D1, D2, D4, D5, D7, and D8 are used to get the result of A0 after performing five DNA OR and one DNA AND operation. Whereas D3 to D8 is needed to get the outcome of A1 after conducting another five DNA OR and one DNA AND operation. The Block Diagram of Multi-Valued DNA 9-to-2 Encoder is depicted in Figure 17.16.

17.4.2   Working Principle of Multiple-Valued DNA Encoder

The working principle of Multiple-Valued DNA Encoders are explained in this section.

1. **Multi-Valued DNA 3-to-1 Encoder**

    (a) To get Output A = **ACCTAG** , the multi-valued 3-to-1 Encoder D0 will be activated, whereas other input lines remain closed, i.e. **ACC-TAG**

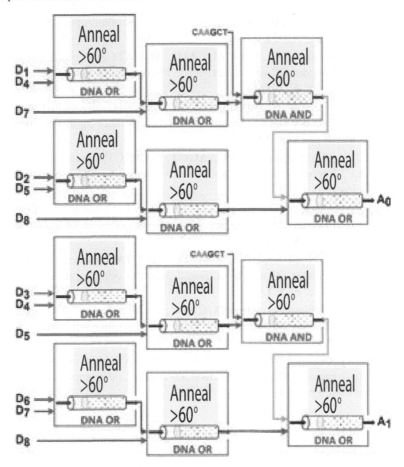

FIGURE 17.16
Circuit Architecture of Multi-Valued DNA 9-to-2 Encoder

    (b) To get Output A = **CAAGCT,** the multi-valued 3-to-1 Encoder D1 will be activated, whereas other input lines remain closed, i.e. **ACCTAG**

    (c) To get Output A = **TGGATC,** the multi-valued 3-to-1 Encoder D2 will be activated, whereas other input lines remain closed, i.e. **ACCTAG**

2. **Multi-Valued DNA 9-to-1 Multiplexer**

    (a) To get Output A1, A0 = **ACCTAG, ACCTAG,** the multi-valued 9-to-2 Encoder D0 will be activated, whereas other input lines remain closed, i.e. **ACCTAG**

    (b) To get Output A1, A0 = **ACCTAG, CAAGCT,** the multi-valued

9-to-2 Encoder D1 will be activated, whereas other input lines remain closed, i.e. **ACCTAG**

(c) To get Output A1, A0 = **ACCTAG, TGGATC,** the multi-valued 9-to-2 Encoder D2 will be activated, whereas other input lines remain closed, i.e. **ACCTAG**

(d) To get Output A1, A0 = **CAAGCT, ACCTAG,** the multi-valued 9-to-2 Encoder D3 will be activated, whereas other input lines remain closed, i.e. **ACCTAG**

(e) To get Output A1, A0 = **CAAGCT, CAAGCT,** the multi-valued 9-to-2 Encoder D4 will be activated, whereas other input lines remain closed, i.e. **ACCTAG**

(f) To get Output A1, A0 = **CAAGCT, TGGATC,** the multi-valued 9-to-2 Encoder D5 will be activated, whereas other input lines remain closed, i.e. **ACCTAG**

(g) To get Output A1, A0 = **TGGATC , ACCTAG,** the multi-valued 9-to-2 Encoder D6 will be activated, whereas other input line remain closed, i.e. **ACCTAG**

(h) To get Output A1, A0 = **TGGATC, CAAGCT,** the multi-valued 9-to-2 Encoder D7 will be activated, whereas other input line remain closed, i.e. **ACCTAG**

(i) To get Output A1, A0 = **TGGATC , TGGATC,** the multi-valued 9-to-2 Encoder D8 will be activated, whereas other input line remain closed, i.e. **ACCTAG**

17.5   Multiple-Valued DNA Decoder

The Multi-Valued DNA Decoder is a combinational logic circuit with n input lines and 3^n output lines, as shown in Figure 17.17. Essentially, it converts a ternary-molecular DNA sequence to its equivalent decimal value.

A 2-to-9 DNA Decoder is built using two Multi-Valued 1-to-3 DNA Decoders and nine Multi-Valued DNA AND (D0-D8) operations (Figure 17.18). Table 17.7 contains the truth table for 2-to-9 DNA Decoders.

Figure 17.18 shows the block diagram multi-valued DNA 2-to-9 decoder. Here two inputs are A and B will be decoded into six inputs. Nine multi-valued DNA AND operations are needed here to contruct Multi-Valued DNA 2-to-9 Decoder.

Figure 17.19 shows the block diagram multi-valued DNA n-to-3 ^{n} Decoder. In the case of n input DNA decoders, a total of n 1-to-3 decoders are required to design an n-to-3^n DNA decoder. Then it will be able to get from D_0 to $D_3{}^{n.}$

TABLE 17.7
Truth Table of Multi-Valued DNA 2-to-9 Decoder

Input		Output									
A	B	D8	D7	D6	D5	D4	D3	D2	D1	D0
ACCTAG	ACCTAG	ACCTAG	ACCTAG	ACCTAG	ACCTAG	ACCTAG	ACCTAG	ACCTAG	ACCTAG	TGGATC
ACCTAG	CAAGCT	ACCTAG	ACCTAG	ACCTAG	ACCTAG	ACCTAG	ACCTAG	ACCTAG	TGGATC	ACCTAG
ACCTAG	TGGATC	ACCTAG	ACCTAG	ACCTAG	ACCTAG	ACCTAG	ACCTAG	TGGATC	ACCTAG	ACCTAG
CAAGCT	ACCTAG	ACCTAG	ACCTAG	ACCTAG	ACCTAG	ACCTAG	TGGATC	ACCTAG	ACCTAG	ACCTAG
CAAGCT	CAAGCT	ACCTAG	ACCTAG	ACCTAG	ACCTAG	TGGATC	ACCTAG	ACCTAG	ACCTAG	ACCTAG
CAAGCT	TGGATC	ACCTAG	ACCTAG	ACCTAG	TGGATC	ACCTAG	ACCTAG	ACCTAG	ACCTAG	ACCTAG
TGGATC	ACCTAG	ACCTAG	ACCTAG	TGGATC	ACCTAG	ACCTAG	ACCTAG	ACCTAG	ACCTAG	ACCTAG
TGGATC	CAAGCT	ACCTAG	TGGATC	ACCTAG	ACCTAG	ACCTAG	ACCTAG	ACCTAG	ACCTAG	ACCTAG
TGGATC	TGGATC	TGGATC	ACCTAG	ACCTAG	ACCTAG	ACCTAG	ACCTAG	ACCTAG	ACCTAG	ACCTAG

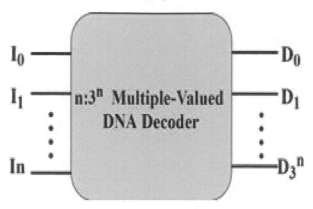

FIGURE 17.17
Multi-Valued DNA n-to- 3^n Decoder

17.5.1 Circuit Architecture of Multiple-Valued DNA Decoder

The construction of the Multi-Valued DNA 2-to-9 decoder circuit is shown in Figure 17.20. In this circuit, the input, A and B, at first, pass through two 1-to-3 Decoders separately so that inserted DNA molecular sequence, i.e. **ACCTAG, CAAGCT** or **TGGATC,** which represent "1", "2" and "3", respectively, can be identified using the output of 1-to-3 decoders. Nine DNA AND operation is used in 2-to-9 DNA Decoder where each input of these AND operations come from the various combinations of the output of 1-to-3 decoder. Based on the value of A and B only one line of each 1-to-3 decoder generates sequence **TGGATC** along the line whereas others provide **ACCTAG.** Thus only one DNA AND operation among D0 to D8, gives the result **TGGATC** where other outputs, D0-D8 remain inactive. i.e. **ACCTAG.**

17.5.2 Working Principle of Multiple-Valued DNA Decoder

The working principle of Multiple-Valued DNA Decoders are explained in this section.

1. For input sequences A, B =**ACCTAG, ACCTAG**, the multi-valued 1-to-3 decoder will perform (A0 && B0) and the D0 line will be open. As a result, the output sequence of D0 equal **TGGATC** and remaining lines, D1 to D8 will remain closed**, i.e. ACCTAG.**

2. For input sequences A, B =**ACCTAG, CAAGCT**, the multi-valued 1-to-3 decoder will perform (A0 && B1) and the D1 line will be open. As a result, the output sequence of D1 equal **TGGATC** and remaining lines, D0, D2 to D8 will remain closed**, i.e. ACCTAG.**

3. For input sequences A, B =**ACCTAG, TGGATC**, the multi-valued 1-to-3 decoder will perform (A0 && B2) and the D2 line will be open. As a

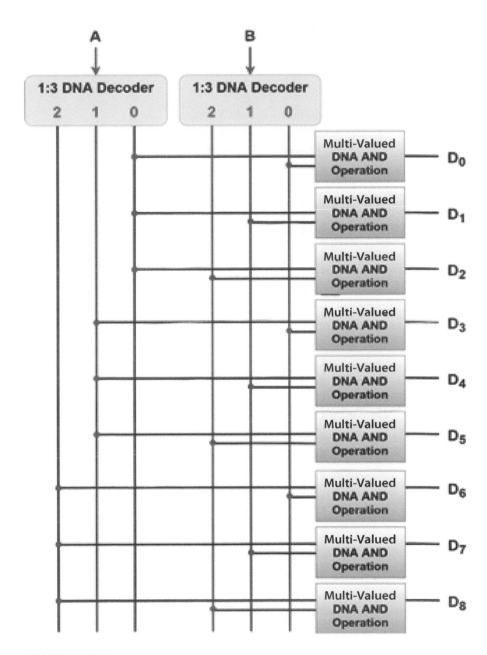

FIGURE 17.18
Block Diagram Multi-Valued DNA 2-to-9 Decoder

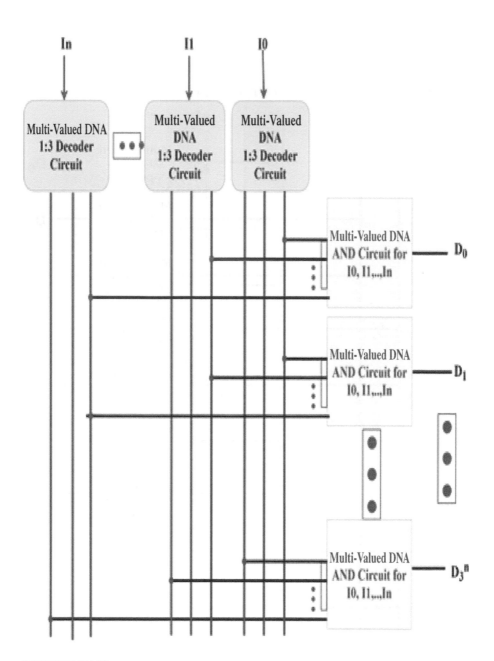

FIGURE 17.19
Block Diagram Multi-Valued DNA n-to-3 ^{n} Decoder

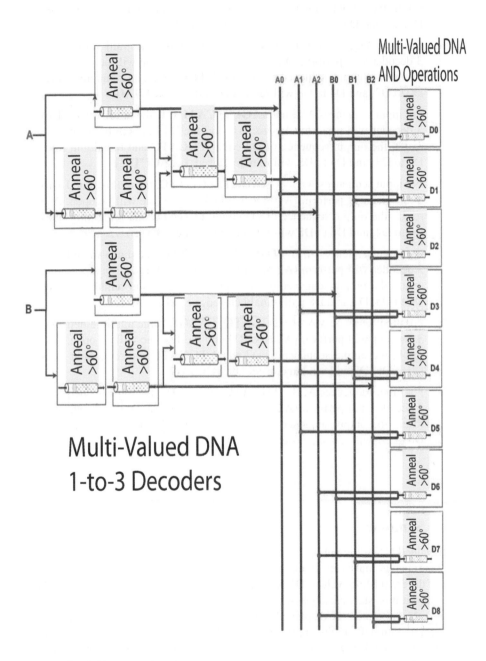

FIGURE 17.20
Circuit Architecture of Multi-Valued DNA 2-to-9 Decoder

result, the output sequence of D2 equal **TGGATC** and remaining lines, D0, D1 and D3 to D8 will remain closed, **i.e. ACCTAG**.

4. For input sequences A, B =**CAAGCT, ACCTAG**, the multi-valued 1-to-3 decoder will perform (A1 && B0) and the D3 line will be open. As a result, the output sequence of D3 equal **TGGATC** and remaining lines, D0 to D2, and D4 to D8 will remain closed, **i.e. ACCTAG**.

5. For input sequences A, B =**CAAGCT, CAAGCT**, the multi-valued 1-to-3 decoder will perform (A1 && B1) and the D4 line will be open. As a result, the output sequence of D4 equal **TGGATC** and remaining lines, D0 to D3, and D5 to D8 will remain closed, **i.e. ACCTAG**.

6. For input sequences A, B =**CAAGCT, TGGATC**, the multi-valued 1-to-3 decoder will perform (A1 && B2) and the D5 line will be open. As a result, the output sequence of D5 equal **TGGATC** and remaining lines, D0 to D4, and D6 to D8 will remain closed, **i.e. ACCTAG**.

7. For input sequences A, B =**TGGATC, ACCTAG**, the multi-valued 1-to-3 decoder will perform (A2 && B0) and the D6 line will be open. As a result, the output sequence of D6 equal TGGATC and remaining lines, D0 to D5, D7, and D8 will remain closed, **i.e. ACCTAG**.

8. For input sequences A, B =**TGGATC, CAAGCT**, the multi-valued 1-to-3 decoder will perform (A2 && B1) and the D7 line will be open. As a result, the output sequence of D7 equal **TGGATC** and remaining lines, D0 to D6, D8 will remain closed, **i.e. ACCTAG**.

9. For input sequences A, B =**TGGATC, TGGATC**, the multi-valued 1-to-3 decoder will perform (A2 && B2) and the D8 line will be open. As a result, the output sequence of D8 equal TGGATC and remaining lines, D0 to D7 will remain closed, **i.e. ACCTAG**.

17.6   Summary

Multi-Valued DNA computing is gaining popularity as a result of the tread technology. This chapter makes an attempt to provide architectural concepts for the development of different Multi-Valued DNA combinational circuits. Each combinational circuit's algorithms, such as adder, subtractor, multiplexer, demultiplexer, encoder, decoder, and comparator, are displayed. The architecture designs for high-performance DNA combinational circuits are discussed in this chapter. Some combinational circuits for 2 molecular DNA sequence and N-molecular DNA sequence, as well as their design approaches and operating principles are also shown. As it is known to all, multi-valued DNA computers require a lot of heat to complete the reactions, as a result, the quantity of heat required by these Multi-Valued DNA combinational circuits is shown, providing a clear image of their efficiency. A speed calculation was also performed to see how much quicker these Multi-Valued DNA combinational circuits would be if they were perfectly implemented. All of these articles have been implemented hypothetically.

Bibliography

[1] Dhande, A. P., & Ingole, V. T. (2005, March). Design and implementation of 2 bit ternary ALU slice. In Proc. Int. Conf. IEEE-Sci. Electron., Technol. Inf. Telecommun (Vol. 17).

[2] Hallworth, R. P., & Heath, F. G. (1962). Semiconductor circuits for ternary logic. Proceedings of the IEE-Part C: Monographs, 109(15), 219-225.

[3] Giri, S., & Saraswathi, M. N. (2012). Implementation of combinational circuits using ternary multiplexer. International Journal of Computational Engineering Research, 2(2), 457-463.

[4] Mandal, S. B., Chakrabarti, A., & Sur-Kolay, S. (2011, May). Synthesis techniques for ternary quantum logic. In 2011 41st IEEE International Symposium on Multiple-Valued Logic (pp. 218-223). IEEE.

18

Multiple-Valued Quantum-DNA Combinational Circuits

18.1 Introduction

As discussed in the previous chapters, biomolecules and biomolecular processes are meant to apply computational algorithms in DNA computing. Quantum computing involves performing calculations at a scale where quantum mechanical effects are significant. Both of these new computing paradigms have been considered as potential successors to solid-state computers. Both of these qualities could be captured by combining DNA and quantum computers. DNA computers could self-assemble quantum logic circuits from DNA strand-attached gates. Furthermore, quantum computers might be built directly from the physical properties of the DNA molecule. Multi-Valued logic enhances the advantages of this combination. The combination of this quantum computing and DNA computing is called quantum molecular biology. This chapter will describe the combinational circuits in multi-valued quantum-DNA computing.

18.2 Multiple-Valued Quantum-DNA Multiplexer

A balanced multiplexer is an essential component of a digital system that converts multiple inputs to a single output. This circuit is useful for building multi-valued arithmetic circuits. There are three parts of n-to-1 multiplexer. There are $3n$ inputs, 1 output, and an n selector line. In a multi-valued Quantum-DNA n-to-1 Multiplexer, all the inputs perform some quantum operations then the output of those quantum operations conduct the remaining operations in DNA computing to generate the final output of the Quantum-DNA Multiplexer.

Table 18.1 and Figure 18.1, respectively, show the truth table and associative block diagram of Multi-Valued Quantum-DNA 3-to-1 Multiplexer

For Multi-Valued Quantum-DNA 3-to-1 Multiplexer, one 1-to-3 Quantum decoder is needed to get the output of selection line S. Additionally, three DNA AND and one DNA OR operation is required as shown in Figure 18.1. To match the speed between Quantum and DNA operation a Quantum cache Memory is used. Data

DOI: 10.1201/9781003381938-18

TABLE 18.1

Truth Table of Multi-Valued 3-to-1 Quantum-DNA Multiplexer

Inputs	Outputs
|S>	Y
|0>	I0
|1>	I1
|2>	I2

conversion circuits are used to convert qubits into DNA sequences. A heat transfer circuit is used here to transfer ecvessive heat produced from quantum part to DNA part, because DNA part needs extra heat to conduct computations.

When total inputs are 9 and the selection line is 2, as shaown in Table 18.2. Table 18.2 shows the 9-to-1 Quantum-DNA Multiplexer Truth Table.

In Multi-Valued Quantum-DNA 9-to-1 Multiplexer, the quantum AND operations will take all inputs and the outputs of these will be passing through the quantum cache memory and data conversion circuit. Finally, these outputs from the quantumpart will be the input to the DNA OR operation. DNA OR operation will produce the final output. Figure 18.2 gives the block diagram of Multi-Valued Quantum-DNA 9-to-1 Multiplexer.

18.2.1 Circuit Architecture of Multiple-Valued Quantum-DNA Multiplexer

Figure 18.3 shows the construction of Multi-Valued Quantum-DNA 3-to-1 multiplexer circuit. In this circuit, all the inputs |I0> - |I2> connect to three different DNA AND gates after performing data conversion, NMR relaxation, via Quantum cache memory in which another input comes from the Selection line |S>.

TABLE 18.2

Truth Table of Multi-Valued 9-to-1 Quantum-DNA Multiplexer

Inputs		Outputs
|S1>	|S0>	Y
|0>	|0>	I0
|0>	|1>	I1
|0>	|2>	I2
|1>	|0>	I3
|1>	|1>	I4
|1>	|2>	I5
|2>	|0>	I6
|2>	|1>	I7
|2>	|2>	I8

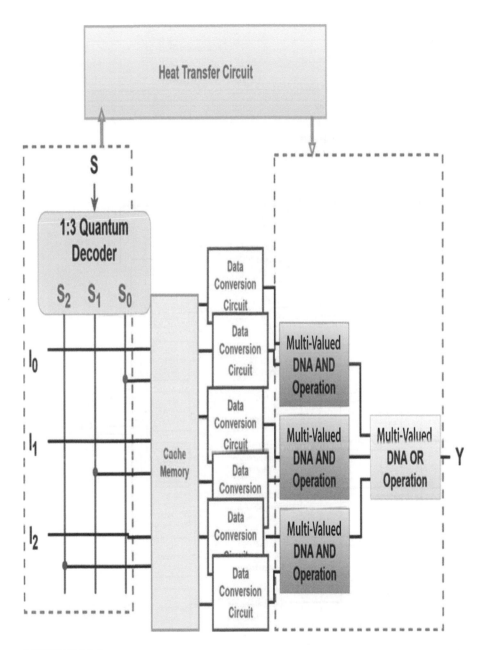

FIGURE 18.1
Block Diagram of Multi-Valued Quantum-DNA 3-to-1 Multiplexer

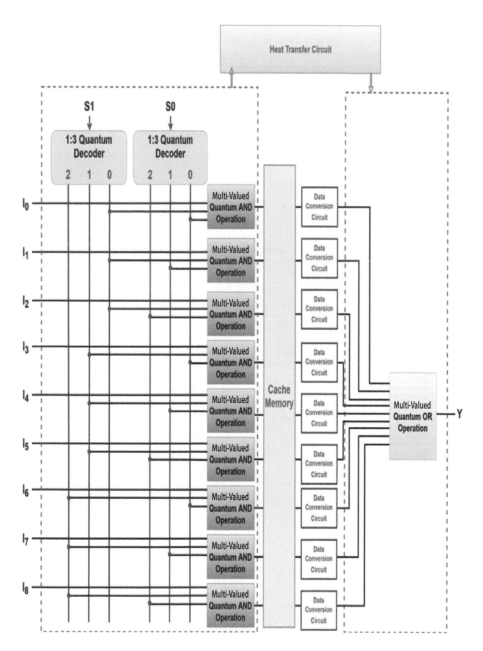

FIGURE 18.2
Block Diagram of Multi-Valued Quantum-DNA 9-to-1 Multiplexer

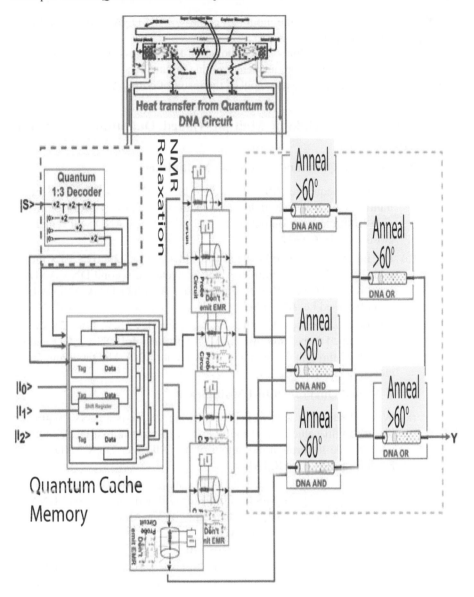

FIGURE 18.3
Circuit Architecture of Multi-Valued Quantum-DNA 3-to-1 Multiplexer

To select one particular input, qubit |0>, |1> or |2> is provided through |S> based on which one of the three output lines of the Quantum 1-to-3 Decoder gets activated and passes qubit |2> through it. Other output lines of the decoder provide qubit |0>. Thus only one DNA AND operation is open to take the input to the output. Finally, the DNA OR operation passes the input that comes via the DNA AND operation to the output of the multiplexer.

The architecture of the Multi-Valued Quantum-DNA 9-to-1 Multiplexer is also the same (Figure 18.4). However, as there are nine inputs, |I0> - |I8>, two selection lines are required to select only one input as the result of output Y.

18.2.2 Working Principle of Multiple-Valued Quantum-DNA Multiplexer

1. **Multi-Valued Quantum-DNA 3-to-1 Multiplexer**

   (a) For input sequences |I0>, |I1>, |I2>, |S> = **|2>, |2>, |2>, |0>** the multi-valued 3-to-1 multiplexer will activate first DNA AND and provide I0 through the output of Y . All other DNA AND will remain closed i.e. **ACCTAG**.

   (b) For input sequences |I0>, |I1>, |I2>, |S> = **|2>, |2>, |2>, |1>** the multi-valued 3-to-1 multiplexer will activate the second DNA AND and provide I1 through the output of Y . All other DNA AND will remain closed i.e. **ACCTAG**.

   (c) For input sequences |I0>, |I1>, |I2>, |S> = **|2>, |2>, |2>, |2>** the multi-valued 3-to-1 multiplexer will activate third DNA AND and provide I2 through the output of Y . All other DNA AND will remain closed i.e. **ACCTAG**.

2. **Multi-Valued Quantum 9-to-1 Multiplexer**

   (a) When all the input sequences |I0>- |I9> are |2>, and |S0>,|S1> = **|0>, |0>** the multi-valued 9-to-1 multiplexer will activate first DNA AND and provide I0 through the output of Y . All other DNA AND will remain closed i.e. **ACCTAG**.

   (b) When all the input sequences |I0>- |I9> are |2>, and |S0>,|S1> = **|0>, |1>** the multi-valued 9-to-1 multiplexer will activate a second DNA AND and provide I1 through the output of Y . All other DNA AND will remain closed i.e. **ACCTAG**.

   (c) When all the input sequences |I0>- |I9> are |2>, and |S0>,|S1> = **|0>, |2>** the multi-valued 9-to-1 multiplexer will activate third DNA AND and provide I2 through the output of Y . All other DNA AND will remain closed i.e. **ACCTAG**.

   (d) When all the input sequences |I0>- |I9> are |2>, and |S0>,|S1> = **|1>, |0>** the multi-valued 9-to-1 multiplexer will activate fourth DNA AND and provide I3 through the output of Y . All other DNA AND will remain closed i.e. **ACCTAG**.

   (e) When all the input sequences |I0>- |I9> are |2>, and |S0>,|S1> = **|1>, |1>** the multi-valued 9-to-1 multiplexer will activate fifth DNA AND and provide I4 through the output of Y . All other DNA AND will remain closed i.e. **ACCTAG**.

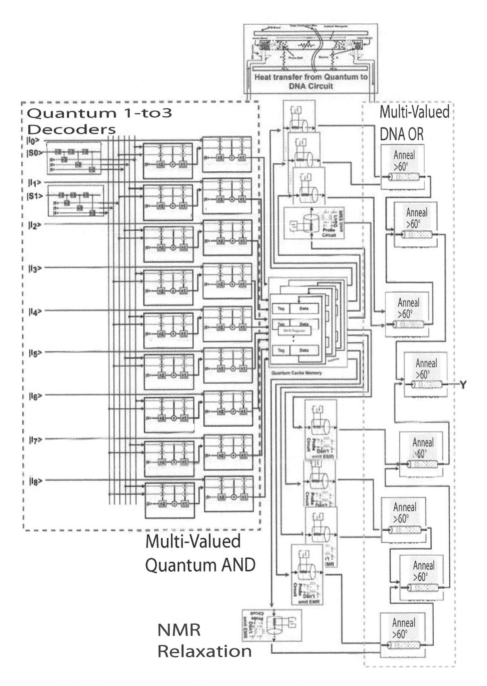

FIGURE 18.4
Circuit Architecture of Multi-Valued Quantum-DNA 9-to-1 Multiplexer

(f) When all the input sequences |I0>- |I9> are |2>, and |S0>,|S1> = |1>, |2> the multi-valued 9-to-1 multiplexer will activate sixth DNA AND and provide I5 through the output of Y . All other DNA AND will remain closed i.e. **ACCTAG**.

(g) When all the input sequences |I0>- |I9> are |2>, and |S0>,|S1> = |2>, |0> the multi-valued 9-to-1 multiplexer will activate seventh DNA AND and provide I6 through the output of Y . All other DNA AND will remain closed i.e. **ACCTAG**.

(h) When all the input sequences |I0>- |I9> are |2>, and |S0>,|S1> = |2>, |1> the multi-valued 9-to-1 multiplexer will activate eighth DNA AND and provide I7 through the output of Y . All other DNA AND will remain closed i.e. **ACCTAG**.

(i) When all the input sequences |I0>- |I9> are |2>, and |S0>,|S1> = |2>, |2> the multi-valued 9-to-1 multiplexer will activate the ninth DNA AND and provide I8 through the output of Y . All other DNA AND will remain closed i.e. **ACCTAG**.

18.3   Multiple-Valued Quantum-DNA Demultiplexer

Demultiplexer circuits are among the most important circuits in the design of complex hardware. It maps a single input to multiple outputs. Thereby, a 1-to-n Quantum-DNA Demultiplexer consists of three main parts: one input, n selection line, and 3^n outputs. In a multi-valued Quantum-DNA 1-to-n Demultiplexer, all the inputs perform some Quantum operation then the output of those conduct the remaining operation in DNA to generate the final output of the Quantum-DNA Demultiplexer.

The truth table of Multi-Valued Quantum-DNA 1-to-3 Demultiplexer is shown in Table 18.3. Here only the selected output equals the input |I> and all other outputs are 0. To represent qubit |0>, |1>, and |2>, in the molecular sequence **ACCTAG**, **CAAGCT** and **TGGATC** are used, respectively.

Figure 18.5 depicts the Quantum-DNA 1-to-3 demultiplexer's block diagram. Depending on the combination chosen, one of the decoder outputs becomes |2> while the others remain |0>.

TABLE 18.3

Truth Table of Multi-Valued Quantum-DNA
1-to-3 Demultiplexer

Inputs	Outputs		
S	$|Y_2>$	$|Y_1>$	$|Y_0>$
|0>	ACCTAG	ACCTAG	I
|1>	ACCTAG	I	ACCTAG
|2>	I	ACCTAG	ACCTAG

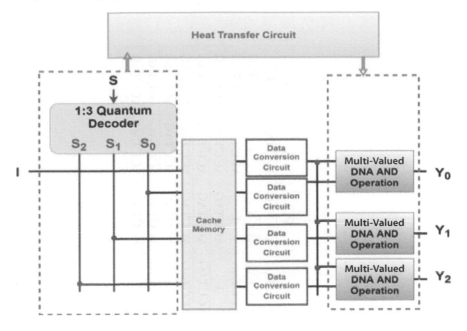

FIGURE 18.5
Block Diagram of Multi-Valued Quantum-DNA 1-to-3 Demultiplexer

Table 18.6 shows the truth table of multi-valued Quantum-DNA 1-to-9 Demultiplexer. Here two selection lines (S1 and S2) are needed to select one output line among nine.

Figure 18.6 gives the block diagram of Multi-Valued Quantum-DNA 1-to-9 Demultiplexer. Here nine output lines produced by nine multi-valued DNA AND operations. Input I and two selection lines are decoded into six lines by two quantum decoders.

18.3.1 Circuit Architecture of Multiple-Valued Quantum-DNA Demultiplexer

The construction of the Multi-Valued Quantum-DNA 1-to-3 Demultiplexer circuit is shown in Figure 18.7. In this circuit, the input, |I> is directly connected to all the DNA AND operations as one input after going through an NMR relaxation to convert qubits into the molecular sequence. The various outputs of selection line |S> contribute as another input of these DNA AND operations after passing through NMR as well. The result of |S> decides which output line must be activated.

To select one particular output, qubit |0>, |1> or |2> is provided through |S> based on which one of the three output lines of the Quantum 1-to-3 Decoder gets activated and passes molecular sequence **TGGATC** through after NMR. Other

TABLE 18.4

Truth Table of Multi-Valued Quantum-DNA 1-to-9 Demultiplexer

Inputs		Outputs								
S1	S0	Y_8	Y_7	Y_6	Y_5	Y_4	Y_3	Y_2	Y_1	Y_0
|0>	|0>	ACCTAG	ACCTAG	ACCTAG	ACCTAG	ACCTAG	ACCTAG	ACCTAG	ACCTAG	I
|0>	|1>	ACCTAG	ACCTAG	ACCTAG	ACCTAG	ACCTAG	ACCTAG	ACCTAG	I	ACCTAG
|0>	|2>	ACCTAG	ACCTAG	ACCTAG	ACCTAG	ACCTAG	ACCTAG	I	ACCTAG	ACCTAG
|1>	|0>	ACCTAG	ACCTAG	ACCTAG	ACCTAG	ACCTAG	I	ACCTAG	ACCTAG	ACCTAG
|1>	|1>	ACCTAG	ACCTAG	ACCTAG	ACCTAG	I	ACCTAG	ACCTAG	ACCTAG	ACCTAG
|1>	|2>	ACCTAG	ACCTAG	ACCTAG	I	ACCTAG	ACCTAG	ACCTAG	ACCTAG	ACCTAG
|2>	|0>	ACCTAG	ACCTAG	I	ACCTAG	ACCTAG	ACCTAG	ACCTAG	ACCTAG	ACCTAG
|2>	|1>	ACCTAG	I	ACCTAG	ACCTAG	ACCTAG	ACCTAG	ACCTAG	ACCTAG	ACCTAG
|2>	|2>	I	ACCTAG	ACCTAG	ACCTAG	ACCTAG	ACCTAG	ACCTAG	ACCTAG	ACCTAG

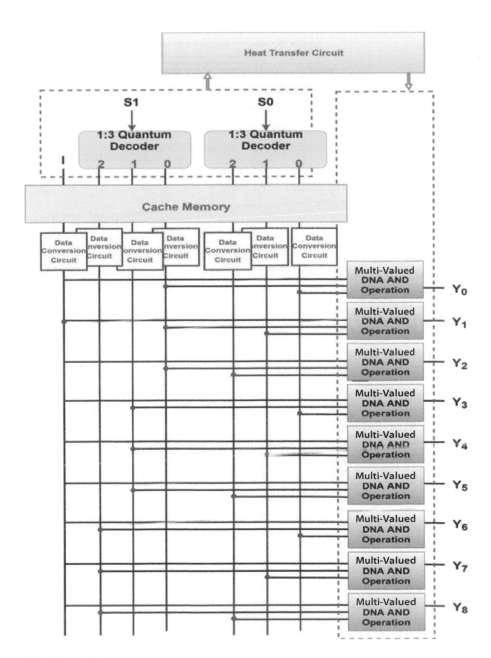

FIGURE 18.6
Block Diagram of Multi-Valued Quantum-DNA 1-to-9 Demultiplexer

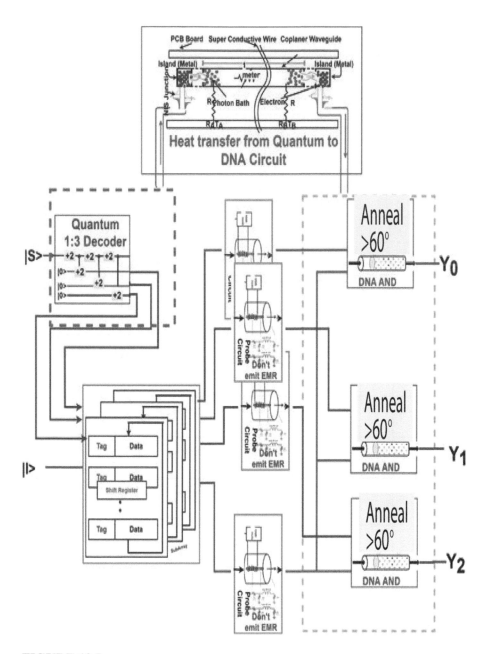

FIGURE 18.7
Circuit Architecture of Multi-Valued Quantum-DNA 1-to-3 Demultiplexer

output lines of the decoder provide molecular sequence **ACCTAG**. Thus only one DNA AND operation is open to select one output among Y0 to Y2.

The architecture of the Multi-Valued Quantum-DNA 1-to-9 Demultiplexer also applies the same procedure (Figure 18.8). However, as there are nine outputs, Y0 - Y8, two selection lines are required to activate only one output among Y0 - Y8, where other outputs remain inactive. i.e. **ACCTAG**.

18.3.2 Working Principle of Multiple-Valued Quantum-DNA Demultiplexer

1. **Multi-Valued Quantum 1-to-3 Demultiplexer**

   (a) For input sequences $|I>$, $|S> = |2>$, $|0>$ the multi-valued 1-to-3 Demultiplexer will activate first DNA AND and provide "2", **TGGATC**, through the output of Y0. All other outputs Y0 - Y2 will remain close as remaining DNA AND operation will generate "0", **ACCTAG**.

   (b) For input sequences $|I>$, $|S> = |2>$, $|1>$ the multi-valued 1-to-3 Demultiplexer will activate second DNA AND and provide "2", **TGGATC**, through the output of Y1. All other outputs Y0 - Y2 will remain close as remaining DNA AND operation will generate "0", **ACCTAG**.

   (c) For input sequences $|I>$, $|S> = |2>$, $|2>$ the multi-valued 1-to-3 Demultiplexer will activate third DNA AND and provide "2", **TGGATC**, through the output of Y2. All other outputs Y0 - Y2 will remain close as remaining DNA AND operation will generate "0", **ACCTAG**.

2. **Multi-Valued Quantum 1-to-9 Demultiplexer**

   (a) When the input qubit $|I>$ is $|2>$, and $|S0>$,$|S1> = |0>$, $|0>$ the multi-valued 1-to-3 Demultiplexer will activate first DNA AND and provide "2", **TGGATC**, through the output of Y0 . All other outputs Y0 - Y8 will remain close as remaining DNA AND operation will generate "0", **ACCTAG**.

   (b) When the input qubit $|I>$ is $|2>$, and $|S0>$,$|S1> = |0>$, $|1>$ the multi-valued 1-to-3 Demultiplexer will activate second DNA AND and provide "2", **TGGATC**, through the output of Y1 . All other outputs Y0 - Y8 will remain close as remaining DNA AND operation will generate "0", **ACCTAG**.

   (c) When the input qubit $|I>$ is $|2>$, and $|S0>$,$|S1> = |0>$, $|2>$ the multi-valued 1-to-3 Demultiplexer will activate third DNA AND and provide "2", **TGGATC**, through the output of Y2 . All other outputs Y0 - Y8 will remain close as remaining DNA AND operation will generate "0", **ACCTAG**.

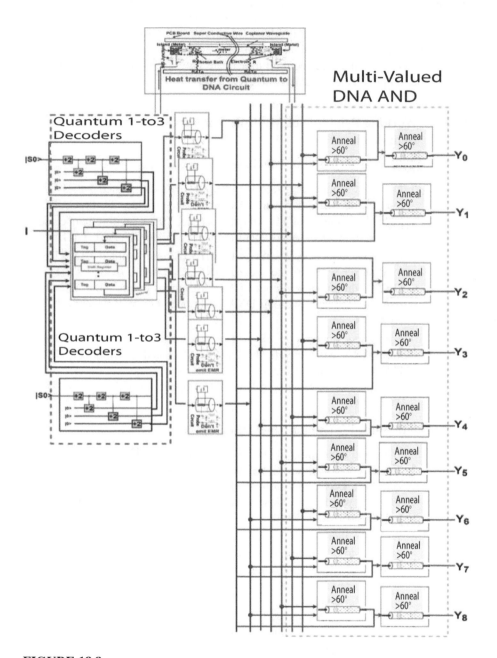

FIGURE 18.8
Circuit Architecture of Multi-Valued Quantum-DNA 1-to-9 Demultiplexer

(d) When the input qubit |I> is |2>, and |S0>,|S1> = **|1>, |0>** the multi-valued 1-to-3 Demultiplexer will activate fourth DNA AND and provide "2", **TGGATC**, through the output of Y3 . All other outputs Y0 - Y8 will remain close as remaining DNA AND operation will generate "0", **ACCTAG**.

(e) When the input qubit |I> is |2>, and |S0>,|S1> = **|1>, |1>** the multi-valued 1-to-3 Demultiplexer will activate fifth DNA AND and provide "2", **TGGATC**, through the output of Y4 . All other outputs Y0 - Y8 will remain close as remaining DNA AND operation will generate "0", **ACCTAG**.

(f) When the input qubit |I> is |2>, and |S0>,|S1> = **|1>, |2>** the multi-valued 1-to-3 Demultiplexer will activate sixth DNA AND and provide "2", **TGGATC**, through the output of Y5 . All other outputs Y0 - Y8 will remain close as remaining DNA AND operation will generate "0", **ACCTAG**.

(g) When the input qubit |I> is |2>, and |S0>,|S1> = **|2>, |0>** the multi-valued 1-to-3 Demultiplexer will activate seventh DNA AND and provide "2", **TGGATC**, through the output of Y6 . All other outputs Y0 - Y8 will remain close as remaining DNA AND operation will generate "0", **ACCTAG**.

(h) When the input qubit |I> is |2>, and |S0>,|S1> = **|2>, |1>** the multi-valued 1-to-3 Demultiplexer will activate the eighth DNA AND and provide "2", **TGGATC**, through the output of Y7 . All other outputs Y0 - Y8 will remain close as remaining DNA AND operation will generate "0", **ACCTAG**.

(i) When the input qubit |I> is |2>, and |S0>,|S1> = **|2>, |2>** the multi-valued 1-to-3 Demultiplexer will activate ninth DNA AND and provide "2", **TGGATC**, through the output of Y8 . All other outputs Y0 - Y8 will remain close as remaining DNA AND operation will generate "0", **ACCTAG**.

18.4   Multiple-Valued Quantum-DNA Encoder

The Multi-Valued Encoder is a combinational logic circuit with 3^n input lines and n output lines that functions as a multi-input and multi-output device. An encoder is a critical component of a digital computer.

Table 18.5 represents the truth table of 3-to-1 Multi-Valued Quantum-DNA Encoder where |D0>, |D1> and |D2> are the input lines and A is the output line.

Figure 18.9 gives the idea of how a 3-to-1 Multi-Valued Quantum-DNA Encoder is designed. As given in the diagram, 3-to-1 Multi-Valued Quantum-DNA Encoder needs two Quantum AND, one Quantum NOT and one DNA OR operation.

TABLE 18.5
Truth Table of Multi-Valued 3-to-1
Quantum-DNA Encoder

Inputs			Outputs
|D2>	|D1>	|D0>	A
|0>	|0>	|2>	ACCTAG
|0>	|2>	|0>	CAAGCT
|2>	|0>	|0>	TGGATC

For multi-valued quantum-DNA 9-to-2 Encoder, Table 18.6 shows the truth table
where |D0> to |D8> are the nine input lines and A0 and A1 are the outputs. The block
diagram of the 9-to-2 Multi-Valued Quantum-DNA Encoder is displayed in Figure
18.10.

18.4.1 Circuit Architecture of Multiple-Valued Quantum-DNA Encoder

Figure 18.11 depicts the architecture of Multi-Valued Quantum-DNA 3-to-1 En-
coder. As when an input line is on, it will generate qubit |2> whereas others will
be closed at |0>. Hence, each input line passes through some Quantum operations.

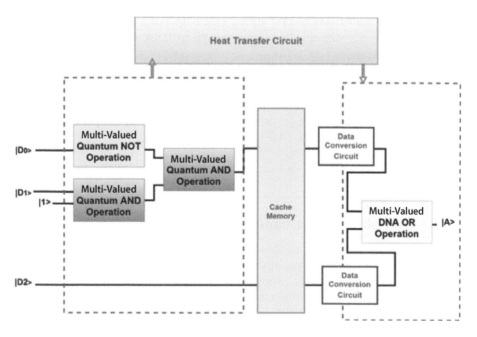

FIGURE 18.9
Block Diagram of Multi-Valued Quantum-DNA 3-to-1 Encoder

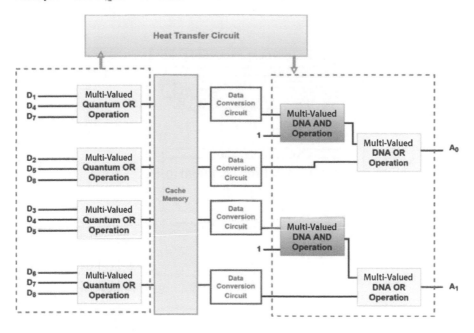

FIGURE 18.10
Block Diagram of Multi-Valued Quantum-DNA 9-to-2 Encoder

First, input line |D0> passes through a Quantum NOT operation to perform a Standard Ternary Inversion (STI). i.e., qubit |0> will become |2> or vice versa and |D1> conduct a Quantum AND operation with qubit |1> as the other input so that it can generate the outcome as |1> whenever |D1> is on. Additionally, |D2> is directly connected to the DNA OR operation as it produces output '**TGGATC**', which represents

TABLE 18.6
Truth Table of Multi-Valued 9-to-2 Quantum-DNA Encoder

Inputs									Outputs	
|D8>	|D7>	|D6>	|D5>	|D4>	|D3>	|D2>	|D1>	|D0>	A1	A0
|0>	|0>	|0>	|0>	|0>	|0>	|0>	|0>	|2>	ACCTAG	ACCTAG
|0>	|0>	|0>	|0>	|0>	|0>	|0>	|2>	|0>	ACCTAG	CAAGCT
|0>	|0>	|0>	|0>	|0>	|0>	|2>	|0>	|0>	ACCTAG	TGGATC
|0>	|0>	|0>	|0>	|0>	|2>	|0>	|0>	|0>	CAAGCT	ACCTAG
|0>	|0>	|0>	|0>	|2>	|0>	|0>	|0>	|0>	CAAGCT	CAAGCT
|0>	|0>	|0>	|2>	|0>	|0>	|0>	|0>	|0>	CAAGCT	TGGATC
|0>	|0>	|2>	|0>	|0>	|0>	|0>	|0>	|0>	TGGATC	ACCTAG
|0>	|2>	|0>	|0>	|0>	|0>	|0>	|0>	|0>	TGGATC	CAAGCT
|2>	|0>	|0>	|0>	|0>	|0>	|0>	|0>	|0>	TGGATC	TGGATC

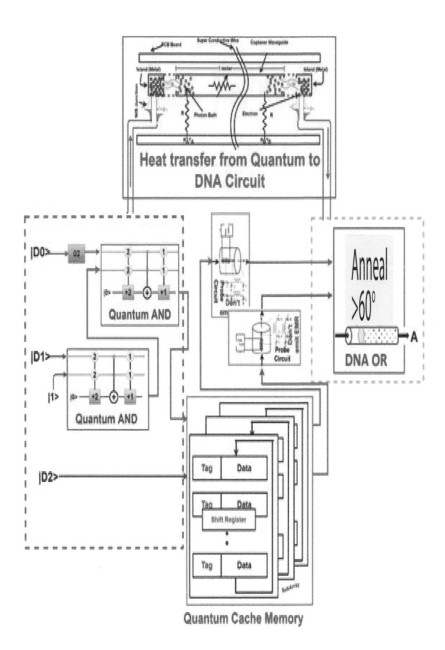

FIGURE 18.11
Circuit Architecture of Multi-Valued Quantum-DNA 3-to-1 Encoder

'2'. DNA OR executes at last to get the output of |A>. However, before conducting the DNA OR operation, Quantum qubits are converted to their equivalent DNA molecular sequences by using NMR relaxation.

For a Multi-Valued Quantum-DNA 9-to-2 Encoder (Figure 18.12), the input of |D1>, |D2>, |D4>, |D5>, |D7>, and |D8> are used to get the result of A0 after performing four Quantum OR, one DNA OR and one DNA AND operation. Whereas |D3>, |D4>, |D5>, |D6>, |D7>, and |D8> is needed to get the outcome of |A1> after conducting another four Quantum OR, one DNA OR, and one DNA AND operation.

For each case, a Quantum cache memory is placed between Quantum operation and DNA operation, so that the qubits can be held temporarily to match the speed.

18.4.2 Working Principle of Multiple-Valued Quantum-DNA Encoder

1. **Multi-Valued Quantum-DNA 3-to-1 Encoder**

   (a) To get Output |A> = |0>, the multi-valued 3-to-1 Encoder D0 will be activated, whereas other input lines remain closed, i.e. **ACCTAG**

   (b) To get Output |A> = |1>, the multi-valued 3-to-1 Encoder D1 will be activated, whereas other input lines remain closed, i.e. **ACCTAG**

   (c) To get Output |A> = |2>, the multi-valued 3-to-1 Encoder D2 will be activated, whereas other input lines remain closed, i.e. **ACCTAG**

2. **Multi-Valued Quantum-DNA 9-to-2 Encoder**

   (a) To get Output |A1>, |A0> = |0>, |0>, the multi-valued 9-to-2 Encoder D0 will be activated, whereas other input line remain closed, i.e. **ACCTAG**

   (b) To get Output |A1>, |A0> = |0>, |1>, the multi-valued 9-to-2 Encoder D1 will be activated, whereas other input line remain closed, i.e. **ACCTAG**

   (c) To get Output |A1>, |A0> = |0>, |2>, the multi-valued 9-to-2 Encoder D2 will be activated, whereas other input line remain closed, i.e. **ACCTAG**

   (d) To get Output |A1>, |A0> = |1>, |0>, the multi-valued 9-to-2 Encoder D3 will be activated, whereas other input line remain closed, i.e. **ACCTAG**

   (e) To get Output |A1>, |A0> = |1>, |1>, the multi-valued 9-to-2 Encoder D4 will be activated, whereas other input line remain closed, i.e. **ACCTAG**

   (f) To get Output |A1>, |A0> = |1>, |2>, the multi-valued 9-to-2 Encoder D5 will be activated, whereas other input line remain closed, i.e. **ACCTAG**

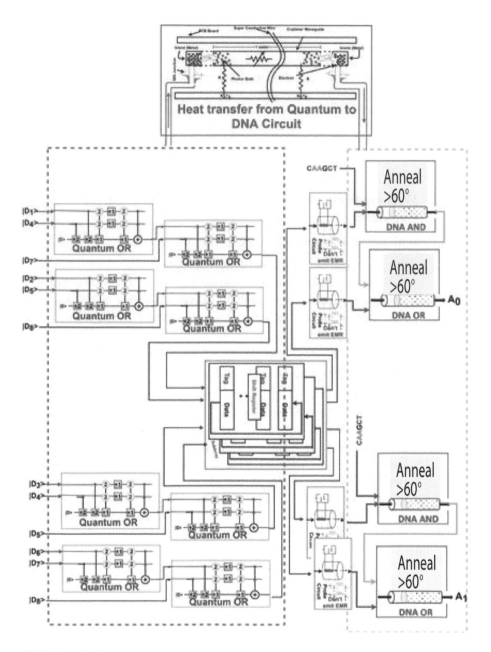

FIGURE 18.12
Circuit Architecture of Multi-Valued Quantum-DNA 9-to-2 Encoder

(g) To get Output |A1>, |A0> = **|2>, |0>,** the multi-valued 9-to-2 Encoder D6 will be activated, whereas other input line remain closed, i.e. **ACCTAG**

(h) To get Output |A1>, |A0> = **|2>, |1>,** the multi-valued 9-to-2 Encoder D7 will be activated, whereas other input line remain closed, i.e. **ACCTAG**

(i) To get Output |A1>, |A0> = **|2>, |2>,** the multi-valued 9-to-2 Encoder D8 will be activated, whereas other input line remain closed, i.e. **ACCTAG**

18.5 Multiple-Valued Quantum-DNA Decoder

The Multi-Valued Quantum-DNA Decoder is a combinational logic circuit with n input lines and 3^n output lines. Essentially, it converts a ternary qubit to its equivalent decimal value.

A Quantum-DNA decoder first performs some Quantum operation on its inputs, then proceeds to the DNA operations after converting the qubit to its corresponding DNA molecule. Table 18.7 contains the truth table for multi-valued quantum-DNA 2-to-9 Decoders.

A 2-to-9 Quantum-DNA Decoder is built using two Multi-Valued 1-to-3 Quantum Decoders and nine Multi-Valued DNA AND (D0-D8) operations. Figure 18.13 depicts the block diagram of multi-valued quantum-dna 2-to-9 decoder.

18.5.1 Circuit Architecture of Multiple-Valued Quantum-DNA Decoder

The construction of the Multi-Valued Quantum-DNA 2-to-9 decoder circuit is shown in Figure 18.14. In this circuit, the input, |A> and |B>, at first, pass through two 1-to-3 Decoders separately so that inserted qubit, i.e., |0>, |1>, or |2> can be identified using the output of 1-to-3 decoders.

Nine DNA AND operation is used in Quantum-DNA 2-to-9 Decoder where each input of these AND operations come from the various combinations of the output of 1-to-3 decoder. However, as the output of Quantum operation will act as an input of DNA operation, thus each qubit is converted to its corresponding DNA molecular sequence by using the NMR relaxation technique. Before the conversion Quantum outputs are stored in a Quantum Cache memory shortly to match the speed to operations of different kinds.

Finally, based on the value of |A> and |B>, only one line of each 1-to-3 decoder generates sequence **TGGATC** along the line whereas others provide **ACCTAG**. Thus, only one DNA AND operation among D0 to D8, gives the result **TGGATC**, where other outputs, D0-D8 remain inactive. i.e., **ACCTAG**.

TABLE 18.7
Truth Table of Multi-Valued Quantum-DNA 2-to-9 Decoder

Input		Output								
|A>	|B>	D8	D7	D6	D5	D4	D3	D2	D1	D0
|0>	|0>	ACCTAG	ACCTAG	ACCTAG	ACCTAG	ACCTAG	ACCTAG	ACCTAG	ACCTAG	TGGATC
|0>	|1>	ACCTAG	ACCTAG	ACCTAG	ACCTAG	ACCTAG	ACCTAG	ACCTAG	TGGATC	ACCTAG
|0>	|2>	ACCTAG	ACCTAG	ACCTAG	ACCTAG	ACCTAG	ACCTAG	TGGATC	ACCTAG	ACCTAG
|1>	|0>	ACCTAG	ACCTAG	ACCTAG	ACCTAG	ACCTAG	TGGATC	ACCTAG	ACCTAG	ACCTAG
|1>	|1>	ACCTAG	ACCTAG	ACCTAG	ACCTAG	TGGATC	ACCTAG	ACCTAG	ACCTAG	ACCTAG
|1>	|2>	ACCTAG	ACCTAG	ACCTAG	TGGATC	ACCTAG	ACCTAG	ACCTAG	ACCTAG	ACCTAG
|2>	|0>	ACCTAG	ACCTAG	TGGATC	ACCTAG	ACCTAG	ACCTAG	ACCTAG	ACCTAG	ACCTAG
|2>	|1>	ACCTAG	TGGATC	ACCTAG	ACCTAG	ACCTAG	ACCTAG	ACCTAG	ACCTAG	ACCTAG
|2>	|2>	TGGATC	ACCTAG	ACCTAG	ACCTAG	ACCTAG	ACCTAG	ACCTAG	ACCTAG	ACCTAG

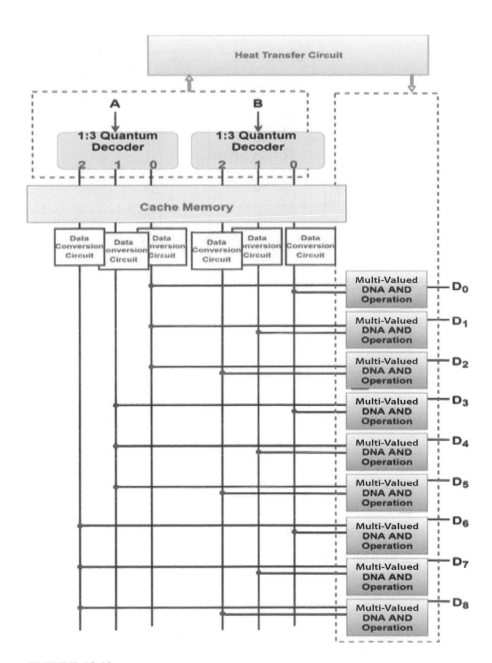

FIGURE 18.13
Block Diagram of Multi-Valued Quantum-DNA 2-to-9 Decoder

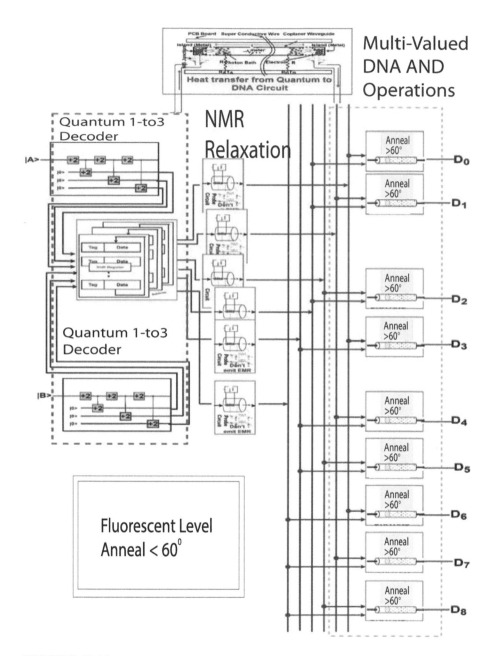

FIGURE 18.14
Circuit Architecture of Multi-Valued Quantum-DNA 2-to-9 Decoder

18.5.2 Working Principle of Multiple-Valued Quantum-DNA Decoder

1. For input sequences |A>, |B> = |0>, |0>, the multi-valued 1-to-3 decoder will perform (A0 && B0) and the D0 line will be open. As a result, the output sequence of D0 equal TGGATC and remaining lines, D1 to D8 will remain closed, **i.e., ACCTAG.**

2. For input sequences |A>, |B> = |0>, |1>, the multi-valued 1-to-3 decoder will perform (A0 && B1) and the D1 line will be open. As a result, the output sequence of D1 equal TGGATC and remaining lines, D0, D2 to D8 will remain closed, **i.e. ACCTAG.**

3. For input sequences |A>, |B> = |0>, |2>, the multi-valued 1-to-3 decoder will perform (A0 && B2) and the D2 line will be open. As a result, the output sequence of D2 equal **TGGATC** and remaining lines, D0, D1 and D3 to D8 will remain closed, **i.e. ACCTAG.**

4. For input sequences |A>, |B> = |1>, |0>, the multi-valued 1-to-3 decoder will perform (A1 && B0) and the D3 line will be open. As a result, the output sequence of D3 equal **TGGATC** and remaining lines, D0 to D2, and D4 to D8 will remain closed, **i.e. ACCTAG.**

5. For input sequences |A>, |B> = |1>, |1> the multi-valued 1-to-3 decoder will perform (A1 && B1) and the D4 line will be open. As a result, the output sequence of D4 equal **TGGATC** and remaining lines, D0 to D3, and D5 to D8 will remain closed, **i.e. ACCTAG.**

6. For input sequences |A>, |B> = |1>, |2>, the multi-valued 1-to-3 decoder will perform (A1 && B2) and the D5 line will be open. As a result, the output sequence of D5 equal **TGGATC** and remaining lines, D0 to D4, and D6 to D8 will remain closed, **i.e. ACCTAG.**

7. For input sequences A, B= |2>, |0>, the multi-valued 1-to-3 decoder will perform (A2 && B0) and the D6 line will be open. As a result, the output sequence of D6 equal **TGGATC** and remaining lines, D0 to D5, D7, D8 will remain closed, **i.e. ACCTAG.**

8. For input sequences |A>, |B> = |2>, |1>, the multi-valued 1-to-3 decoder will perform (A2 && B1) and the D7 line will be open. As a result, the output sequence of D7 equal **TGGATC** and remaining lines, D0 to D6, D8 will remain closed, **i.e. ACCTAG.**

9. For input sequences |A>, |B> = |2>, |2>, the multi-valued 1-to-3 decoder will perform (A2 && B2) and the D8 line will be open. As a result, the output sequence of D8 equal **TGGATC** and remaining lines, D0 to D7 will remain closed, **i.e. ACCTAG.**

18.6 Summary

Another novel concept of Multi-Valued Quantum-DNA computing is presented in this chapter, along with possible architectural concepts for building multiple multi-valued combinational circuits. Each Multi-Valued Quantum-DNA combinational circuit's algorithms, such as adder, subtractor, multiplexer, demultiplexer, encoder, decoder, and comparator, are displayed. This chapter goes into specific architecture designs for Multi-Valued Quantum-DNA combinational circuits with exceptional performance. Combinational circuits for 2 qubits and N-qubit are also illustrated, along with their design methodologies and operating principles. Quantum computers, produce a lot of heat, which leads to qubit disorder. DNA computers, on the other hand, require a lot of heat to carry out the reactions. As a result, the quantity of heat that these Multi-Valued Quantum-DNA combinational circuits may require or generate is exhibited, providing a clear image of their efficiency. A speed calculation was also done to see how much faster they would be if these Multi-Valued Quantum-DNA combinational circuits were perfectly implemented. All of these articles have been implemented hypothetically.

Bibliography

[1] Dhande, A. P., & Ingole, V. T. (2005, March). Design and implementation of 2 bit ternary ALU slice. In Proc. Int. Conf. IEEE-Sci. Electron., Technol. Inf. Telecommun (Vol. 17).

[2] Lin, S., Kim, Y. B., & Lombardi, F. (2009). CNTFET-based design of ternary logic gates and arithmetic circuits. IEEE Transactions on Nanotechnology, 10(2), 217-225.

[3] Heung, A., & Mouftah, H. T. (1985). Depletion/enhancement CMOS for a lower power family of three-valued logic circuits. IEEE Journal of Solid-State Circuits, 20(2), 609-616.

[4] Monfared, A. T., & Haghparast, M. (2017). Design of novel quantum/reversible ternary adder circuits. International Journal of Electronics Letters, 5(2), 149-157.

[5] Zadeh, R. P., & Haghparast, M. (2011). A new reversible/quantum ternary comparator. Australian Journal of Basic and Applied Sciences, 5(12), 2348-2355.

19

Multiple-Valued DNA-Quantum Combinational Circuits

19.1 Introduction

Biomolecules and biomolecular processes, as explained in earlier chapters, are designed to use computational methods in DNA computing. Quantum computing entails carrying out calculations at a scale where quantum mechanical effects are noticeable. Both of these new computing paradigms have been proposed as solid-state computers' successors. Both DNA computing and quantum computing have the potential to outperform ordinary digital computers, but considerable technical hurdles must be overcome first. Because of their coherent superposition of states, quantum computers are more powerful than traditional turing machines. DNA computers can be evolved via biotechnology techniques. Combining DNA and quantum computers could capture both of these properties. With this in mind, this chapter is devoted to DNA-Quantum Computing, which combines two distinct technologies. The overall chapter format is similar to the preceding chapter, with the introduction of DNA-Quantum computing at the beginning. Combining DNA and quantum computers could capture both of these properties. Quantum logic circuits could be self-assembled by DNA computers using DNA strand-attached gates. Quantum computers could also be constructed directly from the physical features of the DNA molecule. This chapter will discuss about multi-valued DNA-Quantum multiplexer, demultiplexer, encoder and decoder.

19.2 Multiple-Valued DNA-Quantum Multiplexer

A balanced multiplexer is an essential component of a digital system that converts multiple inputs to a single output. This circuit is useful for building multi-valued arithmetic circuits. There are three parts to an n-to-1 multiplexer. There are 3^n inputs, 1 output, and n selector line. In a multi-valued DNA-Quantum n-1 Multiplexer, first, all the inputs perform some DNA operation then the output of those conduct the

DOI: 10.1201/9781003381938-19

TABLE 19.1

Truth Table of Multi-Valued DNA-Quantum
3-to-1 Multiplexer

Inputs	Outputs
S	|Y>
ACCTAG	|I0>
CAAGCT	|I1>
TGGATC	|I2>

remaining operation in Quantum to generate the final output of the DNA-Quantum Multiplexer.

Table 19.1 and Figure 19.1, respectively, show the truth table and associative block diagram of Multi-Valued DNA-Quantum 3-to-1 Multiplexer.

For multi-valued DNA-Quantum 3-to-1 Multiplexer, one 1-to-3 DNA decoder is needed to get the output of selection line S. Additionally, three Quantum AND and one Quantum OR operation is required as shown in Figure 19.1. To match the speed between DNA and Quantum operation a DNA cache Memory is used.

When total inputs are 9 and the selection line is 2, then the truth table of multi-valued Quantum 9-to-1 multiplexer will look like as shown in Table 19.2.

Here, among the quantum AND operations, only one AND will provide its input to the output line via quantum OR operation. Figure 19.2 gives the block diagram of the multi-valued quantum 9-to-1 multiplexer. DNA cache memory is used for temporary storage of DNA sequences. Data conversion circuit is used to convert DNA sequences into qubits.

TABLE 19.2

Truth Table of Multi-Valued Quantum
9-to-1 Multiplexer

Inputs		Outputs
S1	S0	|Y>
ACCTAG	ACCTAG	|I0>
ACCTAG	CAAGCT	|I1>
ACCTAG	TGGATC	|I2>
CAAGCT	ACCTAG	|I3>
CAAGCT	CAAGCT	|I4>
CAAGCT	TGGATC	|I5>
TGGATC	ACCTAG	|I6>
TGGATC	CAAGCT	|I7>
TGGATC	TGGATC	|I8>

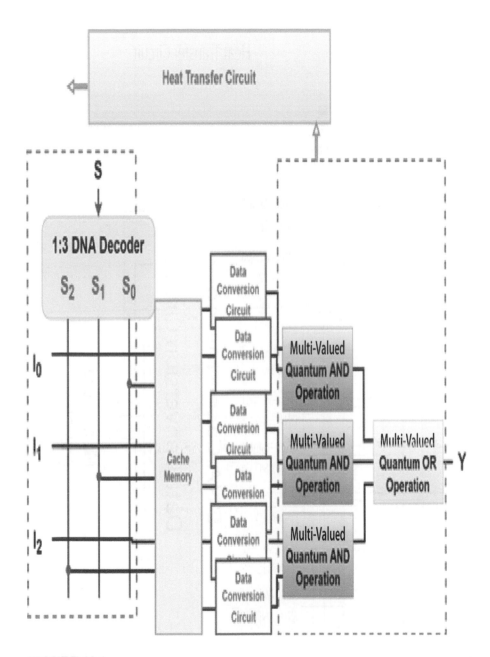

FIGURE 19.1
Block Diagram of Multi-Valued DNA-Quantum 3-to-1 Multiplexer

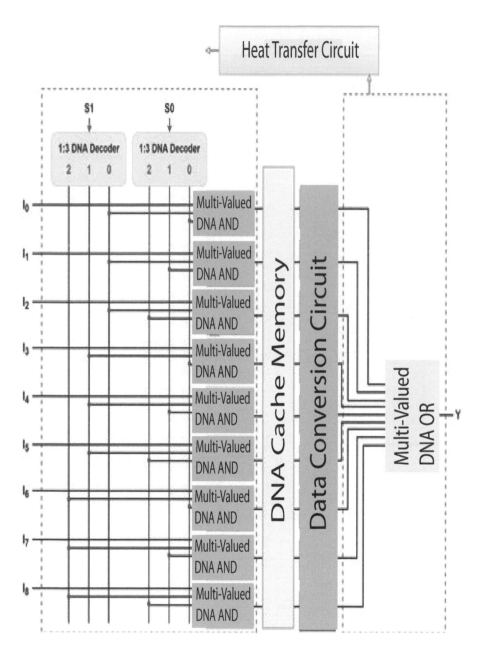

FIGURE 19.2
Block Diagram of Multi-Valued DNA-Quantum 9-to-1 Multiplexer

19.2.1 Circuit Architecture of Multiple-Valued DNA-Quantum Multiplexer

Figure 19.3 shows the construction of a multi-valued DNA-Quantum 3-to-1 multiplexer circuit. In this circuit, all the inputs I0 - I2 connect to three different DNA AND gates after performing data conversion, trap ion, via DNA cache memory in which another input comes from the Selection line S.

To select one particular input, molecular sequence **ACCTAG, CAAGCT** or **TGGATC** is provided through S to represent 1, 2, or 3, respectively. Based on this output of the selection line one of the three output lines of the DNA 1-to-3 Decoder gets activated and passes molecular sequence **TGGATC** through it. Other output lines of the decoder provide molecular sequence **ACCTAG**. Thus, only one Quantum AND operation is open after the trap ion to take the input to the output. Finally, the Quantum OR operation passes the input that comes via the Quantum AND operation to the output of the multiplexer.

The architecture of the Multi-Valued DNA-Quantum 9-to-1 Multiplexer is also the same (Figure 19.4). However, as there are nine inputs (I0 - I8), two selection lines are required to select only one input as the result of output |Y>.

19.2.2 Working Principle of Multiple-Valued DNA-Quantum Multiplexer

1. **Multi-Valued DNA-Quantum 3-to-1 Multiplexer**

   (a) For input sequences I0, I1, I2, S = **TGGATC, TGGATC, TGGATC**, and **ACCTAG** which represent 2, 2, 2, and 0, the multi-valued 3-to-1 multiplexer will activate the first Quantum AND and provide |I0> through the output of |Y> . All other Quantum AND will remain closed |0>.

   (b) For input sequences I0, I1, I2, S = **TGGATC, TGGATC, TGGATC**, and **CAAGCT** which represent 2, 2, 2, and 1, the multi-valued 3-to-1 multiplexer will activate the first Quantum AND and provide |I1> through the output of |Y>. All other Quantum AND will remain closed |0>.

   (c) For input sequences I0, I1, I2, S = **TGGATC, TGGATC, TGGATC**, and **TGGATC** which represent 2, 2, 2, and 2, the multi-valued 3-to-1 multiplexer will activate first Quantum AND and provide |I2> through the output of |Y>. All other Quantum AND will remain closed |0>.

2. **Multi-Valued DNA-Quantum 9-to-1 Multiplexer**

   (a) When all the input sequences I0- I9 are **TGGATC**, and S0, S1 = **ACCTAG, ACCTAG** the multi-valued 9-to-1 multiplexer will activate the first Quantum AND and provide |I0> through the output of |Y>. All other Quantum AND will remain closed |0>.

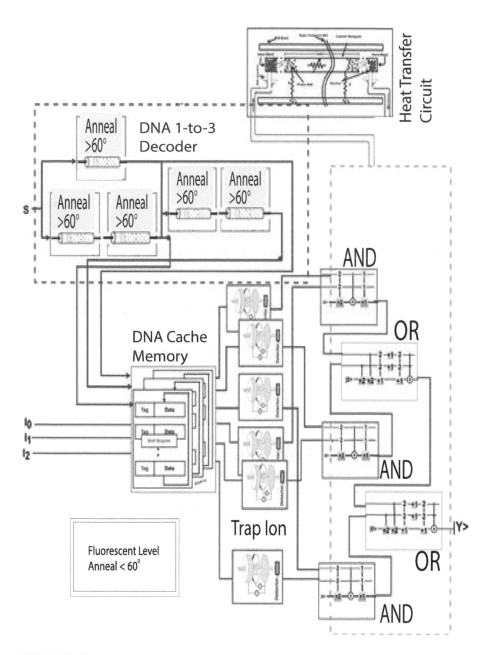

FIGURE 19.3
Circuit Architecture of Multi-Valued DNA-Quantum 3-to-1 Multiplexer

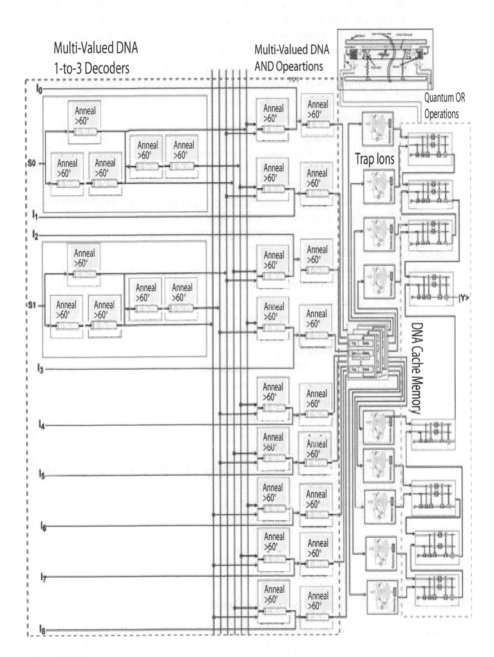

FIGURE 19.4
Circuit Architecture of Multi-Valued DNA-Quantum 9-to-1 Multiplexer

(b) When all the input sequences I0- I9 are **TGGATC**, and S0, S1 = **AC-CTAG, CAAGCT the** multi-valued 9-to-1 multiplexer will activate the second Quantum AND and provide |I1> through the output of |Y> . All other Quantum AND will remain closed **|0>.**

(c) ces I0- I9 are **TGGATC**, and S0, S1 = **ACCTAG** , **TGGATC** the multi-valued 9-to-1 multiplexer will activate the third Quantum AND and provide |I2> through the output of |Y> . All other Quantum AND will remain closed **|0>.**

(d) When all the input sequences I0- I9 are **TGGATC**, and S0, S1 = **CAAGCT, ACCTAG the** multi-valued 9-to-1 multiplexer will activate the fourth Quantum AND and provide |I3> through the output of |Y>. All other Quantum AND will remain closed **|0>.** .

(e) When all the input sequences I0- I9 are **TGGATC**, and S0, S1 = **CAAGCT, CAAGCT the** multi-valued 9-to-1 multiplexer will activate the fifth Quantum AND and provide |I4> through the output of |Y>. All other Quantum AND will remain closed **|0>.**

(f) When all the input sequences I0- I9 are **TGGATC**, and S0, S1 = **CAAGCT, TGGATC** the multi-valued 9-to-1 multiplexer will activate a sixth Quantum AND and provide |I5> through the output of |Y> . All other Quantum AND will remain closed **|0>.**

(g) When all the input sequences I0- I9 are **TGGATC**, and S0, S1 = **TG-GATC, ACCTAG the** multi-valued 9-to-1 multiplexer will activate seventh Quantum AND and provide |I6 > through the output of |Y> . All other Quantum AND will remain closed **|0>.**

(h) When all the input sequences I0- I9 are **TGGATC**, and S0, S1 = **TG-GATC, CAAGCT the** multi-valued 9-to-1 multiplexer will activate eighth Quantum AND and provide |I7> through the output of |Y> . All other Quantum AND will remain closed **|0>.**

(i) When all the input sequences I0- I9 are **TGGATC**, and S0, S1 = **TG-GATC, TGGATC** the multi-valued 9-to-1 multiplexer will activate the ninth Quantum AND and provide |I8> through the output of |Y>. All other Quantum AND will remain closed **|0>.**

19.3   Multiple-Valued DNA-Quantum Demultiplexer

Demultiplexer circuits are among the most important circuits in the design of complex hardware. It maps a single input to multiple outputs. Thereby, a 1-to-N DNA-Quantum Demultiplexer consists of three main parts: one input, an N selection line and 3^N outputs. In a multi-valued DNA-Quantum 1-to-N Demultiplexer, all the inputs perform some DNA operation then the output of those DNA part will conduct the

TABLE 19.3

Truth Table of Multi-Valued DNA-Quantum 1-to-3 Demultiplexer

Inputs	Outputs		
S	$|Y_2>$	$|Y_1>$	$|Y_0>$
ACCTAG	$|0>$	$|0>$	$|I>$
CAAGCT	$|0>$	$|I>$	$|0>$
TGGATC	$|I>$	$|0>$	$|0>$

remaining operation in Quantum part to generate the final output of the DNA-Quantum Demultiplexer.

The truth table of Multi-Valued DNA-Quantum 1-to-3 Demultiplexer is shown in Table 19.3 where only the selected output equals the input I and all other outputs are 0. To represent 0, 1, and 2 in the molecular sequence, **ACCTAG**, **CAAGCT** and **TGGATC** are used, respectively.

Figure 19.5 depicts the block diagram of multi-valued 1-to-3 demultiplexer. Depending on the chosen combination, one of the decoder outputs becomes 2 while the others remain 0.

Here one selection line will be decoded into three by DNA 1-to-3 decoder. Three multi-valued quantum AND operations are needed to produce the final outputs.

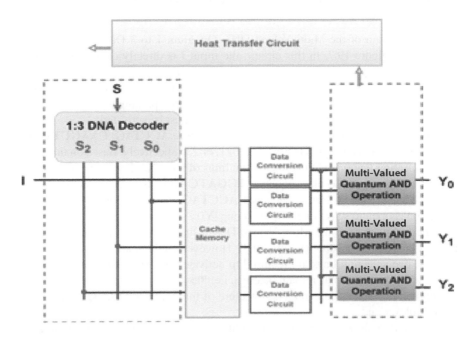

FIGURE 19.5

Block Diagram of Multi-Valued DNA-Quantum 1-to-3 Demultiplexer

TABLE 19.4

Truth Table of Multi-Valued DNA-Quantum 2-to-9 Demultiplexer

Inputs		Outputs								
S1	S0	|Y0>	|Y1>	|Y2>	|Y3>	|Y4>	|Y5>	|Y6>	|Y7>	|Y8>
ACCTAG	ACCTAG	|I>	|0>	|0>	|0>	|0>	|0>	|0>	|0>	|0>
ACCTAG	CAAGCT	|0>	|I>	|0>	|0>	|0>	|0>	|0>	|0>	|0>
ACCTAG	TGGATC	|0>	|0>	|I>	|0>	|0>	|0>	|0>	|0>	|0>
CAAGCT	ACCTAG	|0>	|0>	|0>	|I>	|0>	|0>	|0>	|0>	|0>
CAAGCT	CAAGCT	|0>	|0>	|0>	|0>	|I>	|0>	|0>	|0>	|0>
CAAGCT	TGGATC	|0>	|0>	|0>	|0>	|0>	|I>	|0>	|0>	|0>
TGGATC	ACCTAG	|0>	|0>	|0>	|0>	|0>	|0>	|I>	|0>	|0>
TGGATC	CAAGCT	|0>	|0>	|0>	|0>	|0>	|0>	|0>	|I>	|0>
TGGATC	TGGATC	|0>	|0>	|0>	|0>	|0>	|0>	|0>	|0>	|I>

Table 19.4 shows the truth table of multi-valued DNA-Quantum 2-to-9 Demultiplexer where 2 selection lines are needed to select one output line among nine. Figure 19.6 gives the block diagram of Multi-Valued DNA-Quantum 2-to-9 Demultiplexer.

19.3.1   Circuit Architecture of Multiple-Valued DNA-Quantum Demultiplexer

The construction of the Multi-Valued DNA-Quantum 1-to-3 Demultiplexer circuit is shown in Figure 19.7. In this circuit, the input I is directly connected to all the quantum AND operations as one input after going through a trap ion to convert the molecular sequence into the qubit. The various outputs of selection line S contribute as another input of these quantum AND operations after passing through the trap ion as well. The result of S decides which output line must be activated.

To select one particular input, molecular sequence **ACCTAG**, **CAAGCT** or **TG-GATC** is provided through S to represent 0, 1, or 2, respectively. Based on this output of the selection line one of the three output lines of the DNA 1-to-3 Decoder gets activated and passes molecular sequence **TGGATC** through it. Other output lines of the decoder provide molecular sequence **ACCTAG**. Thus only one Quantum AND operation is open to select one output among |Y0> to |Y2>.

The architecture of the Multi-Valued DNA-Quantum 2-to-9 Demultiplexer also applies the same procedure (Figure 19.8). However, as there are nine outputs, |Y0> – |Y8>, two selection lines are required to activate only one output among |Y0> – |Y8>, where other outputs remain inactive. i.e. **|0>**.

Figure 19.7 shows the circuit architecture of multi-valued DNA-Quantum 1-to-3 demultiplexer and

Figure 19.8 shows the circuit architecture of multi-valued DNA-Quantum 2-to-9 demultiplexer.

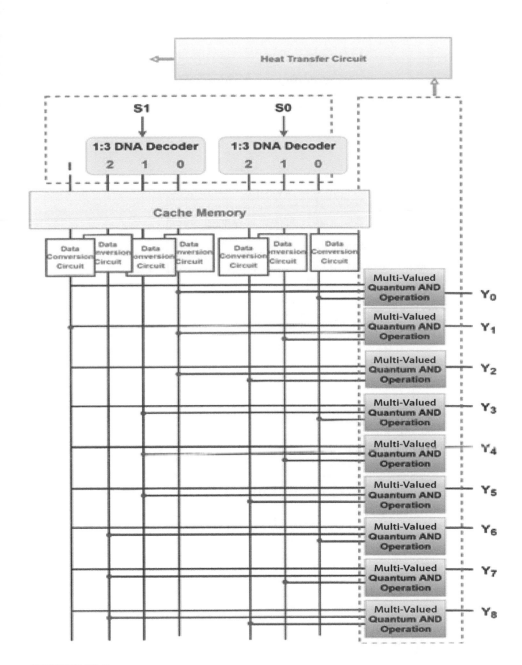

FIGURE 19.6
Block Diagram of Multi-Valued DNA-Quantum 2-to-9 Demultiplexer

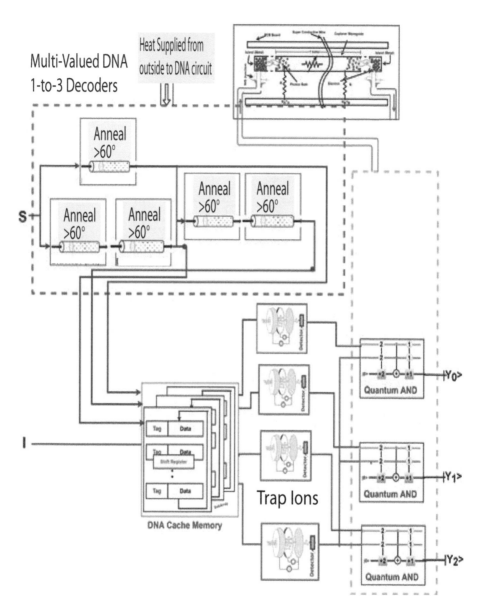

FIGURE 19.7
Circuit Architecture of Multi-Valued DNA-Quantum 1-to-3 Demultiplexer

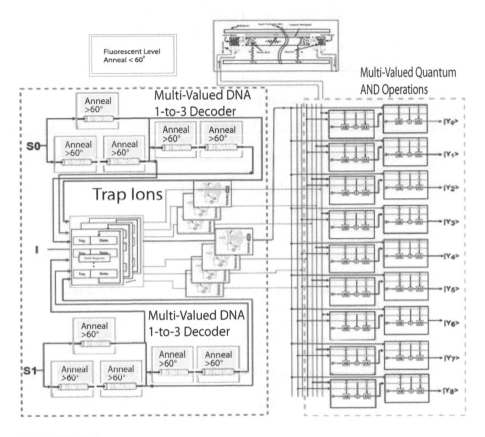

FIGURE 19.8

Circuit Architecture of Multi-Valued DNA-Quantum 2-to-9 Demultiplexer

19.3.2 Working Principle of Multiple-Valued DNA-Quantum Demultiplexer

1. **Multi-Valued DNA 1-to-3 Demultiplexer**

    (a) For input sequences I, S = **TGGATC, ACCTAG** which represent 2, and 0, the multi-valued 1-to-3 Demultiplexer will activate the first Quantum AND and provide |2> through the output of |Y0> to activate it . All other outputs |Y0> - |Y2> will remain close as the remaining Quantum AND operation will generate |**0>**.

(b) For input sequences I, S = **TGGATC, CAAGCT** which represent 2, and 0, the multi-valued 1-to-3 Demultiplexer will activate a second Quantum AND and provide |2> through the output of |Y1> to activate it . All other outputs |Y0> - |Y2> will remain close as the remaining Quantum AND operation will generate **|0>**.

(c) For input sequences I, S = **TGGATC, TGGATC** which represent 2, and 0, the multi-valued 1-to-3 Demultiplexer will activate the third Quantum AND and provide |2> through the output of |Y2> to activate it . All other outputs |Y0> - |Y2> will remain close as the remaining Quantum AND operation will generate **|0>**.

2. **Multi-Valued DNA 1-to-9 Demultiplexer**

(a) When the input qubit, I = **TGGATC**, and S0, S1 = **ACCTAG, AC-CTAG; the** multi-valued 2-to-9 Demultiplexer will activate the first Quantum AND and provide |2> through the output of |Y0> to activate it. All other outputs |Y0> - |Y9> will remain close as the remaining Quantum AND operation will generate **|0>**.

(b) When the input qubit, I = **TGGATC**, and S0, S1 = **ACCTAG, CAAGCT**; the multi-valued 2-to-9 Demultiplexer will activate the second Quantum AND and provide |2> through the output of |Y1> to activate it. All other outputs |Y0> - |Y9> will remain close as the remaining Quantum AND operation will generate **|0>**.

(c) When the input qubit, I = **TGGATC**, and S0, S1 = **ACCTAG, TG-GATC; the** multi-valued 2-to-9 Demultiplexer will activate the third Quantum AND and provide |2> through the output of |Y2> to activate it . All other outputs |Y0> - |Y9> will remain close as the remaining Quantum AND operation will generate **|0>**.

(d) When the input qubit, I = **TGGATC**, and S0, S1 = **CAAGCT, ACC-TAG; the** multi-valued 2-to-9 Demultiplexer will activate the fourth Quantum AND and provide |2> through the output of |Y3> to activate it. All other outputs |Y0> - |Y9> will remain close as the remaining Quantum AND operation will generate **|0>**.

(e) When the input qubit, I = **TGGATC**, and S0, S1 = **CAAGCT, CAAGCT**; the multi-valued 2-to-9 Demultiplexer will activate the fifth Quantum AND and provide |2> through the output of |Y4> to activate it. All other outputs |Y0> - |Y9> will remain close as the remaining Quantum AND operation will generate **|0>**.

(f) When the input qubit, I = **TGGATC**, and S0, S1 = **CAAGCT , TG-GATC;** the multi-valued 2-to-9 Demultiplexer will activate the sixth Quantum AND and provide |2> through the output of |Y5> to activate it . All other outputs |Y0> - |Y9> will remain close as the remaining Quantum AND operation will generate **|0>**.

(g) When the input qubit, I = **TGGATC**, and S0, S1 = **TGGATC, ACC-TAG; the** multi-valued 2-to-9 Demultiplexer will activate the seventh

Quantum AND and provide |2> through the output of |Y6> to activate it . All other outputs |Y0> - |Y9> will remain close as the remaining Quantum AND operation will generate **|0>**.

(h) When the input qubit, I = **TGGATC**, and S0, S1 = **TGGATC, CAAGCT; the** multi-valued 2-to-9 Demultiplexer will activate the eighth Quantum AND and provide |2> through the output of |Y7> to activate it . All other outputs |Y0> - |Y9> will remain close as the remaining Quantum AND operation will generate **|0>**.

(i) When the input qubit, I = **TGGATC**, and S0, S1 = **TGGATC, TG-GATC; the** multi-valued 2-to-9 Demultiplexer will activate ninth Quantum AND and provide |2> through the output of |Y8> to activate it . All other outputs |Y0> - |Y9> will remain close as the remaining Quantum AND operation will generate **|0>**.

19.3.3   Multiple-Valued DNA-Quantum Encoder

The Multi-Valued Encoder is a combinational logic circuit with 3^n input lines and n output lines that functions as a multi-input and multi-output device. An encoder is a critical component of a digital computer.

Table 19.5 represents the truth table of 3-to-1 Multi-Valued DNA-Quantum Encoder where D0, D1, and D2 are the input lines and |A> is the output line. Figure 19.9 shows the block diagram of 3-to-1 Multi-Valued DNA Encoder. As given in the diagram, 3-to-1 Multi-Valued DNA-Quantum Encoder needs two DNA AND, one DNA NOT and one Quantum OR operation.

For multi-valued DNA-Quantum 9-to-2 Encoder, the truth table is shown in Table 19.6 where D0 to D8 are the nine input lines and |A0> and |A1> are the outputs. The block diagram of 9-to-2 Multi-Valued DNA-Quantum Encoder is displayed in Figure 19.10.

19.3.4   Circuit Architecture of Multiple-Valued DNA-Quantum Encoder

Figure 19.11 depicts the architecture of Multi-Valued DNA-Quantum 3-to-1 Encoder. As when an input line is on, it will generate a molecular sequence '**TGGATC**',

TABLE 19.5
Truth Table of Multi-Valued DNA-Quantum 3-to-1 Encoder

Inputs			Outputs
D2	**D1**	**D0**	**|A>**
ACCTAG	**ACCTAG**	**TGGATC**	**|0>**
ACCTAG	**TGGATC**	**ACCTAG**	**|1>**
TGGATC	**ACCTAG**	**ACCTAG**	**|2>**

TABLE 19.6
Truth Table of Multi-Valued DNA-Quantum 9-to-2 Encoder

Inputs									Outputs	
D8	D7	D6	D5	D4	D3	D2	D1	D0	|A1>	|A0>
ACCTAG	ACCTAG	ACCTAG	ACCTAG	ACCTAG	ACCTAG	ACCTAG	ACCTAG	TGGATC	|0>	|0>
ACCTAG	ACCTAG	ACCTAG	ACCTAG	ACCTAG	ACCTAG	ACCTAG	TGGATC	ACCTAG	|0>	|1>
ACCTAG	ACCTAG	ACCTAG	ACCTAG	ACCTAG	ACCTAG	TGGATC	ACCTAG	ACCTAG	|0>	|2>
ACCTAG	ACCTAG	ACCTAG	ACCTAG	ACCTAG	TGGATC	ACCTAG	ACCTAG	ACCTAG	|1>	|0>
ACCTAG	ACCTAG	ACCTAG	ACCTAG	TGGATC	ACCTAG	ACCTAG	ACCTAG	ACCTAG	|1>	|1>
ACCTAG	ACCTAG	ACCTAG	TGGATC	ACCTAG	ACCTAG	ACCTAG	ACCTAG	ACCTAG	|1>	|2>
ACCTAG	ACCTAG	TGGATC	ACCTAG	ACCTAG	ACCTAG	ACCTAG	ACCTAG	ACCTAG	|2>	|0>
ACCTAG	TGGATC	ACCTAG	ACCTAG	ACCTAG	ACCTAG	ACCTAG	ACCTAG	ACCTAG	|2>	|1>
TGGATC	ACCTAG	ACCTAG	ACCTAG	ACCTAG	ACCTAG	ACCTAG	ACCTAG	ACCTAG	|2>	|2>

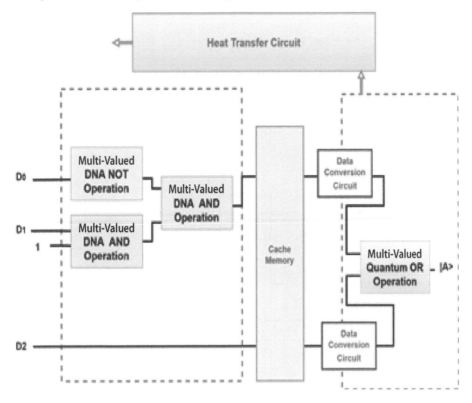

FIGURE 19.9
Block Diagram of Multi-Valued DNA-Quantum 3-to-1 Encoder

which represents '2', whereas others will be closed at '**ACCTAG**', as it represents '0'. Hence, each input line passes through some DNA operations. First, input line D0 passes through a DNA NOT operation to perform a Standard Ternary Inversion (STI), i.e. sequence '**ACCTAG**' will become '**TGGATC**' or vice versa and D1 conduct a DNA AND operation with the molecular sequence '**CAAGCT**', which represents '1', as the other input. Additionally, D2 is directly connected to the Quantum OR operation as it produces output |2>. Quantum OR executes at last to get the output of |A>. However, before conducting the Quantum OR operation, DNA molecular sequences are converted to their equivalent qubit using trap ions.

For a Multi-Valued DNA-Quantum 9-to-2 Encoder (Figure 19.12), the input of D1, D2, D4, D5, D7, and D8 are used to get the result of |A0> after performing four DNA OR, one Quantum OR, and one Quantum AND operation. Whereas D3 to D8 is needed to get the outcome of |A1> after conducting another four DNA OR, one Quantum OR, and one Quantum AND operation.

For each case, a DNA cache memory is placed between DNA operation and Quantum operation, so that the molecular sequence can be held temporarily to match the speed.

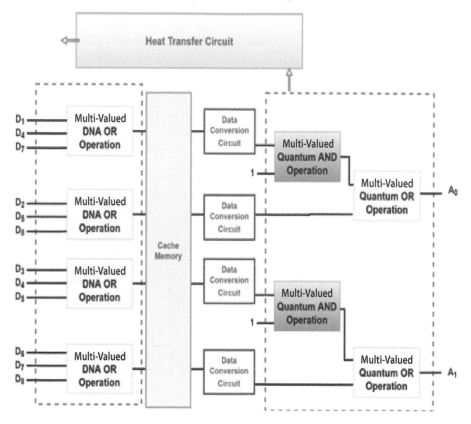

FIGURE 19.10
Block Diagram of Multi-Valued DNA-Quantum 9-to-2 Encoder

19.3.5   Working Principle of Multiple-Valued DNA-Quantum Encoder

1. **Multi-Valued DNA-Quantum 3-to-1 Encoder**

    (a) To get Output A = ACCTAG , the multi-valued 3-to-1 Encoder |D0> will be activated, whereas other input lines remain closed, i.e. **|0>**

    (b) To get Output A = CAAGCT, the multi-valued 3-to-1 Encoder |D1> will be activated, whereas other input lines remain closed, i.e. **|0>**

    (c) To get Output A = TGGATC, the multi-valued 3-to-1 Encoder |D2> will be activated, whereas other input lines remain closed, i.e. **|0>**

2. **Multi-Valued DNA-Quantum 9-to-1 Multiplexer**

    (a) To get Output A1, A0 = **ACCTAG, ACCTAG,** the multi-valued 9-to-2 Encoder |D0> will be activated, whereas other input lines remain closed, i.e. **|0>**

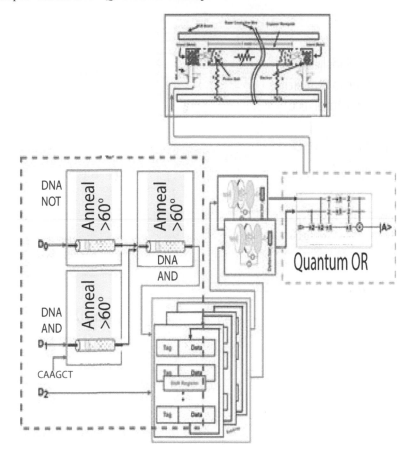

FIGURE 19.11
Circuit Architecture of Multi-Valued DNA-Quantum 3-to-1 Encoder

    (b) To get Output A1, A0 = **ACCTAG, CAAGCT,** the multi-valued 9-to-2 Encoder |D1> will be activated, whereas other input lines remain closed, i.e. **|0>**

    (c) To get Output A1, A0 = **ACCTAG, TGGATC,** the multi-valued 9-to-2 Encoder |D2> will be activated, whereas other input lines remain closed, i.e. **|0>**

    (d) To get Output A1, A0 = **CAAGCT, ACCTAG,** the multi-valued 9-to-2 Encoder |D3> will be activated, whereas other input lines remain closed, i.e. **|0>**

    (e) To get Output A1, A0 = **CAAGCT, CAAGCT,** the multi-valued 9-to-2 Encoder |D4> will be activated, whereas other input lines remain closed, i.e. **|0>**

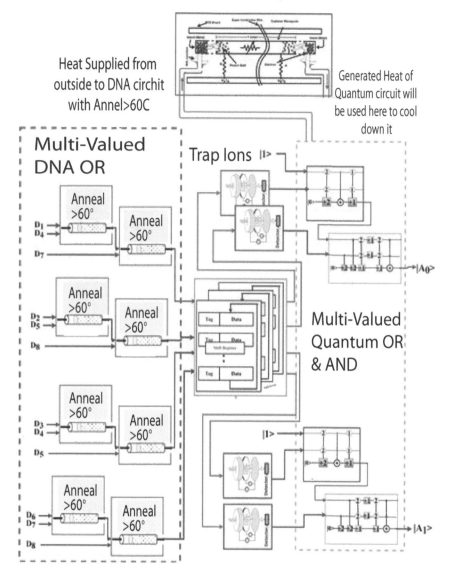

FIGURE 19.12
Circuit Architecture of Multi-Valued DNA-Quantum 9-to-2 Encoder

    (f) To get Output A1, A0 = **CAAGCT, TGGATC,** the multi-valued 9-to-2 Encoder |D5> will be activated, whereas other input lines remain closed, i.e. **|0>**

    (g) To get Output A1, A0 = **TGGATC , ACCTAG,** the multi-valued 9-to-2 Encoder |D6> will be activated, whereas other input lines remain closed, i.e. **|0>**

TABLE 19.7

Truth Table of Multi-Valued DNA-Quantum 2-to-9 Decoder

Input		Output								
A	B	|D8>	|D7>	|D6>	|D5>	|D4>	|D3>	|D2>	|D1>	|D0>
ACCTAG	ACCTAG	|0>	|0>	|0>	|0>	|0>	|0>	|0>	|0>	|2>
ACCTAG	CAAGCT	|0>	|0>	|0>	|0>	|0>	|0>	|0>	|2>	|0>
ACCTAG	TGGATC	|0>	|0>	|0>	|0>	|0>	|0>	|2>	|0>	|0>
CAAGCT	ACCTAG	|0>	|0>	|0>	|0>	|0>	|2>	|0>	|0>	|0>
CAAGCT	CAAGCT	|0>	|0>	|0>	|0>	|2>	|0>	|0>	|0>	|0>
CAAGCT	TGGATC	|0>	|0>	|0>	|2>	|0>	|0>	|0>	|0>	|0>
TGGATC	ACCTAG	|0>	|0>	|2>	|0>	|0>	|0>	|0>	|0>	|0>
TGGATC	CAAGCT	|0>	|2>	|0>	|0>	|0>	|0>	|0>	|0>	|0>
TGGATC	TGGATC	|2>	|0>	|0>	|0>	|0>	|0>	|0>	|0>	|0>

(h) To get Output A1, A0 = **TGGATC, CAAGCT,** the multi-valued 9-to-2 Encoder |D7> will be activated, whereas other input lines remain closed, i.e. **|0>**

(i) To get Output A1, A0 = **TGGATC , TGGATC,** the multi-valued 9 to-2 Encoder |D8> will be activated, whereas other input lines remain closed, i.e. **|0>**

19.4 Multiple-Valued DNA-Quantum Decoder

The Multi-Valued DNA-Quantum Decoder is a combinational logic circuit with n input lines and 3^n output lines. Essentially, it converts a ternary-molecular DNA sequence to its equivalent decimal value.

A DNA-Quantum decoder first performs some DNA operation on its inputs, then proceeds to Quantum operation after converting the DNA molecule to its corresponding qubit. Table 19.7 contains the truth table for 2-to-9 DNA Decoders.

Figure 19.13 shows the block diagram multi-valued DNA-Quantum 2-to-9 Decoder. A 2-to-9 DNA Decoder is built using two Multi-Valued 1-to-3 DNA Decoders and nine Multi-Valued Quantum AND (D0-D8) operations (Figure 19.13).

19.4.1 Circuit Architecture of Multiple-Valued DNA-Quantum Decoder

The construction of the Multi-Valued DNA-Quantum 2-to-9 decoder circuit is shown in Figure 19.14. In this circuit, the input, A and B, at first, pass through two 1-to-3 Decoders separately so that inserted DNA molecular sequence, i.e. **ACCTAG,**

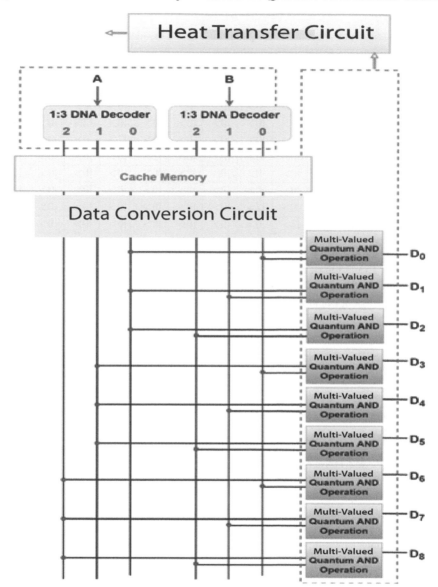

FIGURE 19.13
Block Diagram Multi-Valued DNA-Quantum 2-to-9 Decoder

CAAGCT or **TGGATC,** which represent "0", "1" and "2", respectively, can be identified using the output of 1-to-3 decoders.

Nine Quantum AND operation is used in DNA-Quantum 2-to-9 Decoder where each input of these AND operations come from the various combinations of the output of 1-to-3 decoder. However, as the output of DNA operation will act as an input of Quantum operation, thus each DNA molecular sequence is converted to its corresponding qubit by using the trap ion technique. Before the conversion DNA outputs are stored in a DNA Cache memory shortly to match the speed to different operations.

Finally, based on the value of A and B, only one line of each 1-to-3 decoder generates qubit |2> along the line whereas others provide |0>. Thus only one Quantum AND operation among |D0> to |D8>, gives the result |2>, where other outputs, |D0>-|D8> remain inactive. i.e. |0>.

19.4.2 Working Principle of Multiple-Valued DNA-Quantum Decoder

1. For input sequences A, B =**ACCTAG, ACCTAG**, the multi-valued 1-to-3 decoder will perform (|A0> && |B0>) and the |D0> line will be open. As a result, the output qubit of |D0> equal **|2>** and remaining lines |D1> to |D8> will remain closed **|0>**.

2. For input sequences A, B =**ACCTAG, CAAGCT**, the multi-valued 1-to-3 decoder will perform (|A0> && |B1>) and the |D1> line will be open. As a result, the output qubit of |D1> equal **|2>** and remaining lines, |D0>, |D2> to |D8> will remain closed**|0>**.

3. For input sequences A, B =**ACCTAG, TGGATC** , the multi-valued 1-to-3 decoder will perform (|A0> && |B2>) and the |D2> line will be open. As a result, the output qubit of |D2> equal **|2>** and remaining lines, |D0>, |D1> and |D3> to |D8> will remain closed **|0>**.

4. For input sequences A, B =**CAAGCT, ACCTAG**, the multi-valued 1-to-3 decoder will perform (|A1> && |B0>) and the |D3> line will be open. As a result, the output qubit of |D3> equal **|2>** and remaining lines, |D0> to |D2> and |D4> to |D8> will remain closed **|0>**.

5. For input sequences A, B =**CAAGCT, CAAGCT**, the multi-valued 1-to-3 decoder will perform (|A1> && |B1>) and the |D4> line will be open. As a result, the output qubit of |D4> equal **|2>** and remaining lines, |D0> to |D3> and |D5> to |D8> will remain closed **|0>**.

6. For input sequences A, B =**CAAGCT, TGGATC** , the multi-valued 1-to-3 decoder will perform (|A1> && |B2>) and the |D5> line will be open. As a result, the output qubit of |D5> equal **|2>** and remaining lines, |D0> to |D4> and |D6> to |D8> will remain closed **|0>**.

7. For input sequences A, B =**TGGATC , ACCTAG**, the multi-valued 1-to-3 decoder will perform (|A2> && |B0>) and the |D6> line will be open. As a result, the output qubit of |D6> equal **|2>** and remaining lines, |D0> to |D5> and |D7> to |D8> will remain closed **|0>**.

8. For input sequences A, B =**TGGATC , CAAGCT**, the multi-valued 1-to-3 decoder will perform (|A2> && |B1>) and the |D7> line will be

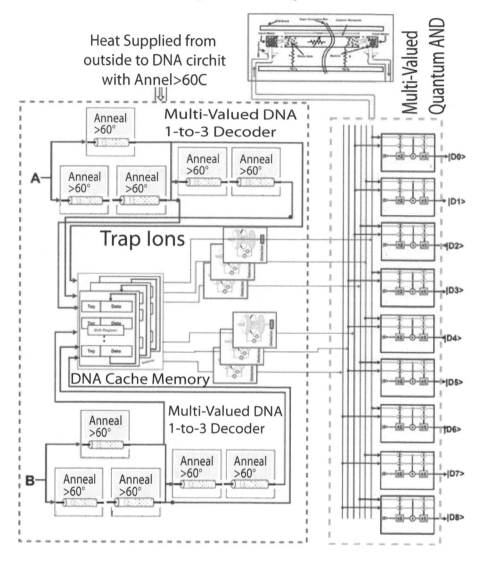

FIGURE 19.14
Circuit Architecture of Multi-Valued DNA-Quantum 2-to-9 Decoder

open.As a result, the output qubit of |D7> equal **|2>** and remaining lines, |D0> to |D6> and |D8> will remain closed **|0>**.

9. For input sequences A, B =**TGGATC** , **TGGATC**, the multi-valued 1-to-3 decoder will perform (|A2> && |B2>) and the |D8> line will be open. As a result, the output qubit of |D8> equal **|2>** and remaining lines, |D0> to |D7> will remain closed **|0>**.

19.5   Summary

In this chapter, a new concept is proposed of Multi-Valued DNA-Quantum computing with the possible architectural ideas for constructing various multi-valued combinational circuits. The algorithms for each Multi-Valued DNA-Quantum combinational circuit, i.e., adder, subtractor, multiplexer, demultiplexer, encoder, decoder, and comparator are shown. This chapter delves into specific architecture designs for high-performance Multi-Valued DNA-Quantum combinational circuits. Some combinational circuits are also shown for 2 qubits and N-qubit, together with their design techniques and functioning principles. Quantum computers generate a lot of heat, which causes disorder among qubits. On the other hand, DNA computers need a lot of heat to execute the reactions. As a result, the amount of heat that can be needed or can be generated by these Multi-Valued DNA-Quantum combinational circuits is displayed, giving a clear picture of their efficiency. A speed calculation has also been performed to indicate how much faster they will be if these Multi-Valued DNA-Quantum combinational circuits are implemented perfectly. All of these articles have been hypothetically implemented.

Bibliography

[1] Mandal, S. B., Chakrabarti, A., & Sur-Kolay, S. (2011, May). Synthesis techniques for ternary quantum logic. In 2011 41st IEEE International Symposium on Multiple-Valued Logic (pp. 218-223). IEEE.

[2] Giri, S., & Saraswathi, M. N. (2012). Implementation of combinational circuits using ternary multiplexer. International Journal of Computational Engineering Research, 2(2), 457-463.

[3] Hallworth, R. P., & Heath, F. G. (1962). Semiconductor circuits for ternary logic. Proceedings of the IEE-Part C: Monographs, 109(15), 219-225.

[4] Dhande, A. P., & Ingole, V. T. (2005, March). Design and implementation of 2 bit ternary ALU slice. In Proc. Int. Conf. IEEE-Sci. Electron., Technol. Inf. Telecommun (Vol. 17).

Final Remarks

Multiple-valued quantum computers provide a higher degree of security, parallel processing capability, and computation speed. Multiple-valued quantum computing, which has become an intriguing option in recent years, can solve many intractable traditional multiple valued computer issues. Multiple-valued DNA (deoxyribose nucleic acid) computing stands out among traditional computer systems because of its parallel processing, large storage capacity, and ability to execute nano-level multiple-valued computing. Traditional computer technologies demand a greater amount of processing power than DNA computing. Logic gates have special qualities in multiple valued DNA computing, such as stability and reusability. The ability of a DNA computer to solve traditional NP problems in polynomial time has drawn a lot of attention. Multiple valued DNA has the potential to store a significant amount of data. As a result, multiple-valued quantum molecular biology may be able to solve the data storage challenge, which is a big problem in the modern age. This book is a great resource for the multiple-valued logic in quantum, DNA, quantum-DNA, and DNA-quantum computing for researchers, students, and academicians. It will quench the thirst of beginners to advanced-level readers.

Multiple-valued quantum molecular biology is completely a new and informative approach introduced to get the advantages of both multi-valued quantum and DNA computing. Multi-valued quantum molecular biology means multiple-valued quantum-DNA and multiple-valued DNA-quantum computing can be obtained by merging multiple-valued quantum and multiple-valued DNA computing and have the capacity to execute parallel operations, which is one of the finest advantages of multiple-valued quantum and DNA computing. This book is a great source for the researchers of quantum and DNA computing.

Trap ion, quadrupole trap ion, NMR, NMR relaxation, quantum and DNA cache memory, and heat transfer circuits are used here in the multiple-valued quantum-DNA and DNA-quantum computing. The architecture and detailed design with figures of these technologies are explained in this book, which are very important for the students and researchers to be able to work in this field.

The arithmetic logic unit (ALU) is the heart of a computer and arithmetic operations are the root of all computations. This book describes the arithmetic operations in multi-valued quantum and DNA computing. Combinational logic circuits in quantum, DNA, quantum-DNA, and DNA-quantum are explained in this book, which is important for students and researchers in the future world of nanotechnology.

Index